AF454308

Genomic Analyses and New Breeding Technologies for the Enhancement of Horticultural Plants

Editors
Fei Chen
Hainan University
Sanya, China

Jia-Yu Xue
Nanjing Agricultural
University
Nanjing, China

Editorial Office
MDPI
St. Alban-Anlage 66
4052 Basel, Switzerland

This is a reprint of articles from the Special Issue published online in the open access journal *Horticulturae* (ISSN 2311-7524) (available at: https://www.mdpi.com/journal/horticulturae/special_issues/Genomic_Breeding_Horticultural).

For citation purposes, cite each article independently as indicated on the article page online and as indicated below:

Lastname, A.A.; Lastname, B.B. Article Title. *Journal Name* **Year**, *Volume Number*, Page Range.

ISBN 978-3-0365-8510-9 (Hbk)
ISBN 978-3-0365-8511-6 (PDF)
doi.org/10.3390/books978-3-0365-8511-6

Cover image courtesy of Fei Chen

Genomic Analyses and New Breeding Technologies for the Enhancement of Horticultural Plants

Editors

Fei Chen
Jia-Yu Xue

Basel • Beijing • Wuhan • Barcelona • Belgrade • Novi Sad • Cluj • Manchester

Contents

About the Editors

Fei Chen

Dr. Fei Chen is a professor of crop science and a professor of electronic information at Hainan University. His research interests are molecular horticulture, tropical plants, genomics and genetics, big data, databases, data visualization, and artificial intelligence. He has published 56 papers, and the number of citations of his papers is 2321 according to Google Scholar. In addition, he serves as the chief executive of Tropical Plants and the associate editor of Horticulture Research.

Jia-Yu Xue

Dr. Jia-Yu Xue is a scientist working on the transition of plants from aquatic to terrestrial environments, the origin and early diversification of angiosperms, and plant–pathogen interactions. He has published 3 book chapters and 59 indexed papers in different journals, of which he published 15 as the first author and 8 as the corresponding author. His h-index is 28, and the number of citations of his papers is 3102 according to Scopus. In addition, he serves as the associate editor for three journals.

 *horticulturae*

Editorial

Genomics and Biotechnology Empower Plant Science Research

Yufan Liang [1], Fei Chen [1,*] and Jia-Yu Xue [2,*]

1   College of Breeding and Multiplication, Hainan University, Sanya 572025, China; 22210901000041@hainanu.edu.cn
2   College of Horticulture, Nanjing Agricultural University, Nanjing 210095, China
*   Correspondence: feichen@hainanu.edu.cn (F.C.); xuejy@njau.edu.cn (J.-Y.X.)

Genomics and biotechnology play crucial roles in biological research, rapidly propelling the field of plant science. They provide the means to delve deeply into the understanding of plant genomics and functions, driving plant improvement and genetic studies, as well as deciphering mechanisms of plant responses to environmental stress. Furthermore, they further our comprehension and application of plant evolution and diversity. These advancements reported in this Special Issue, entitled "Genomic Analyses and New Breeding Technologies for the Enhancement of Horticultural Plants", contribute to the better utilization of plant resources, enhancing crop yield and quality, stress adaptations, as well as the preservation and conservation of plant species.

Genomics and Biotechnologies in Horticulture

Horticultural plants encompass a diverse range of botanical species characterized by unique biological traits. They often necessitate controlled environments for optimal growth, rely on grafting techniques for propagation, and possess the ability for asexual reproduction. These plants hold immense economic and cultural significance for humankind. Currently, more than 200 horticultural plant species, including fruit trees, vegetables, flowers, medicinal plants, and beverage crops, have been subjected to genomic sequencing [1]. Remarkable strides have also been achieved in the fields of genome genotyping, pan-genomics, and the assembly of telomere-to-telomere genomes [2], marking an unprecedented advancement in our understanding of these plants.

Currently, the rapid development and application of multi-omics technologies, such as phenomics, metabolomics, and hormone profiling, as well as various gene editing technologies such as CRISPR-Cas, have greatly facilitated plant gene function research and the molecular breeding of new cultivars.

The Molecular Mechanisms of Horticultural Plant Adaptation to the Environment

Environmental adaptation is a key characteristic of horticultural plants for their growth and reproduction under various environmental conditions. However, the study of horticultural plants lags behind model plants due to their diverse traits. With the decoding of the genomes of many horticultural plants and breakthroughs in various biotechnological approaches, significant progress has been made in the molecular genetics of horticultural plants. Identifying important genes is crucial in the face of various stresses. In this Special Issue, *PebHLH56* in passion fruit was found to be involved in cold stress [3]; SOD in water lilies is associated with temperature stress and heavy metal stress [4]; and lncRNA participates in plant–pathogen interactions in tomatoes [5]. Furthermore, in addressing various disease and pest problems, Chae et al. identified key genomic loci and alleles that play critical roles in pepper disease resistance through quantitative trait locus (QTL) analysis. These research findings will provide references for the development of SNP markers associated with pepper disease resistance QTLs and the breeding of disease-resistant pepper cultivars [6]. Shao et al., exploring the duplication/loss of NLR genes in Asteraceae species, discovered a unique evolutionary pattern of "expansion followed by contraction" in NLR

Citation: Liang, Y.; Chen, F.; Xue, J.-Y. Genomics and Biotechnology Empower Plant Science Research. *Horticulturae* **2023**, *9*, 863. https://doi.org/10.3390/horticulturae9080863

Received: 9 July 2023
Accepted: 17 July 2023
Published: 28 July 2023

genes, thus establishing the NLR gene repertoire in Asteraceae plants and revealing the genetic basis of disease resistance in Asteraceae plants [7].

The Molecular Mechanisms Underlying the Formation of Quality Traits in Horticultural Crops

The formation of quality traits in horticultural crops is a complex process involving the regulation of multiple molecular mechanisms. These mechanisms directly or indirectly influence the crop's appearance, taste, aroma, and nutritional value, among other quality features. The accumulation of sugars in fruits is crucial for quality formation, and sugar accumulation is associated with several molecular mechanisms. Shui et al. [8] investigated lychee fruit quality by applying excessive calcium fertilizer and found that downregulation and expression of the CHS gene family may lead to the reduced accumulation of chalcones, resulting in oxidative damage to fruit flesh and the inhibition of soluble sugar accumulation in the tissue. Xu et al. utilized transcriptomic approaches to study the gene expression profile underlying grape skin color formation and identified key MYB genes involved in skin color transitions [9]. Sun et al. [10] performed cluster analyses on chrysanthemum varieties, dividing them into five categories and summarizing the typical plant configurations of each variety. Finally, they conducted genome-wide association studies (GWASs) to identify potential functional genes. A comprehensive understanding of these mechanisms contributes to optimizing the quality and yield of horticultural crops.

Big Data Tools for Horticulture Research

Genomics, through high-throughput sequencing technologies such as whole-genome sequencing and transcriptome sequencing, enables us to rapidly and comprehensively develop understanding of the composition, structure, and function of plant genomes. This provides a powerful tool for uncovering the functions, genetic variations, and evolution of plant genes. However, genomic data have not been effectively utilized. To address this gap and efficiently utilize big data, Zhou et al. [11] established the first multi-omics database focused on strawberries, storing all available genomic and transcriptomic data. They developed a series of bioinformatics tools, creating an integrated platform that will facilitate genetic and breeding research in strawberries. RNA-Seq analysis is a fundamental transcriptomics research method; Shi et al. [12] proposed a clustering-based correlation analysis method (C-CorA), which is an efficient approach for analyzing the correlation of various types of data across different dimensions. This method can be applied to RNA-Seq data for candidate gene detection in fruit quality research. Jin et al. [13] conducted genome re-sequencing of Gypsophila paniculata, constructed a whole-genome InDel marker system, and established the first genome-wide genetic map for Gypsophila paniculata, providing a complete marker system for molecular studies. This technology enables us to study and improve important traits in plants, such as disease resistance, stress tolerance, and yield.

Genomics and biotechnology have provided abundant data and tools for studying genetic diversity and evolution in plants. By comparing the genome sequences and conducting functional genomics studies of different plant species, we can determine the patterns and mechanisms of plant evolution, infer species relationships, and understand the genomic bases of plant adaptations to different environments. Using biotechnological tools such as gene knockout, gene expression regulation, and transgenic techniques, we can identify and validate the functions of plant genes, thus revealing their roles in biological processes.

Funding: F.C. acknowledges grants from National Natural Science Foundation of China (32172614), Science and Technology special fund of Hainan Province (ZDYF2023XDNY050). J.-Y.X. acknowledges the grant from the Fundamental Research Funds for the Central Universities (no. KYCXJC2022003).

Acknowledgments: We thank all the authors who contributed to this Special Issue.

Conflicts of Interest: The authors declare no conflict of interest.

References

1. Chen, F.; Song, Y.; Li, X.; Chen, J.; Mo, L.; Zhang, X.; Lin, Z.; Zhang, L. Genome sequences of horticultural plants: Past, present and future. *Hortic. Res.* **2019**, *6*, 112. [CrossRef] [PubMed]
2. Zhou, Y.; Zhang, J.; Xiong, X.; Cheng, Z.-M.; Chen, F. De novo assembly of plant complete genomes. *Trop. Plants* **2022**, *1*, 7. [CrossRef]
3. Xu, Y.; Zhou, W.; Ma, F.; Huang, D.; Xing, W.; Wu, B.; Sun, P.; Chen, D.; Xu, B.; Song, S. Characterization of the Passion Fruit (*Passiflora edulis* Sim) bHLH Family in Fruit Development and Abiotic Stress and Functional Analysis of *PebHLH56* in Cold Stress. *Horticulturae* **2023**, *9*, 272. [CrossRef]
4. WKhan, W.U.; Khan, L.U.; Chen, D.; Chen, F. Comparative Analyses of Superoxide Dismutase (SOD) Gene Family and Expression Profiling under Multiple Abiotic Stresses in Water Lilies. *Horticulturae* **2023**, *9*, 781. [CrossRef]
5. Liang, G.; Niu, Y.; Guo, J. Systematic Identification of Long Non-Coding RNAs under Allelopathic Interference of Para-Hydroxybenzoic Acid in *S. lycopersicum*. *Horticulturae* **2022**, *8*, 1134. [CrossRef]
6. Chae, S.-Y.; Lee, K.; Do, J.-W.; Hong, S.-C.; Lee, K.-H.; Cho, M.-C.; Yang, E.-Y.; Yoon, J.-B. QTL Mapping of Resistance to Bacterial Wilt in Pepper Plants (*Capsicum annuum*) Using Genotyping-by-Sequencing (GBS). *Horticulturae* **2022**, *8*, 115. [CrossRef]
7. Li, X.-T.; Zhou, G.-C.; Feng, X.-Y.; Zeng, Z.; Liu, Y.; Shao, Z.-Q. Frequent Gene Duplication/Loss Shapes Distinct Evolutionary Patterns of NLR Genes in Arecaceae Species. *Horticulturae* **2021**, *7*, 539. [CrossRef]
8. Shui, X.; Wang, W.; Ma, W.; Yang, C.; Zhou, K. Mechanism by Which High Foliar Calcium Contents Inhibit Sugar Accumulation in Feizixiao Lychee Pulp. *Horticulturae* **2022**, *8*, 1044. [CrossRef]
9. Wen, W.; Fang, H.; Yue, L.; Khalil-Ur-Rehman, M.; Huang, Y.; Du, Z.; Yang, G.; Xu, Y. RNA-Seq Based Transcriptomic Analysis of Bud Sport Skin Color in Grape Berries. *Horticulturae* **2023**, *9*, 260. [CrossRef]
10. Sun, D.; Zhang, L.; Su, J.; Yu, Q.; Zhang, J.; Fang, W.; Wang, H.; Guan, Z.; Chen, F.; Song, A. Genetic Diversity and Genome-Wide Association Study of Architectural Traits of Spray Cut Chrysanthemum Varieties. *Horticulturae* **2022**, *8*, 458. [CrossRef]
11. Zhou, Y.; Qiao, Y.; Ni, Z.; Du, J.; Xiong, J.; Cheng, Z.; Chen, F. GDS: A Genomic Database for Strawberries (*Fragaria* spp.). *Horticulturae* **2022**, *8*, 41. [CrossRef]
12. Qian, J.; Liu, W.; Shi, Y.; Zhang, M.; Wu, Q.; Chen, K.; Chen, W. C-CorA: A Cluster-Based Method for Correlation Analysis of RNA-Seq Data. *Horticulturae* **2022**, *8*, 124. [CrossRef]
13. Jin, C.; Liu, B.; Ruan, J.; Yang, C.; Li, F. Development of InDel Markers for *Gypsophila paniculata* Based on Genome Resequencing. *Horticulturae* **2022**, *8*, 921. [CrossRef]

 *horticulturae*

Article

Comparative Analyses of Superoxide Dismutase (SOD) Gene Family and Expression Profiling under Multiple Abiotic Stresses in Water Lilies

Wasi Ullah Khan [1,2,3], Latif Ullah Khan [1], Dan Chen [4] and Fei Chen [1,2,3,*]

1 Sanya Nanfan Research Institute, Hainan University, Sanya 572025, China; wasibiotechnologist@gmail.com (W.U.K.)
2 School of Tropical Crops, Hainan University, Haikou 570228, China
3 Hainan Yazhou Bay Seed Laboratory, Sanya 572025, China
4 Hainan Institute of Zhejiang University, Sanya 572025, China
* Correspondence: feichen@hainanu.edu.cn

Abstract: Plants in their natural habitat frequently face different biotic and abiotic stresses, which lead to the production of reactive oxygen species (ROS) that can damage cell membranes, cause peroxidation and deterioration of macromolecules, and ultimately result in cell death. Superoxide dismutase (SOD), a class of metalloenzymes, is primarily found in living organisms and serves as the principal line of defense against ROS. The *SOD* gene family has not yet been characterized in any species of water lily from the genus *Nymphaea*. The present study aims to conduct a genome-wide study to discover *SOD* genes in four representative water lily species. In our present comparative study, we discovered 43 *SOD* genes in the genomes of four water lily species. The phylogenetic investigation results revealed that *SOD* genes from water lily and closely related plant species formed two distinct groups, as determined by their binding domains with high bootstrap values. Enzymatic ion-binding classified the *SOD* gene family into three groups, *FeSOD*, *Cu/ZnSOD*, and *MnSOD*. The analysis of gene structure indicated that the *SOD* gene family exhibited a relatively conserved organization of exons and introns, as well as motif configuration. Moreover, we discovered that the promoters of water lily *SODs* contained five phytohormones, four stress-responsive elements, and numerous light-responsive *cis*-elements. The predicted 3D protein structures revealed water lily *SODs* form conserved protein dimer signatures that were comparable to each other. Finally, the RT-qPCR gene expression analysis of nine *NcSOD* genes revealed their responsiveness to heat, saline, cold, cadmium chloride, and copper sulphate stress. These findings establish a basis for further investigation into the role of the *SOD* gene family in *Nymphaea colorata* and offer potential avenues for genetic enhancement of water lily aquaculture.

Keywords: *NcSOD* genes; sequence analysis; abiotic stresses; expression pattern

Citation: Khan, W.U.; Khan, L.U.; Chen, D.; Chen, F. Comparative Analyses of Superoxide Dismutase (SOD) Gene Family and Expression Profiling under Multiple Abiotic Stresses in Water Lilies. *Horticulturae* 2023, 9, 781. https://doi.org/10.3390/horticulturae9070781

Academic Editor: Hongmei Du

Received: 24 May 2023
Revised: 11 June 2023
Accepted: 14 June 2023
Published: 8 July 2023

1. Introduction

Plants residing in their native environments often encounter a multitude of stress factors, including elevated salinity levels, prolonged drought periods, high temperatures, and the presence of heavy metals. These stressors exert notable influences on the plants' overall growth, development, and productivity [1,2]. Under stress, plants adapt their homeostatic mechanisms by generating an excess of reactive oxygen species (ROS) within their cells. ROS are primarily generated in various parts of the plant cell, including the plasma membrane, peroxisomes, apoplast, cell walls, endoplasmic reticulum, mitochondria, and chloroplasts [3]. These ROS are toxic free radicals that can oxidize proteins, damage cell membranes, and cause harm to DNA when formed in excessive amounts [4,5]. The occurrence of stresses in plants inevitably leads to the production of ROS, such as peroxide radicals (HOO−), hydrogen peroxide (H_2O_2), and singlet oxygen ($^1O_2$). For instance, several potent scavengers of active oxygen have the ability to mitigate environmental stresses by regulating the expression of genes belonging to enzyme reaction families, such as superoxide dismutase (*SOD*), catalase (*CAT*), peroxidase (*POD*), glutathione peroxidase (*GPX*), and peroxidase (*PrxR*) [6–9]. Plants have developed effective and intricate antioxidant defense mechanisms comprising a variety of enzymatic and non-enzymatic antioxidants to manage the harmful effects of ROS. Among various antioxidant enzymes, *SOD*, a group of metalloenzymes, is predominantly present in living creatures. In managing environmental cues, *SOD*s show a vital part in the physio-biochemical processes of plants via acting as the primary defense against ROS [3]. In plants, *SOD* enzymes are encoded by a family of genes that are classified based on their metal cofactors: (*FeSOD*), (*Cu/ZnSOD*), (*MnSOD*), and (*NiSOD*) [10–12]. Among them *NiSOD* is predominantly found in cyanobacteria, streptomyces, and marine organisms, but has not yet been documented in plants [13,14]. Iron and manganese superoxide dismutase are mainly present in lower plants, whereas copper and zinc are found in higher plants [15]. Such *SOD*s are usually dispersed in different cell parts [16]. In the main, *Cu/ZnSOD*s are localized in the cytosol, peroxisomes, and chloroplasts. *MnSOD*s are found inside mitochondria and *FeSOD*s are usually found in the peroxisomes and chloroplasts [17,18].

*SOD*s have been demonstrated in recent studies to secure plants against abiotic stress factors including cold, drought, heat, salinity, ethylene and abscisic acid [19–22]. Various findings have demonstrated that *SOD* genes might be transcribed and induced in many plants in different stress circumstances [23,24]. In recently published articles, the *SOD* gene family under various abiotic and hormones stress situations in *Brassica napus* [25], *Zostera marina* [26], *Salvia miltiorrhiza* [27], and *Hordeum vulgare* [28] were reported. Furthermore, different stress conditions can result in varied expression patterns of diverse forms of *SOD* genes. For example, tomatoes (*Solanum lycopersicum*) exhibit specific patterns of regulation in their *SOD* genes; for instance, under salt stress, *SlSOD1* is a single gene among the nine *SlSOD* genes that shows significant upregulation, while *SlSOD2*, *SlSOD5*, *SlSOD6*, and *SlSOD8* are also regulated. However, in drought conditions, the expression levels of four genes among the nine, namely "*SlSOD2*, *SlSOD5*, *SlSOD6*, and *SlSOD8*," are observed to be high [23]. Moreover, the expression profiles of the identical *SOD* gene type varied in the presence of stress. For instance, the studies revealed that the expression of *MnSOD*s in *Arabidopsis* remained unchanged during oxidative stress, while scientists observed a considerable alteration in the expression of *MnSOD*s in *Zostera marina*, peas (*Pisum sativum*), and wheat (*Triticum aestivum*) during salinity stress [26,29–31]. The findings imply that diverse *SOD* genes unveil distinct expression patterns in reaction to varying environmental stresses. In addition, scientists have revealed that the regulation of *SOD* expression may involve various miRNAs and alternative splicing mechanisms [32,33].

Water lilies are the most significant ornamental waterscape plants in the world. It is a perennial aquatic plant of the order Nymphaeales, genus *Nymphaea* in the family *Nymphaeaceae*. *Nymphaea* (*Nymphaeaceae*), also called flowering plants, are angiosperms with large and showy flowers. There are more than 60 species in the world, mostly distributed in tropical, subtropical, and temperate regions. They have curved or rounded and variously notched waxy-coated leaves on long stalks, usually grow on the water, and surround flowers. Each plant can grow approximately 70 to 80 flowers. The aquaculture of water lily, flowers can also be used as fresh cut flowers, in tea, in dried flower crafts, and in textile production. Water lilies have garnered significant attention from scientists, researchers, and entrepreneurs worldwide due to their immense economic, medicinal, and cultural value. While these plants hold great importance in phylogenetic research, the accessibility to comprehensive genetic and genomic information remains somewhat limited [34]. Since we released the first water lily (*Nymphaea colorata*) genome sequence in 2020 [35], the *SOD* gene family has not yet been discovered in any species of water lily.

To fill in this gap, the present study aims to conduct a comparative genome-wide study to discover *SOD* genes in representative water lily species genomes. Our analysis included the characterization of their phylogenetic connections, conserved motifs, cis-elements, gene structure, expression analysis, protein-protein interaction and 3D structures, in order to decode its structural characteristics and functions under stresses.

2. Materials and Methods

2.1. Retrieval of SOD Gene Family in Water Lily Species

To investigate the *SOD* gene family in water lilies, the Blastp search method was employed, utilizing the *Arabidopsis SOD* sequence as a query to examine the entire genome of each water lily species individually [36]. In our study, we used two methodologies, protein blast and the hidden Markov model (HMM), to detect *SOD* genes in four water lily species in which the genome sequences of two species *Nymphaea colorata*, and *Nymphaea thermarum*, are available online; while the other two, i.e., *Nymphaea minuta*, and *Nymphaea mexicana*, have unpublished genome sequences. For BLASTP, we utilized eight *A. thaliana SOD* amino acid sequences (AT1G08830.1/*AtCSD1*, AT2G28190.1/*AtCSD2*, AT5G18100.1/*AtCSD3*, AT4G25100.1/*AtFSD1*, AT5G51100.1/*AtFSD2*, AT5G23310.1/*AtFSD3*, AT3G10920.1/*AtMSD1*, and AT3G56350.1/*At00MSD2*) as the query, with an e-value set to 1×10^{-5}. We obtained these eight *AtSODs* amino acid sequences from the *Arabidopsis* genome database TAIR (http://www.arabidopsis.org/ accessed on 5 January 2023). To identify conserved domains *SOD_Cu* (PF00080), *SOD_Fe_C* (PF00081) and *SOD _Mn* (PF02777), we performed scans of specific amino acid sequences using the web resources Pfam (Pfamv34.0-19178pSSMs) protein domain database (http://pfam.xfam.org/ accessed on 8 January 2023), and SMART (http://smart.embl-heidelberg.de/ accessed on 9 January 2023) [37].

2.2. Analysis of Physicochemical Features and Subcellular Localization

In order to anticipate the physicochemical characteristics of *SOD* proteins in water lily species, such as amino acid count (A.A), theoretical isoelectric point (pI), molecular weight (kDa), Pfam Domains, Functional annotations, and grand average of hydropathicity (GRAVY), we employed the ProtParam website accessible at http://web.expasy.org/protparam/ accessed on 13 January 2023 [38]. In order to predict the subcellular localization of *SOD* proteins, we employed the WoLF PSORT (https://wolfpsort.hgc.jp/ accessed on 17 January 2023) [39] and ProtComp 9.0 server (http://linux1.softberry.com/ accessed on 18 January 2023) [27].

2.3. Phylogenetic Analysis and Conserved Motifs

We generated a phylogenetic tree to investigate the evolutionary connections of the water lily *SOD* gene family. This was done using protein sequences from *N. colorata*, *N. mexicana*, *N. minuta*, *N. thermarum*, and *A. trichopoda* (*Amborella trichopoda*). First, we aligned the protein sequences using MUSCLE with default parameters [39]. Then, we used the MEGA11 software (https://megasoftware.net/home accessed on 27 January 2023) to make a phylogenetic tree using the Neighbor-Joining algorithms. We assigned confidence levels to each branch of the tree using bootstrap tests (1000). We used the MEME server (https://meme-suite.org/meme/db/motifs accessed on 2 February 2023) with default settings to detect conserved motifs in the protein sequences of water lily *SODs* [40].

2.4. Prediction of the Cis-Regulatory Elements in the Promoter

To examine the potential cis-elements in the promoters of water lily *SODs*, we obtained the 2 Kb sequence upstream of start codons from each species' genome. Then we used the PlantCARE website (http://bioinformatics.psb.ugent.be/webtools/plantcare/html/ accessed on 6 February 2023) [41], to examine the promoter sequence of each gene, and then we created figure using TBtools (V 1.068).

2.5. Examination of the 3D Structures of Water Lily SOD Proteins

The comprehension of a protein's functions requires a detailed understanding of its 3D structure. To this end, we analyzed the predicted 3D structures of four water lily species, using the online servers SOPMA and SWISS MODEL, both available through ExPASy at https://www.expasy.org/ accessed on 15 February 2023. Finally, we applied the UCSF Chimera visualization tool to visualize the 3D structures [42].

2.6. Analysis of SOD Gene Structure of Water Lily

The Gene Structure Display Server (GSDS; http://gsds.cbi.pku.edu.cn/index.php accessed on 19 February 2023) program was used to visualize the organization of exons and introns in the *SOD* genes of water lilies [27].

2.7. Analysis of Potential Protein Interaction

To make the *SOD* protein interaction network, we utilized STRING 11.0 (https://string-db.org/cgi/input.pl accessed on 4 March 2023) tool for this purpose [43].

2.8. Expression Profiling of NcSOD Genes in Pollen and Ovule

The expression pattern of the *NcSOD* gene family was obtained from our own RNA-seq raw data (unpublished). All 9 *NcSOD* genes' expression levels were explored in 1 day mature pollen, and 0, 1, 2, and 3 days mature ovule. The expression heatmap was constructed using TBtools (V 1.068, https://github.com/CJ-Chen/TBtools/ accessed on 9 March 2023), in which the color bar from light yellow to a dark red exhibited less to high levels of expression, and light blue to dark blue shows less or no expression of *NcSOD* genes.

2.9. Plant Materials and Abiotic Stresses

To examine how *NcSOD* members respond to different abiotic stresses, we grew *N. colorata* mature plants in water tub filled with tap water under an open environment. The plants were then subjected to various stress treatments, including 250 mM NaCl, 200 μM CuSO$_4$, and 2.5 mM CdCl$_2$, as well as being treated with cold stress at 8 °C and heat stress at 42 °C. Each treatment was performed with three independent biological replicates, and each sample was collected from at least five individual plants. Leaves from the plantlets were collected at 0, 2, 4, and 6 h for the salt, heat, cold, and heavy metal stress experiments. After collection, all samples were instantly frozen in liquid nitrogen and preserved at −80 °C until total RNA isolation.

2.10. RNA Isolation and Real-Time Quantitative PCR Expression Analysis

RNA extraction was performed using the RNAprep Pure Plant Kit (TIANGEN, Beijing, China). The concentration of the samples was determined using a NanoDrop 2000 C spectrophotometer (Thermo Fisher Scientific, Waltham, MA, USA). The genomic DNA, was removed by DNase I treatment, followed by cDNA synthesis using the QuantiTect Reverse Transcription Kit (Qiagen, Shanghai, China). The RT-qPCR expression was performed on the Roche LightCycler 96 PCR system, following the recommended guidelines for the ChamQTM SYBR RT-qPCR Master Mix (Vazyme Biotech Co., Ltd., Sanya, China). For each RT-qPCR, the expression level of the actin gene in *N. colorata* was employed to standardize the RNA samples. Three biological replicates for each sample were employed for RT-qPCR, analysis with actin as internal control. Gene-specific primers for *NcSODs* and Nc-actin in the RT-qPCR system were designed using the online NCBI Primer-BLAST Program and their specificity was confirmed using the Oligo Calculator online tool (http://mcb.berkeley.edu/labs/krantz/tools/oligocalc.html accessed on 5 April 2023). The primers were synthesized by NANSHAN BIOTECH, (Sanya), and listed in (Table S1). The $2^{-\Delta\Delta CT}$ method was used to analyze the RT-qPCR gene expression data [44].

3. Results

3.1. Genome-Wide Analysis of SOD Gene Family in Four Water Lily Species

In this comparative study, we discovered 43 *SOD* genes in the genomes of four water lily species. Protein sequences of eight *A. thaliana* (*AtSODs*) were used as queries and removed the repetitive redundant sequences (Table S2), and 9–15 genes were obtained for each species, for example three diploid water lilies *N. colorata* (9 *SODs*), *N. thermarum* (10 *SODs*), *N. minuta* (9 *SODs*), and a tetraploid water lily *N. mexicana* (15 *SODs*) (Table 1; Table S3). After conducting domain scrutiny, we identified 15 proteins with a *Cu/Zn-SODs* domain (Pfam; 00080), 19 with a *Fe-SODs* domain (Pfam; 00081), and 9 with a *Mn-SODs* domain (Pfam; 02777) in water lily species. These results are consistent among all species and contain all *SOD* genes and domains with sub-families.

Table 1. Characteristics of the *SOD* genes from four water lily species.

Plant Species	Transcript ID	Gene Name	Pfam Domains	Protein Length (A.A)	Functional Annotations	MW (kDa)	pI	Subcellular-Localization	GRAVY
Nymphaea colorata	Nycol.F01435	NcCSD1	PF00080	161	Cu/Zn-SOD	16.337	7.19	Cytoplasmic	−0.121
	Nycol.E01211	NcCSD2	PF00080	170	Cu/Zn-SOD	17.134	5.72	Cytoplasmic	−0.108
	Nycol.L00920	NcCSD3	PF00080	223	Cu/Zn-SOD	22.902	5.96	Chloroplast	0.025
	Nycol.A03620	NcFSD1	PF00081	198	Fe-SOD	23.178	6.23	Mitochondrial	−0.409
	Nycol.B01638	NcFSD2	PF00081	275	Fe-SOD	31.321	9.27	Chloroplast	−0.328
	Nycol.D01220	NcFSD3	PF00081	239	Fe-SOD	26.823	5.66	Chloroplast	−0.437
	Nycol.D01221	NcFSD4	PF00081	308	Fe-SOD	35.007	5.31	Chloroplast	−0.618
	Nycol.J01291	NcFSD5	PF00081	249	Fe-SOD	27.667	7.86	Chloroplast	−0.335
	Nycol.L00218	NcMnSD1	PF02777	260	Mn-SOD	29.073	7.71	Mitochondrial	−0.332
Nymphaea mexicana	NM3G27.30	NMCSD1	PF00080	135	Cu/Zn-SOD	13.478	5.45	Cytoplasmic	−0.246
	NM27G140.23	NMCSD2	PF00080	177	Cu/Zn-SOD	18.221	6.82	Cytoplasmic	−0.071
	NM26G14.21	NMCSD3	PF00080	117	Cu/Zn-SOD	12.166	6.41	Cytoplasmic	0.156
	NM16G174.41	NMCSD4	PF00080	135	Cu/Zn-SOD	13.478	5.45	Cytoplasmic	−0.246
	NM15G96.16	NMCSD5	PF00080	253	Cu/Zn-SOD	26.721	4.68	Cytoplasmic	−0.091
	NM15G40.65	NMFSD1	PF00081	231	Fe-SOD	25.785	6.8	Mitochondrial	−0.298
	NM9G74.16	NMFSD2	PF00081	259	Fe-SOD	29.11	6.39	Chloroplast	−0.324
	NM5G64.29	NMFSD3	PF00081	259	Fe-SOD	29.12	7.02	Chloroplast	−0.306
	NM6G47.44	NMFSD4	PF00081	306	Fe-SOD	34.951	5.39	Chloroplast	−0.588
	NM5G249.61	NMFSD5	PF00081	297	Fe-SOD	33.521	5.34	Chloroplast	−0.607
	NM14G120.45	NMMnSD1	PF02777	259	Mn-SOD	28.748	8.7	Mitochondrial	−0.284
	NM8G53.44	NMMnSD2	PF02777	271	Mn-SOD	31.044	9.05	Chloroplast	−0.302
	NM7G151.45	NMMnSD3	PF02777	247	Mn-SOD	28.178	7.09	Chloroplast	−0.323
	NM5G249.14	NMMnSD4	PF02777	236	Mn-SOD	26.48	5.38	Cytoplasmic	−0.408
	NM6G47.40	NMMnSD5	PF02777	240	Mn-SOD	26.914	5.39	Cytoplasmic	−0.414
Nymphaea minuta	Nmin13g00218	NminCSD1	PF00080	160	Cu/Zn-SOD	16.397	6.86	Cytoplasmic	−0.15
	Nmin06g00456	NminCSD2	PF00080	122	Cu/Zn-SOD	12.78	8.88	Cytoplasmic	−0.115
	Nmin08g00862	NminCSD3	PF00080	320	Cu/Zn-SOD	33.427	5.13	Cytoplasmic	0.104
	Nmin08g01659	NminFSD1	PF00081	231	Fe-SOD	25.827	6.8	Mitochondrial	−0.348
	Nmin00g05403	NminFSD2	PF00081	237	Fe-SOD	26.373	7.17	Cytoplasmic	−0.278
	Nmin02g01006	NminFSD3	PF00081	259	Fe-SOD	29.041	7.71	Mitochondrial	−0.303
	Nmin05g01512	NminFSD4	PF00081	362	Fe-SOD	40.905	5.15	Cytoplasmic	−0.552
	Nmin01g00323	NminMnSD1	PF00081	230	Mn-SOD	24.608	7.96	Cytoplasmic	−0.155
	Nmin05g01511	NminMnSD2	PF02777	238	Mn-SOD	26.783	5.51	Cytoplasmic	−0.421
Nymphaea thermarum	KAF3774564.1	NtCSD1	PF00080	161	Cu/Zn-SOD	16.75	5.36	Cytoplasmic	0.05
	KAF3777549.1	NtCSD2	PF00080	267	Cu/Zn-SOD	28.061	9.47	Mitochondrial	−0.023
	KAF3779999.1	NtCSD3	PF00080	267	Cu/Zn-SOD	28.053	4.86	Cytoplasmic	−0.122
	KAF3781552.1	NtCSD4	PF00080	158	Cu/Zn-SOD	17.099	5.51	Cytoplasmic	−0.124
	KAF3786936.1	NtMnSD1	PF02777	248	Mn-SOD	27.721	7.86	Mitochondrial	−0.352
	KAF3791217.1	NtFSD1	PF00081	274	Fe-SOD	31.403	9.3	Mitochondrial	−0.343
	KAF3782520.1	NtFSD2	PF00081	259	Fe-SOD	28.995	7.71	Mitochondrial	−0.285
	KAF3793027.1	NtFSD3	PF00081	306	Fe-SOD	34.987	5.39	Mitochondrial	−0.635
	KAF3793028.1	NtFSD4	PF00081	238	Fe-SOD	26.841	5.66	Cytoplasmic	−0.415
	KAF3779312.1	NtFSD5	PF00081	351	Fe-SOD	39.341	10.15	Cytoplasmic	−0.468

The biochemical and physiological characteristics of all *SODs* were investigated by calculating various parameters (Table 1). Protein length in the four representative water lily species ranged from 117 to 362 amino acids. Consequently, the molecular weight was found to be from 12.166 to 40.905 kDa. Most of the species investigated in this study displayed acidic properties, exhibiting pI values among 4.68 to 10.15. Furthermore, the predicted GRAVY values of *SOD* proteins were negative, showing that they are hydrophilic.

According to the subcellular localization prediction, the majority of water lily *CSDs* were found in the cytoplasm, while only a few localized in the chloroplast. Water lily *FSDs* and *MnSDs* were specifically located in the chloroplast and mitochondria, respectively (Table 1).

Furthermore, an NCBI domain search was conducted to perform domain-based analysis on all *SOD* proteins to acquire data and TBtools was then used to make the domain structures. As a result of this analysis, the presence of the *SOD* family was verified on all chosen protein sequences (Figure 1).

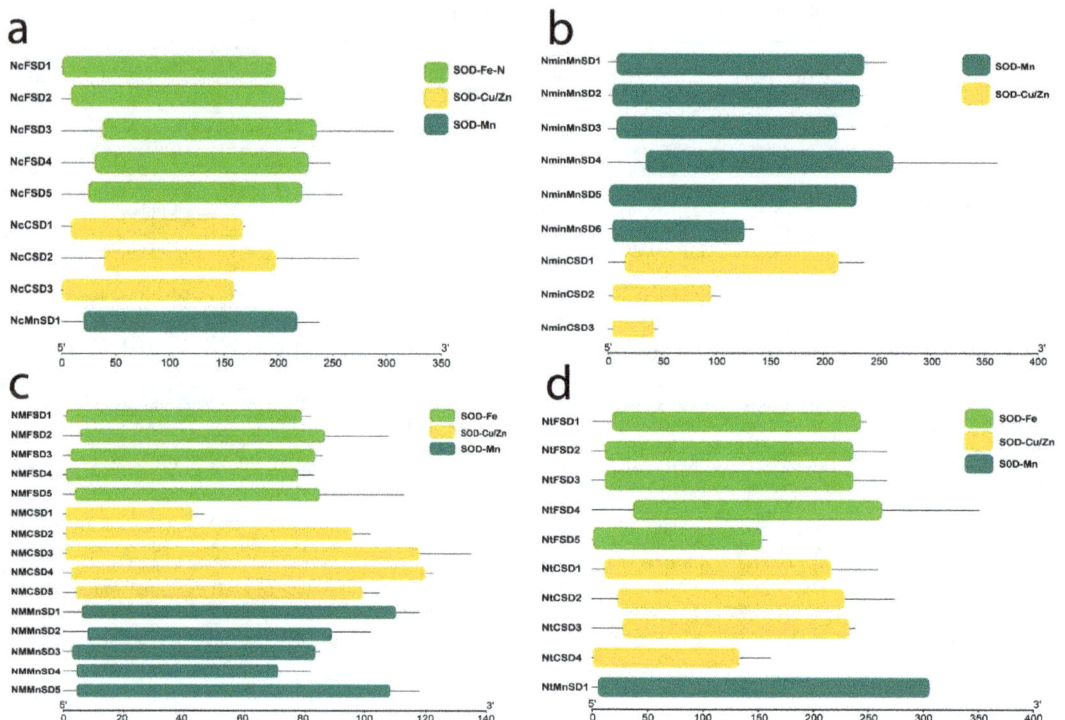

Figure 1. Symbolic *SOD* domain structures of four water lily species: (**a**) *Nymphaea colorata*; (**b**) *Nymphaea minuta*; (**c**) *Nymphaea mexicana*; and (**d**) *Nymphaea thermarum*. Among four species only *Nymphaea minuta* contains Mn and Cu/Zn, while others constitute all three subfamilies of *SOD*.

3.2. Phylogenetic Relationships and Conserved Motif Analysis in Representative Water Lily Species

To uncover the evolutionary relations among *SODs* in various plant species, a phylogenetic tree was created using the complete protein sequences. The present research investigated the evolutionary relationships among genes of *NcSODs*, *NtSODs*, *NminSODs*, *NMSODs*, and *AmtSODs*. By considering the domains (*Fe-SODs*, *Cu/Zn-SODs*, and *Mn-SODs*) and analyzing a phylogenetic tree, a total of 50 *SODs* were classified into two main groups (Figure 2a). To assess the structural diversity of water lily *SOD* proteins and predict their functions, we utilized the MEME software to analyze their full-length protein sequences and identify conserved motifs. By examining conserved motifs in the *SOD* family, this analysis confirmed the categorization and evolutionary connections among *SOD* genes within the water lily species. The investigation revealed that 10 conserved motifs were present in all species (Figure 2b). The detailed information about the identified motifs, including their names, sequences, and widths is displayed in Table S4. The number of conserved motifs in *SOD* proteins ranged from two to six, and their distribution aligned with the groups. The *Cu-SOD* and *Fe-SOD* groups had only one motif in two genes (Figure 2b). Furthermore, *MnSODs* and *FeSODs* were classified in the similar group and subcluster, whereas *Cu/ZnSODs* were placed in a distinct group. Interestingly, motifs 1, 2, 4, and 9 were estimated to be particular to *Cu/Zn-SOD* whereas motifs 3, 5, 6, and 10 were exclusive to the *MnSODs* and *FeSODs* groups. In conclusion, the reliability of group arrangements was strongly supported by analyzing conserved motif patterns and phylogenetic relationships between water lily species. This suggests that water lily *SODs* proteins possess highly conserved amino acid residues within groups. Consequently, it is reasonable to infer that proteins with analogous structures and motifs may have similar efficient roles.

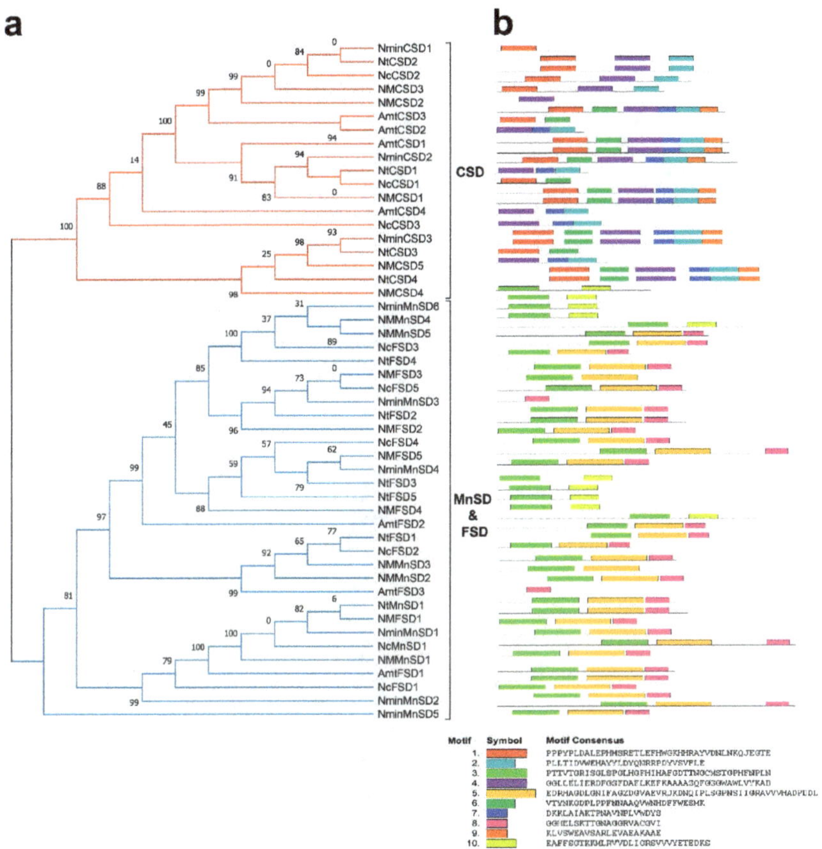

Figure 2. Classification of *SOD* genes according to their subfamilies: (**a**) A neighbor-joining phylogenetic tree; (**b**) Conserved motif analysis. The motifs supported the two subfamilies which are mentioned in tree. Reliability of group arrangements was strongly supported by analyzing conserved motif patterns and phylogenetic relationships between water lily species. Different types of motifs represented by differently colored boxes.

3.3. Analyses of Cis-Elements in Water Lily SOD Gene Promoters

Retrieving cis-regulatory elements from the promoter regions of water lily *SODs* enabled the differentiation of gene functions and regulatory roles. By using the PlantCARE database, an analysis was conducted on the 2 kb upstream region from the start codon of individual *SOD* gene. Based on current findings, the cis-elements were classified into three categories: light related, stress related, and hormone response elements (Figure 3).

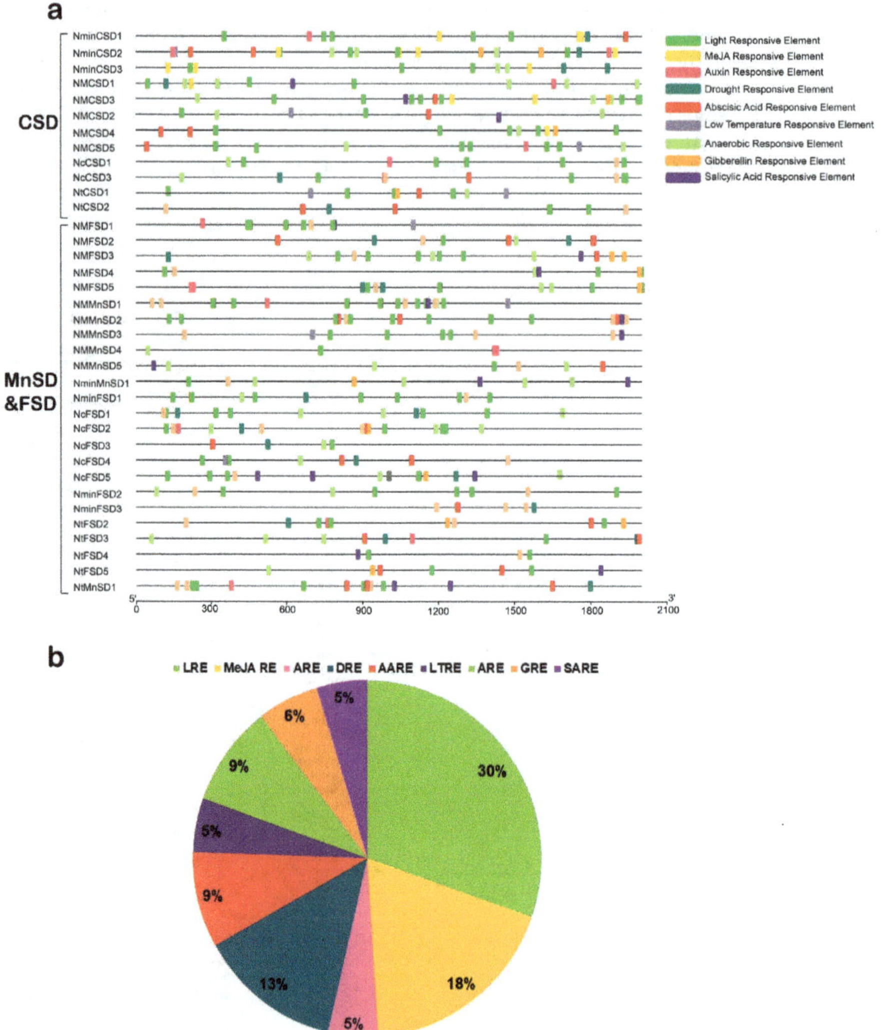

Figure 3. Exploration of abiotic-stress related cis-regulatory components in the water lily *SOD* promoter regions. (**a**) Various hormone-associated and stress-responsive elements are explored. (**b**) The size of pie chart corresponds to the ratio of the respective promoter element. Cis-elements that share functional similarity are represented by the same colors.

In this study, five phytohormone-associated elements (Salicylic acid (SA), Gibberellin, Auxin, Abscisic acid (ABA), and Methyl jasmonate (MeJA), were detected, including ABRE, TCA-components, TGACG-and CGTCA-motif, TATC-box, P-box, etc. Moreover, four stress-responsive components (drought, low-temperature, light, and anaerobic) were recognized, such as LTR, ARE, TCT-motif, LAMP-element, MBS, etc (Figure 3). Generally many light reactive components were detected to be extensively dispersed in same group species, and demonstrating the important part of water lily *SODs* in response to light stress. Comparative analysis of the findings revealed that the *SOD* promoter cis-elements in water lily species can exhibit a significant response to abiotic stresses and can play a role in regulating plant growth and development and stress response.

3.4. 3D Structure Analysis of Water Lily SOD Proteins

The examination of a protein's structure holds a great importance in understanding its function. We used SWISS MODEL and SOPMA online tools with the default search options, to predict the 3D structures of proteins. This study involved the prediction of three-dimensional models for four water lily proteins. The generated models were downloaded for the purpose of visualizing the 3D structures. The helices are represented by yellow, while the sheets or strands are represented by green (Figure 4). Proteins that fall under specific groups exhibit related structural evenness. The *MnSD* and *FSD* subfamilies share a nearly identical structure, with an equal amount of helices and sheets. Similarly, proteins in the *CSD* subfamily also possess an analogous structure.

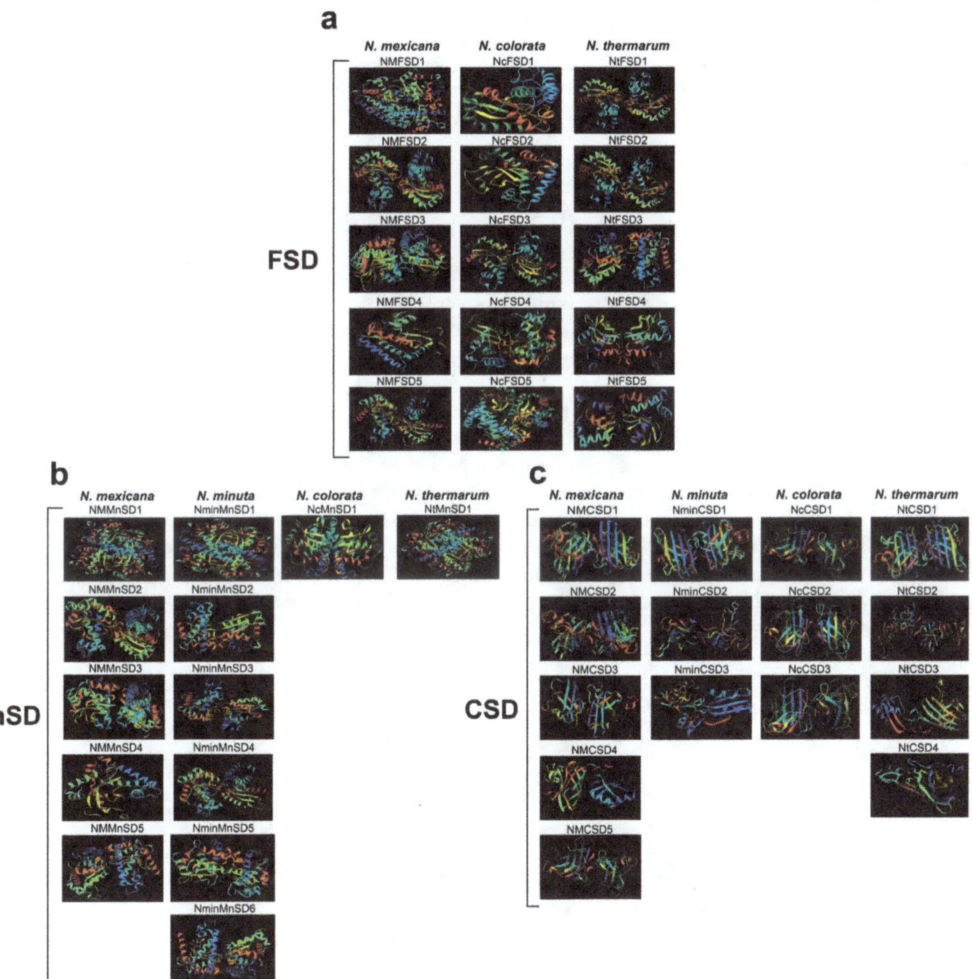

Figure 4. The 3D structures of four water lilies' (*Nymphaea colorata*, *Nymphaea thermarum*, *Nymphaea minuta*, and *Nymphaea mexicana*) *SOD* proteins categorized based on their sub families. (**a**) Represents the 3D structures of *FSD* subfamily; (**b**) Represents the 3D structures of *MnSD* subfamily; (**c**) Represents the 3D structures of *CSD* subfamily in all species. The final models are displayed, with diverse colors representing various sheets, domains, and helices. Note: In group (**a**) the *Nymphaea minuta* has no *FSD* subfamily.

3.5. Analysis of Exon-Intron Structure of NcSOD

The analysis of exon-intron structure of *NcSOD* genes was performed to elucidate the structural characteristics of species (Figure 5). *NcSOD* genes displayed varied exon-intron organizational patterns, with introns ranging from 5 (*NcFSD5*) to 9 (*NcFSD4*). The number of exons in *NcSOD* differs from 1 (*NcFSD1*) to 9 (*NcFSD2*). In one *NcFSD1* gene, introns are absent, and there is only one exon. The gene structure investigation revealed that the *SOD* gene family displayed a relatively conserved exon/intron organization.

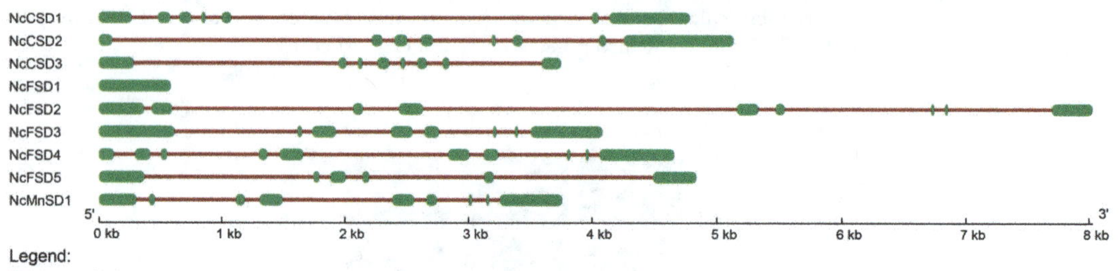

Figure 5. Gene structure of *NcSOD* shows conserved exon/intron organization.

3.6. Expression Examination of NcSOD Genes in Reproductive Organs

The *SOD* gene family has a crucial role in plant growth, development, and response to stress. In order to investigate their specific biological functions in *N. colorata*, we observed the expression patterns of the 9 *NcSOD* genes in pollen and ovules using our own unpublished RNA-seq raw data. Under normal growth conditions, not all predicted genes in the *N. colorata SOD* family were expressed. Our analysis revealed that *NcFSD3*, *NcFSD5*, and *NcMnSD1* were highly expressed in ovules at 0, 1, 2, and 3 days, while showing relatively lower expression in pollen on day 1 (Figure 6; Table S5). *NcCSD1* and *NcCSD2* were moderately expressed in ovules and pollen throughout all days. *NcFSD2* and *NcFSD4* showed a moderate expression in ovule but exhibited no expression in pollen. Both *NcFSD1* and *NcCSD3* showed no expression levels in both ovules and pollen. Generally, results exhibited that genes from all three subfamilies, i.e., Fe, Mn and Cu, play essential roles in *N. colorata* reproduction, growth, and development.

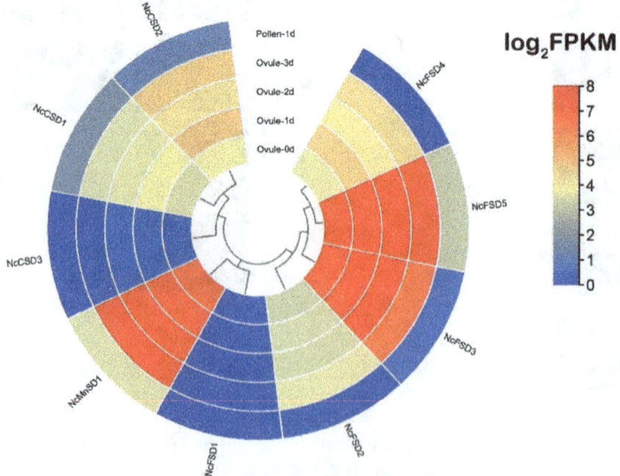

Figure 6. Expression of the *NcSOD* genes was analyzed in pollen and ovule samples at four different time-points: 0 d, 1 d, 2 d, and 3 d. The expression bar from light blue to dark blue shows less or no expression of *NcSOD* genes. The light yellow to a dark red color exhibited less to high level of expression of these genes.

3.7. Potential NcSOD Protein–Protein Interaction

The potential *NcSOD* protein–protein interaction was analyzed via "STRING"11.0 (https://string-db.org/cgi/input.pl accessed on 25 April 2023). As shown in Figure 7, among nine *NcSOD* genes, seven *SOD* proteins participate in strong interaction networks. Interestingly, we observed that different proteins co-regulate each other to respond to stress conditions. For example, *NcCSD3, NcFSD1* and *NcFSD4* are upregulated after 2 h under cold stress (Figure 8c). In water lilies, they potentially exert a regulatory function by forming protein complexes to improve cold tolerance and cope with various stresses.

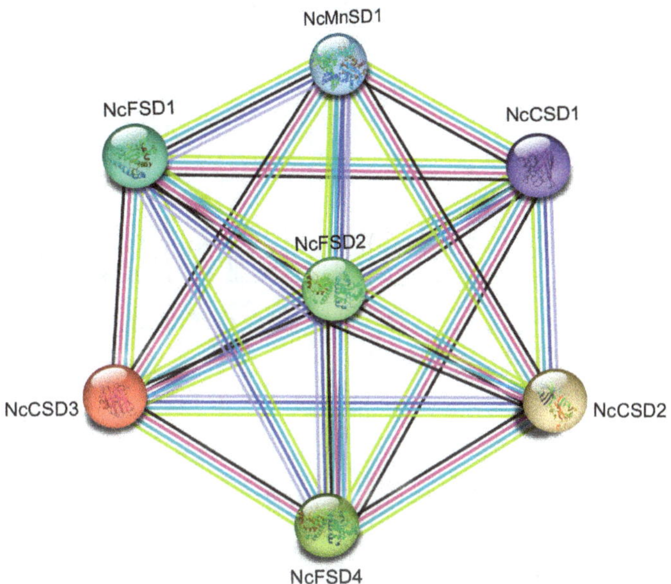

Figure 7. Protein interaction linkage among the seven *SOD* genes from *Nymphaea colorata*. Different colored lines show the interaction of the genes.

3.8. Real-Time Quantitative PCR (RT-qPCR) Analysis of NcSOD Genes under Abiotic Stresses

In order to know the function of *SODs*, we employed RT-qPCR to examine the expression patterns of the *SOD* gene under various stress conditions like salinity, heat, cold, and heavy metals (copper sulphate and cadmium chloride). Substantial variations were perceived in the expression levels of the *NcSOD* genes across various treatments, indicating a complex and dynamic nature of their expression patterns.

Salt treatment strongly induced the expression of all *NcCSDs*, peaking at 2 and 4 h. Our study found high expression of *NcCSDs* at 6 h, suggesting its involvement in salt response in *N. colorata*. Additionally, *NcMnSD1, NcFSD1, NcFSD2,* and *NcFSD5* were strongly induced and highly expressed under salt stress, implying their potential participation in the salt stress response (Figure 8a).

During the heat stress condition, the levels of expression of all *NcSOD* genes were upregulated at both 2 h and 4 h, with the exception of *NcFSD4*, which initially showed a decrease at 2 h and then an increase at 4 h. Following the 6 h treatment, the expression levels of the various genes showed variation (Figure 8b).

Under the cold treatment, distinct expression profiles were observed among all *NcSODs*. *NcCSD3, NcFSD1,* and *NcFSD4* exhibited upregulated expression at almost all time points, and reached their maximum expression at 2 and at 4 h, while *NcCSD1* and *NcCSD2* expression was slightly low. On the other hand, the remaining members showed downregulated expression (Figure 8c).

NcSOD genes showed a positive response against heavy metals. In response to the copper sulphate treatment, the expressions of *NcSOD* genes exhibited variations at different time points. *NcCSD3* and *NcFSD1* displayed high expression at 2, 4 and 6 h, while other genes were low-expressed and different levels of expression were recorded. Furthermore, the highest expression levels for all genes were observed at the 6 h treatment (Figure 8d). During the cadmium chloride treatment, all *NcSOD* genes were upregulated at 2 h. Notably; *NcCSD3* exhibited consistently high expression levels across all time points, as shown in the figure. At the 4 h and 6 h treatment, all genes experienced a gradual increase and demonstrated robust expression levels (Figure 8e). These findings can enhance our comprehension of *NcSOD* genes across various environmental conditions.

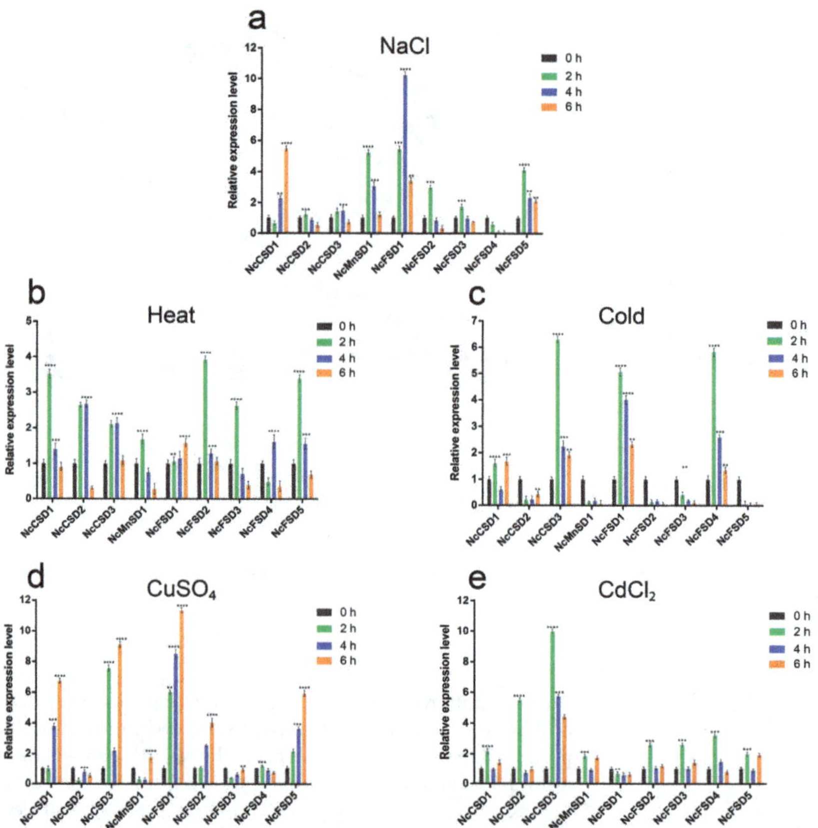

Figure 8. RT-qPCR analysis of the expression patterns of *NcSOD* genes in the leaves under various abiotic stresses: (**a**) salt; (**b**) heat; (**c**) cold; (**d**) $CuSO_4$; and (**e**) $CdCl_2$ stress (0 (CK), 2, 4, and 6 h). Data presented as means, $\pm$ standard error, $n = 3$; statistically significant differences are exhibited by asterisks ($p \leq 0.05$), according to the LSD test.

4. Discussion

Water lilies, with their significant ornamental, economic, medicinal, and cultural value, face challenges stemming from various abiotic stressors. However, through a combination of scientific research, technological innovations, and sustainable practices, we can optimize the growth and production of water lilies while preserving their aesthetic and functional benefits [45]. *SODs* have been demonstrated in recent studies to secure plants against abiotic stress factors including cold, drought, heat, salinity, ethylene, and abscisic acid [20–25].

In last several years, various plant species have been found to contain *SOD* family genes. For example, the aquatic sea grass (*Zostera marina*) has five *SOD* genes [26], *Medicago truncatula* [17] and barley (*Hordeum vulgare*) contain seven genes [28], sorghum (*Sorghum bicolor*) has eight genes [9], tomato (*Solanum lycopersicum*) has nine genes, and grapevine (*Vitis vinifera*) has ten genes [46]. Thus, we explored this family in four representative water lily species and checked the expression analysis in *Nymphaea colorata*.

In the present study, 43 *SOD* genes were identified in four water lily species, including 19 *Fe-SODs*, 15 *Cu/Zn-SODs*, and 9 *Mn-SODs* in all species (Table 1). The genes were classified into three major groups according to their binding domain (Figure 1). The number of genes in various water lily species was similar to that in cucumbers (*Cucumis sativus*) (9) and grapes (*Vitis vinifera*) (10), but fewer genes than the polyploidy crops cotton (*Gossypium hirustum*) (18) and wheat (*Triticum aestivum*) (26). However, the number of genes that encode Fe, Cu/Zn, and *Mn-SOD* differ among various species. For instance, *N. colorata* has three *Cu/Zn-SODs*, five *Fe-SODs*, and one *Mn-SOD*. The variation in *SOD* family member number could be due to the changes in genome sizes among species.

Previous research has shown that *Cu/ZnSODs* are consistently acidic, while *Fe-MnSODs* can be acidic or basic [42]. Most of the species investigated in this study displayed acidic properties. The results of subcellular localization of *SOD* proteins revealed that *Cu/Zn-SODs* are likely to be expressed in the cytoplasm, but *Mn-SODs* and *Fe-SODs* are expressed in mitochondria and chloroplasts, respectively, consistent with previous studies on *SODs* [18]. These distinct cellular locations enable *Fe-SODs*, *Cu/Zn-SODs*, and *Mn-SODs* to collaborate with one another to maintain the balance of free radicals in cells by functioning in different cellular parts.

Previous studies have indicated that the majority of cytoplasmic and chloroplast *SODs* comprise seven introns [13]. However, in our study it was revealed that *NcSOD* had a variable amount of exons ranging from 1 to 9. Furthermore, the number of introns in *NcSOD* varied from 5 to 9 (Figure 5). Notably, Figure 5 indicates that *NcFSD1* comprises only one exon and lacks introns. The variability in the gene structure of *SODs* may arise from the mechanisms involving the insertion or deletion of exons and introns [47].

Various research studies have demonstrated that *SOD* genes from distinct species are divided into three subfamilies [12]. In our study, we examined the evolutionary connections of *SOD* proteins in *N. colorata*, *N. thermarum*, *N. minuta*, *N. mexicana*, and *A. trichopoda* which categorized within three subfamilies (Figure 2a): *Fe*, *Cu/Zn*, and *Mn-SOD*. Within the phylogenetic tree, the three subfamilies were classified into two distinct groups: *Cu/ZnSODs* and *Fe-MnSODs*. *FeSODs* and *MnSODs* were clustered together, and a high bootstrap value separated them. The water lily *SODs* exhibited a strong clustering relationship with closely related species, while showing less affinity with outgroup. This suggests that this gene family has undergone relatively conserved evolution. The presence of specific domains suggests a basis for classifying these genes and the possibility of shared ancestral genes. In cotton (*Gossypium hirustum*), the *MSD* and *FSD* families were found to have originated from a common ancestor, while the *CSD* subfamily developed independently. As a result, the two major groups expanded separately, as reported by Wang [48].

The analysis of promoters unveiled the existence of three main kinds of cis-components related to light, abiotic stress, and hormones response. Additionally, there were cis-elements associated with tissue-specific expression and developmental processes. Significant quantities of light-responsive cis-components were identified within *SOD* gene promoters, indicating the potential involvement of *SODs* in the abiotic stress response. Numerous investigations indicated the participation of *SOD* genes in the abiotic stress response across diverse plant species, including maize (*Zea mays*), *Pennisetum glaucum*, *Dendrobium catenatum*, and *Arabidopsis* [36,49–51]. Moreover, *SOD* gene promoters were found to contain a range of cis-elements linked to abiotic stress responses, including ARE, ABRE, MBS, ERE, Box-4, and TC-rich repeats. These cis-elements potentially contribute to the regulation of gene expression under diverse stress conditions. Among plant species like *Arabidopsis*, banana (*Musa paradisiaca*), rice (*Oryza sativa*), tomato (*Solanum lycopersicum*), poplar (*Populus angusti-folia*), and cotton (*Gossypium herbaceum*), the majority of *SOD* genes exhibit inducibility in response to various abiotic stresses [4,36,52–55]. Under various abiotic and hormones stress situations the *SOD* gene family were recently identified by many researchers in various different types of plants like in *Brassica napus* [25], *Zostera marina* [26], *Salvia miltiorrhiza* [27], and *Hordeum vulgare* [28].

The 3D structures of water lily *SOD* proteins remain relatively conserved, similar to conserved domains, gene structure, and phylogeny. The findings indicate that water lily *SODs* genes potentially perform diverse functions across various tissues and genotypes. These results supported earlier anticipated three dimensional structures of *SODs* in *G. arboretum* [54], sorghum (*Sorghum bicolor*) [9], rice (*Oryza sative*) [52], and in *Gossypium raimondii*. The preceding investigation demonstrated that metal ion binding active sites and the formation of conserved disulfide bonds within individual subunits contribute to protein stability, specificity, and dimerization [56].

To determine the specific expression profiles of *NcSOD* genes during various stages of development, we utilized RNA-seq data from ovules and pollen at various developmental stages. By analyzing the RNA-sequencing data from *N. colorata*, and examined the expressions of the 9 *NcSOD* genes in pollen and ovules at different days post-anthesis. Our analysis revealed that *NcFSD3*, *NcFSD5*, and *NcMnSD1* were highly expressed in ovules at 0, 1, 2, and 3 days, while showing relatively lower expression in pollen on day 1 (Figure 6). While *NcFSD1* and *NcCSD3* showed no expression levels in both ovules and pollen that are in agreement with earlier findings [46].

The RT-qPCR analyses offer valuable insights into the potential role of *NcSODs* in reaction to diverse stresses. Our research revealed significant changes in the expression levels of nine *NcSODs* in varied stress environments, suggesting their crucial regulatory role in response to stress and possible functional interconnections. Overexpressing *Cu/ZnSODs* improved salinity stress resistance in *Triticum aestivum*, *Oryza sativa*, *Puccinellia tenuiflora*, and *Arabidopsis* [57,58]. Salt treatment strongly induced the expression of all *NcCSDs*, peaking at 2 h and 4 h. Our study found high expression of *NcCSD* at 6 h, suggesting its involvement in salt response in *Nymphaea colorata*. Additionally, *NcMnSD1*, *NcFSD1*, *NcFSD2*, and *NcFSD5* were strongly induced and highly expressed under salt stress, implying their potential participation in the salt stress response, similar to *NcCSDs*, in *N. colorata* (Figure 8a). Particularly, most *NcSOD* genes exhibited upregulation throughout heat treatment, with some displaying analogous expression patterns (Figure 8b). During cold treatment, distinct expression profiles were observed among all *NcSODs*. *NcCSD3*, *NcFSD1*, and *NcFSD4* exhibited upregulated expression at almost all time points, and reached their maximum expression at 2 h and at 4 h, while *NcCSD1* and *NcCSD2* were slightly expressed. On the other hand, the remaining members showed down-regulated expression (Figure 8c). These conclusions are consistent with previous findings, which reported a notable increase in *SOD* activity in rapeseed (*Brassica napus*) under cold stress conditions [59]. *NcSOD* genes showed a positive response against heavy metals. In response to the copper sulphate treatment, the expressions of *NcSOD* genes exhibited variations at different time points (Figure 8d). During the cadmium chloride treatment, all *NcSOD* genes were upregulated at 2 h. Notably, *NcCSD3* exhibited consistently high expression levels across all time points, as showed in the (Figure 8e). Nevertheless, certain genes within the nine *NcSODs* exhibited a pattern of initially increasing and subsequently decreasing expressions in response to both heavy metal treatments. Similar results were also reported in reaction to heavy metals treatment in several plants [27]. However, experimental verification is still needed to fully elucidate the roles of *NcSODs* regulatory networks, and their interaction mechanism, under different abiotic stresses.

5. Conclusions

In conclusion, this study conducted a comprehensive genome-wide analysis of the *SOD* gene family in four representative water lily species, resulting in the identification of 43 water lily *SODs*. The gathered information, encompassing exon-intron structure, cis-components, protein features, phylogenetic relations, and expression profiles of *N. colorata*, has shed light on the significant roles played by *NcSOD* genes in responding to salt, heat, cold, and heavy metal stresses. Findings of this systematic investigation provide a valuable resource for future functional research on *NcSOD* proteins in biological processes and lay a solid foundation for stress-resistant breeding of *N. colorata*. To further deepen our understanding of *NcSODs'* functions, our future studies will focus on gene engineering and comprehensive analysis, integrating genomics, transcriptomics, proteomics, and metabolomics.

Supplementary Materials: The following are available online at https://www.mdpi.com/article/10.3390/horticulturae9070781/s1, Table S1: the list of primer was used for gene expression analysis by RT-qPCR, Table S2: the protein sequences of SOD family genes in *Arabidopsis thaliana*, Table S3: the protein sequences of SOD family genes in four representative water lilies, Table S4: the information of identified 10 motifs in water lily SOD proteins, Table S5: The transcriptome data of *Nymphaea colorata* from ovules and pollen at various developmental stages.

Author Contributions: F.C. and W.U.K. conceived and designed this project. W.U.K. performed the analyses. L.U.K. and D.C. participated in the data analyses. W.U.K. carried out the experimental study and wrote the draft manuscript. F.C. and W.U.K. checked and revised the manuscript. All authors have read and agreed to the published version of the manuscript.

Funding: This work is supported by a grant from National Natural Science Foundation of China (32172614), a grant from Hainan Province Science and Technology Special Fund (ZDYF2023XDNY050), a fund from collaborative Innovation Center of Nanfan and Tropical High Efficient Agriculture, XTCX2022NYB04, and a start-up fund from Hainan Institute of Zhejiang University.

Data Availability Statement: The original contributions presented in the study are included in the article/Supplementary Materials, further inquiries can be directed to the corresponding author.

Acknowledgments: The authors thank the editor and anonymous reviewers for their valuable and insightful comments on this manuscript.

Conflicts of Interest: The authors declare no conflict of interest.

References

1. Mittler, R.; Blumwald, E. Genetic Engineering for Modern Agriculture: Challenges and Perspectives. *Annu. Rev. Plant Biol.* **2010**, *61*, 443–462. [CrossRef] [PubMed]
2. Cramer, G.R.; Urano, K.; Delrot, S.; Pezzotti, M.; Shinozaki, K. Effects of abiotic stress on plants: A systems biology perspective. *BMC Plant Biol.* **2011**, *11*, 163. [CrossRef] [PubMed]
3. Hasanuzzaman, M.; Bhuyan, M.H.M.B.; Zulfiqar, F.; Raza, A.; Mohsin, S.M.; Mahmud, J.A.; Fujita, M.; Fotopoulos, V. Reactive Oxygen Species and Antioxidant Defense in Plants under Abiotic Stress: Revisiting the Crucial Role of a Universal Defense Regulator. *Antioxidants* **2020**, *9*, 681. [CrossRef]
4. Lee, S.-H.; Ahsan, N.; Lee, K.-W.; Kim, D.-H.; Lee, D.-G.; Kwak, S.-S.; Kwon, S.-Y.; Kim, T.-H.; Lee, B.-H. Simultaneous overexpression of both CuZn superoxide dismutase and ascorbate peroxidase in transgenic tall fescue plants confers increased tolerance to a wide range of abiotic stresses. *J. Plant Physiol.* **2007**, *164*, 1626–1638. [CrossRef]
5. Karuppanapandian, T.; Wang, H.W.; Prabakaran, N.; Jeyalakshmi, K.; Kwon, M.; Manoharan, K.; Kim, W. 2,4-dichlorophenoxyacetic acid-induced leaf senescence in mung bean (*Vigna radiata* L. Wilczek) and senescence inhibition by co-treatment with silver nanoparticles. *Plant Physiol. Biochem.* **2011**, *49*, 168–177. [CrossRef]
6. Mittler, R.; Vanderauwera, S.; Gollery, M.; Van Breusegem, F. Reactive oxygen gene network of plants. *Trends Plant Sci.* **2004**, *9*, 490–498. [CrossRef] [PubMed]
7. Ahmad, P.; Umar, S.; Sharma, S. Mechanism of Free Radical Scavenging and Role of Phytohormones in Plants Under Abiotic Stresses. *Plant Adapt. Phytoremediation* **2010**, 99–118. [CrossRef]
8. Bafana, A.; Dutt, S.; Kumar, S.; Ahuja, P.S. Superoxide dismutase: An industrial perspective. *Crit. Rev. Biotechnol.* **2011**, *31*, 65–76. [CrossRef] [PubMed]
9. Filiz, E.; Tombuloğlu, H. Genome-Wide Distribution of Superoxide Dismutase (SOD) Gene Families in Sorghum Bicolor. *Turkish J. Biol.* **2015**, *39*, 49–59. [CrossRef]

10. Hodgson, E.K.; Fridovich, I. Reversal of the superoxide dismutase reaction. *Biochem. Biophys. Res. Commun.* **1973**, *54*, 270–274. [CrossRef]

11. Brawn, K.; Fridovich, I. Superoxide Radical and Superoxide Dismutases: Threat and Defense. *Autoxid. Food Biol. Syst.* **1980**, 429–446. [CrossRef]

12. Fink, R.C.; Scandalios, J.G. Molecular Evolution and Structure–Function Relationships of the Superoxide Dismutase Gene Families in Angiosperms and Their Relationship to Other Eukaryotic and Prokaryotic Superoxide Dismutases. *Arch. Biochem. Biophys.* **2002**, *399*, 19–36. [CrossRef] [PubMed]

13. Abreu, I.A.; Cabelli, D.E. Superoxide dismutases—A review of the metal-associated mechanistic variations. *Biochim. Biophys. Acta (BBA)—Proteins Proteom.* **2010**, *1804*, 263–274. [CrossRef] [PubMed]

14. Dupont, C.L.; Neupane, K.; Shearer, J.; Palenik, B. Diversity, function and evolution of genes coding for putative Ni-containing superoxide dismutases. *Environ. Microbiol.* **2008**, *10*, 1831–1843. [CrossRef] [PubMed]

15. Xia, M.; Wang, W.; Yuan, R.; Den, F.; Shen, F.F. Superoxide Dismutase and Its Research in Plant Stress-Tolerance. *Mol. Plant Breed.* **2015**, *13*, 2633–2646.

16. Zeng, X.-C.; Liu, Z.-G.; Shi, P.-H.; Xu, Y.-Z.; Sun, J.; Fang, Y.; Yang, G.; Wu, J.-Y.; Kong, D.-J.; Sun, W.-C. Cloning and Expression Analysis of Copper and Zinc Superoxide Dismutase (Cu/Zn-SOD) Gene from *Brassica campestris* L. *Acta Agron. Sin.* **2014**, *40*, 636–643. [CrossRef]

17. Song, J.; Zeng, L.; Chen, R.; Wang, Y.; Zhou, Y. In Silico Identification and Expression Analysis of Superoxide Dismutase (SOD) Gene Family in Medicago Truncatula. *3 Biotech* **2018**, *8*, 348. [CrossRef]

18. Perry, J.J.P.; Shin, D.S.; Getzoff, E.D.; Tainer, J.A. The structural biochemistry of the superoxide dismutases. *Biochim. Biophys. Acta (BBA)—Proteins Proteom.* **2010**, *1804*, 245–262. [CrossRef]

19. Wang, B.; Lüttge, U.; Ratajczak, R. Specific regulation of SOD isoforms by NaCl and osmotic stress in leaves of the C3 halophyte *Suaeda salsa* L. *J. Plant Physiol.* **2004**, *161*, 285–293. [CrossRef]

20. Pilon, M.; Ravet, K.; Tapken, W. The Biogenesis and Physiological Function of Chloroplast Superoxide Dismutases. *Biochim. Biophys. Acta (BBA)—Bioenerg.* **2011**, *1807*, 989–998. [CrossRef]

21. Krouma, A.; Drevon, J.-J.; Abdelly, C. Genotypic variation of N2-fixing common bean (*Phaseolus vulgaris* L.) in response to iron deficiency. *J. Plant Physiol.* **2006**, *163*, 1094–1100. [CrossRef] [PubMed]

22. Feng, X.; Lai, Z.; Lin, Y.; Lai, G.; Lian, C. Genome-wide identification and characterization of the superoxide dismutase gene family in *Musa acuminata* cv. Tianbaojiao (AAA group). *BMC Genom.* **2015**, *16*, 1–16. [CrossRef] [PubMed]

23. Feng, K.; Yu, J.; Cheng, Y.; Ruan, M.; Wang, R.; Ye, Q.; Zhou, G.; Li, Z.; Yao, Z.; Yang, Y.; et al. The SOD Gene Family in Tomato: Identification, Phylogenetic Relationships, and Expression Patterns. *Front. Plant Sci.* **2016**, *7*, 1279. [CrossRef]

24. Yan, J.-J.; Zhang, L.; Wang, R.-Q.; Xie, B.; Li, X.; Chen, R.-L.; Guo, L.-X.; Xie, B.-G. The Sequence Characteristics and Expression Models Reveal Superoxide Dismutase Involved in Cold Response and Fruiting Body Development in *Volvariella volvacea*. *Int. J. Mol. Sci.* **2015**, *17*, 34. [CrossRef]

25. Su, W.; Raza, A.; Gao, A.; Jia, Z.; Zhang, Y.; Hussain, M.A.; Mehmood, S.S.; Cheng, Y.; Lv, Y.; Zou, X. Genome-Wide Analysis and Expression Profile of Superoxide Dismutase (SOD) Gene Family in Rapeseed (*Brassica napus* L.) under Different Hormones and Abiotic Stress Conditions. *Antioxidants* **2021**, *10*, 1182. [CrossRef] [PubMed]

26. Zang, Y.; Chen, J.; Li, R.; Shang, S.; Tang, X. Genome-wide analysis of the superoxide dismutase (SOD) gene family in *Zostera marina* and expression profile analysis under temperature stress. *PeerJ* **2020**, *8*, e9063. [CrossRef] [PubMed]

27. Han, L.-M.; Hua, W.-P.; Cao, X.-Y.; Yan, J.-A.; Chen, C.; Wang, Z.-Z. Genome-wide identification and expression analysis of the superoxide dismutase (SOD) gene family in *Salvia miltiorrhiza*. *Gene* **2020**, *742*, 144603. [CrossRef] [PubMed]

28. Zhang, X.; Zhang, L.; Chen, Y.; Wang, S.; Fang, Y.; Zhang, X.; Wu, Y.; Xue, D. Genome-wide identification of the SOD gene family and expression analysis under drought and salt stress in barley. *Plant Growth Regul.* **2021**, *94*, 49–60. [CrossRef]

29. Gomez, J.M.; Hernandez, J.A.; Jimenez, A.; Del Rio, L.A.; Sevilla, F. Differential Response of Antioxidative Enzymes of Chloroplasts and Mitochondria to Long-term NaCl Stress of Pea Plants. *Free. Radic. Res.* **1999**, *31*, 11–18. [CrossRef]

30. Wu, G.; Wilen, R.W.; Robertson, A.J.; Gusta, L.V. Isolation, Chromosomal Localization, and Differential Expression of Mitochondrial Manganese Superoxide Dismutase and Chloroplastic Copper/Zinc Superoxide Dismutase Genes in Wheat. *Plant Physiol.* **1999**, *120*, 513–520. [CrossRef]

31. Baek, K.-H.; Skinner, D.Z. Alteration of antioxidant enzyme gene expression during cold acclimation of near-isogenic wheat lines. *Plant Sci.* **2003**, *165*, 1221–1227. [CrossRef]

32. Srivastava, V.; Srivastava, M.K.; Chibani, K.; Nilsson, R.; Rouhier, N.; Melzer, M.; Wingsle, G. Alternative Splicing Studies of the Reactive Oxygen Species Gene Network in Populus Reveal Two Isoforms of High-Isoelectric-Point Superoxide Dismutase. *Plant Physiol.* **2009**, *149*, 1848–1859. [CrossRef] [PubMed]

33. Lu, Y.; Feng, Z.; Bian, L.; Xie, H.; Liang, J. miR398 regulation in rice of the responses to abiotic and biotic stresses depends on CSD1 and CSD2 expression. *Funct. Plant Biol.* **2010**, *38*, 44–53. [CrossRef] [PubMed]

34. Diao, Y.; Chen, L.; Yang, G.; Zhou, M.; Song, Y.; Hu, Z.; Liu, J.Y. Nuclear DNA C-values in 12 species in Nymphaeales. *Caryologia* **2006**, *59*, 25–30. [CrossRef]

35. Zhang, L.; Chen, F.; Zhang, X.; Li, Z.; Zhao, Y.; Lohaus, R.; Chang, X.; Dong, W.; Ho, S.Y.W.; Liu, X.; et al. The water lily genome and the early evolution of flowering plants. *Nature* **2020**, *577*, 79–84. [CrossRef]

36. Kliebenstein, D.J.; Monde, R.-A.; Last, R.L. Superoxide Dismutase in Arabidopsis: An Eclectic Enzyme Family with Disparate Regulation and Protein Localization. *Plant Physiol.* **1998**, *118*, 637–650. [CrossRef] [PubMed]
37. Schultz, J.; Milpetz, F.; Bork, P.; Ponting, C.P. SMART, a simple modular architecture research tool: Identification of signaling domains. *Proc. Natl. Acad. Sci. USA* **1998**, *95*, 5857–5864. [CrossRef] [PubMed]
38. Gasteiger, E.; Gattiker, A.; Hoogland, C.; Ivanyi, I.; Appel, R.D.; Bairoch, A. ExPASy: The Proteomics Server for in-Depth Protein Knowledge and Analysis. *Nucleic Acids Res.* **2003**, *31*, 3784–3788. [CrossRef]
39. Edgar, R. MUSCLE: Multiple sequence alignment with improved accuracy and speed. In Proceedings of the 2004 IEEE Computational Systems Bioinformatics Conference, CSB 2004, Stanford, CA, USA, 19 August 2004; pp. 728–729. [CrossRef]
40. Bailey, T.L.; Boden, M.; Buske, F.A.; Frith, M.; Grant, C.E.; Clementi, L.; Ren, J.; Li, W.W.; Noble, W.S. MEME Suite: Tools for Motif Discovery and Searching. *Nucleic Acid Res.* **2009**, *37*, 202–208. [CrossRef]
41. Lescot, M.; Déhais, P.; Thijs, G.; Marchal, K.; Moreau, Y.; Van de Peer, Y.; Rouzé, P.; Rombauts, S. PlantCARE, a database of plant cis-acting regulatory elements and a portal to tools for in silico analysis of promoter sequences. *Nucleic Acids Res.* **2002**, *30*, 325–327. [CrossRef]
42. Zhou, Y.; Hu, L.; Wu, H.; Jiang, L.; Liu, S. Genome-Wide Identification and Transcriptional Expression Analysis of Cucumber Superoxide Dismutase (SOD) Family in Response to Various Abiotic Stresses. *Int. J. Genom.* **2017**, *2017*, 7243973. [CrossRef] [PubMed]
43. Szklarczyk, D.; Gable, A.L.; Lyon, D.; Junge, A.; Wyder, S.; Huerta-Cepas, J.; Simonovic, M.; Doncheva, N.T.; Morris, J.H.; Bork, P. STRING V11: Protein–Protein Association Networks with Increased Coverage, Supporting Functional Discovery in Genome-Wide Experimental Datasets. *Nucleic Acids Res.* **2019**, *47*, D607–D613. [CrossRef]
44. Livak, K.J.; Schmittgen, T.D. Analysis of Relative Gene Expression Data Using Real-Time Quantitative PCR and the $2^{-\Delta\Delta CT}$ Method. *Methods* **2001**, *25*, 402–408. [CrossRef]
45. Xiong, X.; Zhang, J.; Yang, Y.; Chen, Y.; Su, Q.; Zhao, Y.; Wang, J.; Xia, Z.; Wang, L.; Zhang, L.; et al. Water lily research: Past, present, and future. *Trop. Plants* **2023**, *2*, 1–8. [CrossRef]
46. Hu, X.; Hao, C.; Cheng, Z.-M.; Zhong, Y. Genome-Wide Identification, Characterization, and Expression Analysis of the Grapevine Superoxide Dismutase (SOD) Family. *Int. J. Genom.* **2019**, *2019*, 7350414. [CrossRef]
47. Xu, G.; Guo, C.; Shan, H.; Kong, H. Divergence of duplicate genes in exon–intron structure. *Proc. Natl. Acad. Sci. USA* **2012**, *109*, 1187–1192. [CrossRef] [PubMed]
48. Wang, W.; Zhang, X.; Deng, F.; Yuan, R.; Shen, F. Genome-wide characterization and expression analyses of superoxide dismutase (SOD) genes in Gossypium hirsutum. *BMC Genom.* **2017**, *18*, 1–25. [CrossRef] [PubMed]
49. Divya, K.; Kavi Kishor, P.B.; Bhatnagar-Mathur, P.; Singam, P.; Sharma, K.K.; Vadez, V.; Reddy, P.S. Isolation and functional characterization of three abiotic stress-inducible (Apx, Dhn and Hsc70) promoters from pearl millet (*Pennisetum glaucum* L.). *Mol. Biol. Rep.* **2019**, *46*, 6039–6052. [CrossRef] [PubMed]
50. Huang, H.; Wang, H.; Tong, Y.; Wang, Y. Insights into the Superoxide Dismutase Gene Family and Its Roles in *Dendrobium catenatum* under Abiotic Stresses. *Plants* **2020**, *9*, 1452. [CrossRef] [PubMed]
51. Liu, J.; Xu, L.; Shang, J.; Hu, X.; Yu, H.; Wu, H.; Lv, W.; Zhao, Y. Genome-wide analysis of the maize superoxide dismutase (SOD) gene family reveals important roles in drought and salt responses. *Genet. Mol. Biol.* **2021**, *44*, 1–12. [CrossRef]
52. Dehury, B.; Sarma, K.; Sarmah, R.; Sahu, J.; Sahoo, S.; Sahu, M.; Sen, P.; Modi, M.K.; Sharma, G.D.; Choudhury, M.D.; et al. In silico analyses of superoxide dismutases (SODs) of rice (*Oryza sativa* L.). *J. Plant Biochem. Biotechnol.* **2013**, *22*, 150–156. [CrossRef]
53. Nath, K.; Kumar, S.; Poudyal, R.S.; Yang, Y.N.; Timilsina, R.; Park, Y.S.; Nath, J.; Chauhan, P.S.; Pant, B.; Lee, C.-H. Developmental stage-dependent differential gene expression of superoxide dismutase isoenzymes and their localization and physical interaction network in rice (*Oryza sativa* L.). *Genes Genom.* **2014**, *36*, 45–55. [CrossRef]
54. Wang, W.; Xia, M.; Chen, J.; Deng, F.; Yuan, R.; Zhang, X.; Shen, F. Genome-wide analysis of superoxide dismutase gene family in Gossypium raimondii and G. arboreum. *Plant Gene* **2016**, *6*, 18–29. [CrossRef]
55. Verma, D.; Lakhanpal, N.; Singh, K. Genome-wide identification and characterization of abiotic-stress responsive SOD (superoxide dismutase) gene family in *Brassica juncea* and *B. rapa*. *BMC Genom.* **2019**, *20*, 1–18. [CrossRef] [PubMed]
56. Fisher, C.L.; Cabelli, D.E.; Tainer, J.A.; Hallewell, R.A.; Getzoff, E.D. The role of arginine 143 in the electrostatics and mechanism of Cu, Zn superoxide dismutase: Computational and experimental evaluation by mutational analysis. *Proteins Struct. Funct. Bioinform.* **1994**, *19*, 24–34. [CrossRef] [PubMed]
57. Wang, M.; Zhao, X.; Xiao, Z.; Yin, X.; Xing, T.; Xia, G. A wheat superoxide dismutase gene TaSOD2 enhances salt resistance through modulating redox homeostasis by promoting NADPH oxidase activity. *Plant Mol. Biol.* **2016**, *91*, 115–130. [CrossRef]
58. Guan, Q.; Liao, X.; He, M.; Li, X.; Wang, Z.; Ma, H.; Yu, S.; Liu, S. Tolerance Analysis of Chloroplast OsCu/Zn-SOD Overexpressing Rice under NaCl and NaHCO3 Stress. *PLoS ONE* **2017**, *12*, e0186052. [CrossRef]
59. He, H.; Lei, Y.; Yi, Z.; Raza, A.; Zeng, L.; Yan, L.; Xiaoyu, D.; Yong, C.; Xiling, Z. Study on the mechanism of exogenous serotonin improving cold tolerance of rapeseed (*Brassica napus* L.) seedlings. *Plant Growth Regul.* **2021**, *94*, 161–170. [CrossRef]

Article

Characterization of the Passion Fruit (*Passiflora edulis* Sim) bHLH Family in Fruit Development and Abiotic Stress and Functional Analysis of *PebHLH56* in Cold Stress

Yi Xu [1,2], Weidong Zhou [3], Funing Ma [1,2], Dongmei Huang [1], Wenting Xing [1], Bin Wu [1], Peiguang Sun [1], Di Chen [1], Binqiang Xu [1] and Shun Song [1,2,*]

[1] State Key Laboratory of Biological Breeding for Tropical Crops, Haikou Experimental Station, Sanya Research Institute, Chinese Academy of Tropical Agricultural Sciences, Haikou 571101, China
[2] Hainan Yazhou Bay Seed Laboratory, Sanya 572000, China
[3] College of Agronomy and Biotechnology, Yunnan Agricultural University, Kunming 650201, China
[*] Correspondence: songs@catas.cn

Abstract: Abiotic stress is the focus of research on passion fruit characters because of its damage to the industry. Basic helix-loop-helix (bHLH) is one of the Transcription factors (TFs) which can act in an anti-abiotic stress role through diverse biological processes. However, no systemic analysis of the passion fruit bHLH (PebHLH) family was reported. In this study, 117 PebHLH members were identified from the genome of passion fruit, related to plant stress resistance and development by prediction of protein interaction. Furthermore, the transcriptome sequencing results showed that the *PebHLHs* responded to different abiotic stresses. At different ripening stages of passion fruit, the expression level of most *PebHLHs* in the immature stage (T1) was higher than that in the mature stage (T2 and T3). Eight *PebHLHs* with differentially expressed under different stress treatments and different ripening stages were selected and verified by qRT-PCR. In this research, the expression of one member, *PebHLH56,* was induced under cold stress. Further, the promoter of *PebHLH56* was fused to β-Galactosidase (GUS) to generate the expression vector that was transformed into *Arabidopsis*. It showed that *PebHLH56* could significantly respond to cold stress. This study provided new insights into the regulatory functions of *PebHLH* genes during fruit maturity stages and abiotic stress, thereby improving the understanding of the characteristics and evolution of the *PebHLH* gene family.

Keywords: bHLH; passion fruit; abiotic stress; cold; genome-wide analysis

Citation: Xu, Y.; Zhou, W.; Ma, F.; Huang, D.; Xing, W.; Wu, B.; Sun, P.; Chen, D.; Xu, B.; Song, S. Characterization of the Passion Fruit (*Passiflora edulis* Sim) bHLH Family in Fruit Development and Abiotic Stress and Functional Analysis of *PebHLH56* in Cold Stress. *Horticulturae* **2023**, *9*, 272. https://doi.org/10.3390/horticulturae9020272

Academic Editor: Jia-Long Yao

Received: 6 January 2023
Revised: 11 February 2023
Accepted: 13 February 2023
Published: 17 February 2023

1. Introduction

Transcription factors (TFs) play important roles in regulating growth and responding to external environmental stress in plants [1,2], which can regulate gene expression by binding to cis-promoter elements, thereby exerting regulatory roles in morphogenesis and so on [3,4]. TF genes, such as bHLH and Myeloblastosis (MYB), account for a large proportion of almost all plant genomes and are widely involved in plant development, stress response, and other physiological processes by regulating their target gene [5,6].

Basic/helix-loop-helix (bHLH) is a ubiquitous transcription factor family [7], which forms the second largest TF superfamily in plants [8]. The bHLHs have highly conserved alkaline/helix-loop-helix domains with approximately 50–60 amino acid residues [9]. This domain contains two functional regions: the basic region and the helix-loop-helix (HLH) region. The basic region contains approximately 15 amino acids and is located at the N-terminus, which is a critical region for binding to the cis-acting element E-box (5′-CANNTG-3′) and determining whether the bHLH transcription factor binds to the promoter region of genes [10,11]. The HLH region contains two α-helices connected by a relatively poorly conserved loop distributed at the C-terminus, and this structure is essential for bHLH transcription factors to form homologous or heterodimers [10,12,13].

With the release of more high-quality genomes, many bHLH families in plants have been identified, such as *Carthamus tinctorius* (41) [14], Chinese jujube (92) [15], pineapple (121) [16], pepper (122) [17], potato (124) [18], peanut (132) [19], Jilin ginseng (137) [20], Brachypodium distachyon (146) [21], common bean (155) [22], tomato (159) [23], rice (167) [24], apple (188) [25], maize (208) [26], Chinese cabbage (230) [27], MOSO bamboo (448) [28], wheat (571) [29].

In current research, bHLHs are involved in responding to light [30], hormone signals [31], regulating anthocyanin biosynthesis [32], epidermal cell fate determination [33], and seed germination [34]. At present, the relationship between the bHLH gene and abiotic stress has attracted more and more attention. The bHLH has been shown to play a crucial role in plant resistance to abiotic stresses, such as abnormal temperature, drought, and high salinity. *FtbHLH3* of Tartary buckwheat participates in abiotic stress in response to changes in the polyethylene glycol (PEG) and the abscisic acid (ABA) [35], *FtbHLH2* can improve cold tolerance in plants [36]. *TabHLH39* augments the tolerance of transgenic *Arabidopsis* seedlings to drought, high salt, and low-temperature stress [37]. Overexpression of *AtbHLH92* can significantly improve salt and drought tolerance [38]. The *AtICE1/2* in *Arabidopsis* and their homologs *PtrbHLH* in Poncirus trifoliata can adjust plant cold tolerance and regulate peroxidase to break down hydrogen peroxide [39].

In addition, some studies have reported that *bHLH* genes are related to fruit development. In transgenic tomatoes, *SlbHLH22* is highly induced with the fruit color changed from green to orange, which is achieved by promoting the production of ethylene. Meanwhile, *SlbHLH22* was upregulated by using the exogenous ACC, IAA, ABA, and ethephon [40]. The research from Tan [41] showed that *CmbHLH32* was highly expressed in early developmental fruits. Overexpression of *CmbHLH32* can promote the early ripening of melon fruits, and the transgenic melons ripened earlier than the wild type. Papaya *CpbHLH1* and *CpbHLH2* promote fruit development by regulating genes related to carotenoid biosynthesis [42].

Passion fruit is a rare tropical fruit of the *Passiflora* genus (Passifloraceae), native to South America [43]. *Passiflora* has fresh and ornamental types [44]. *Passiflora* is nutritious and contains more than 100 different fruits in the pulp. At present, the planting area of East Asian countries such as Vietnam and China is growing rapidly, with a growth cycle of 4–6 months [45]. Because of its rich flavor, it has become more and more popular. Similar to other fruit, drought, high salinity, and cold and high-temperature stress seriously affect the development of passion fruit, resulting in a decline in fruit quality and yield. Due to the unpredictability of global climate change in recent years, passion fruit growing areas have been frequently affected by cold injury, which has also caused huge economic losses, resulting in more than 30% yield reduction and fruit stunting. Chilling injury is the most difficult to predict and control among the four abiotic stresses (drought, salt, cold and high temperature). Therefore, the identification of functional genes related to stress resistance and their utilization for variety improvement is of great importance for passion fruit cultivation.

Here, we identified the PebHLH family members in the genome of passion fruit and analyzed the members' biological information. Moreover, the expression patterns of *PebHLH* members were obtained by transcriptome sequencing and qRT-PCR at fruit developmental stages and under typical abiotic stress. More importantly, the *PebHLH* genes that were highly expressed and significantly induced by abiotic stress (drought, high salt, cold, and high temperature) were selected and overexpressed in *Arabidopsis* for functional validation. This provided a good foundation and reserved important information for studying the resistance mechanism in passion fruit and utilizing it for genetic improvement.

2. Materials and Methods

2.1. Identification of bHLH Members in Passion Fruit

The genome sequences of passion fruit were obtained from Phytozome V12.1. The HMM file of the NAM domain (PF00011) was downloaded from the PFAM database. In addition, analysis was performed using the bHLH with the highest comparison value to identify credible *PebHLHs*. The identification of PebHLH proteins used two methods described above were integrated and resolved to remove redundancy. AtbHLH protein sequences were obtained from Plant TFDB software (http://planttfdb.gao-lab.org/ accessed on 15 June 2022). The full-length protein sequences of PebHLHs and AtbHLHs were aligned by online ClustalX2. Moreover, the phylogenetic tree was constructed using MEGA (Version 7.0) [46]. A bootstrap test of 1000 repetitions was performed. Finally, PebHLH protein motifs were achieved using the MEME tool to compare the common domain of *PebHLHs*. Through the above procedures, the PebHLH members were finally obtained.

2.2. Gene Structure, and Chromosomal Locations of PebHLHs

The molecular weight (MW), protein isoelectric point (PI), and molecular formulas of all PebHLH members were reckoned using the online ProtParam. NetPhos (Version 3.1 Server) was used to predict the protein of PebHLH phosphate sites. The WoLF PSORT was used to perform Subcellular localization prediction. The gene structure maps, phylogenetic trees, combinations of motifs and gene structures, visualization of chromosomal localization, and collinearity relationships were obtained using TBtools [47].

2.3. Cis-Acting Elements, Protein Interaction Network and Gene Collinearity of PebHLHs

All *PebHLH* gene transcription start site DNA sequences of the genomes of 2000 bp upstream were imported to PlantCARE and used to analyze the sequence of the promoter region [48]. The Orthovenn2 was used to analyze the orthologous pairs between PebHLHs and AtbHLHs. The regulatory networks of *PebHLHs* and other proteins were identified using the AraNet (Version 2.0). The STRING database and the predicted regulatory network were evinced by Cytoscape software. The PebHLH gene duplication events between different species were analyzed by Multicollinearity Scanning Toolkit (MCScanX) [47].

2.4. Plant Materials, Transcriptome Sequencing, RNA Isolation and Reverse Transcription, Heat Map and qRT-PCR

Healthy passion fruit seedlings about two months old (30 cm in height) in the soil were chosen. The seedlings were planted in incubators and treated with cold, high temperature, high salt, and drought stresses. The control was placed in incubators (28 °C, 70% relative humidity, 12 h light/12 h dark cycle, 200 µmol m^{-2} s^{-1} light intensity). In addition, after fruit ripening, the pericarp turns purplish red. The three stages (T1, T2, and T3) are the time of 7d before ripening, ripening, and 7d after ripening, respectively [45]. The experimental material consisted of three biological replicated samples. The plant materials were used for transcriptome sequencing. The expression data of *PebHLHs* at four stress and three fruit ripening stages are shown in Tables S2 and S4. TBtools software was used for transcriptional analysis of *PebHLHs*, whose Z-score normalized FPKM values were used to produce heat maps [47]. The plant RNA isolation kit was used in *Arabidopsis* transformed with pCAMBIA1304-*PebHLH56*. The Biomic Biotechnology company (Beijing, China) was entrusted with sequencing services. The Primer sequences of *PebHLHs* were designed using the Primer5 software. The expression of *PebHLHs* was detected by qRT-PCR analysis. Relative expression levels were calculated using the $2^{-\Delta\Delta Ct}$ method and normalized to the *PebHLHs*.

2.5. Cloning the Promoter of PebHLH56 and Vector Construction

A 2000 bp DNA sequence before the start codon of the PebHLH56 was amplified by PCR and cloned into the pMD19-T vector. The promoter of PebHLH56 was assessed by DNA-MAN.

Expression vectors were constructed to examine whether *PebHLH56* responds to cold stress. The *PebHLH56* promoter PCR fragment was cloned into the pCAMBIA1304 vector digested with NcoI/HinIII, which was called pCAMBIA1304-*PebHLH56p*. The vector was transferred into the EHA105 strain (*Agrobacterium*).

2.6. Cold Stress Treatment of Transgenic Lines

The Agrobacterium transformed with pCAMBIA1304-*PebHLH56*p were shaken at 28 °C in YEB medium containing Kan and Rif antibiotics, respectively, and added to an *Arabidopsis* transgenic infiltration solution (1/2 MS, 50 g/L) to OD 600 = 0.8–1.0. Inflorescence impregnation was used to transform *Arabidopsis thaliana*. Eight transgenic lines have been obtained. Then three plants from each of the two lines in the T2 generation were cultured for cold stress treatment [8]. The 14-day-old transgenic *Arabidopsis* was treated in an incubator at 4 °C for 0, 24, and 36 h, respectively [3].

2.7. Detection of GUS Activity

The transgenic *Arabidopsis* with pCAMBIA1304-*PebHLH56p* under normal and cold stress were GUS stained. For GUS staining, seedlings of two lines were incubated in X-Gluc solution for 24 h at 37 °C [45,49]. GUS enzyme activity was determined by 4-methylumbel ureylglucuronide fluorometry [50].

3. Data Analysis and Results

3.1. Identification of bHLH Family of Passion Fruit

There were 117 *PebHLH* members identified from the passion fruit genome by the methods of HMM, protein BLAST, and MEME analyses. The characteristics of the *PebHLH* (Table S1) were analyzed, and the length of the *PebHLH* CDS ranged from 276 bp of *PebHLH3* to 3006 bp of *PebHLH73*. The identified *PebHLHs* encoded proteins ranged from 91 amino acids of *PebHLH3* to 1001 amino acids of *PebHLH73*. In addition, all the *PebHLHs* were showed containing three phosphorylation sites (Tyr, Thr, and Ser). The subcellular location prediction showed that the PebHLHs were distributed in the periplasmic, cytoplasmic, extracellular, and outer membranes. The molecular weight ranged from 10.43 Da to 114.67 Da.

3.2. Evolutionary Analysis of the bHLH Genes

The phylogenetic tree was constructed using the full-length sequences of the PebHLH and AtbHLH proteins (Figure 1). The PebHLHs could be divided into eighteen groups according to the evolutionary relationship. In groups 1 to 18, the largest number of *PebHLHs* was in group 1, with 11 members, and the smallest number was two in group 9. The number of PebHLH members in the remaining groups was between 4–10. The known *AtbHLHs* were then used to infer the hypothetical homologous members of *PebHLHs*. Among group 1–6, *PebHLH11/101/80/31/42/88/98/41/18/108/54/46/13/66/44/14* exhibited the closest relationship with At3G50330.1, At3G19500.1, At2G31730.1, At2G43010.5, At1G73830.1, At1G05805.1, At1G51140.1, At4G29100.1, At2G43140.2, At1G27660.1, At1G29950.2, At5G65320.1, respectively. Among groups 7–12, *PebHLH85/112/62/102/38/97/92/100/47/61/19/52/45* was the most homogeneous gene of At4G30980.1, At4G00480.2, At5G51790.2, At2G22760.1, At2G31280, 1At3G23210.1, At2G16910.1, At5G57150.2, At1G32640.1, At2G24260.2, At4G36060.2, At3G19860.3, At5G56960.2, respectively. Among group 13–18, *PebHLH57/43/91/5/12/69/99/ 26/114/84/22/95*, was the best orthologous match of At1G18400.1, At4G21340.1, At5G08130.3, At3G24140.1, At1G71200.1, At4G30410.1, At1G25330.1, At5G62610.1, At2G46510.1, At5G50915.1, At5G61270.2, respectively.

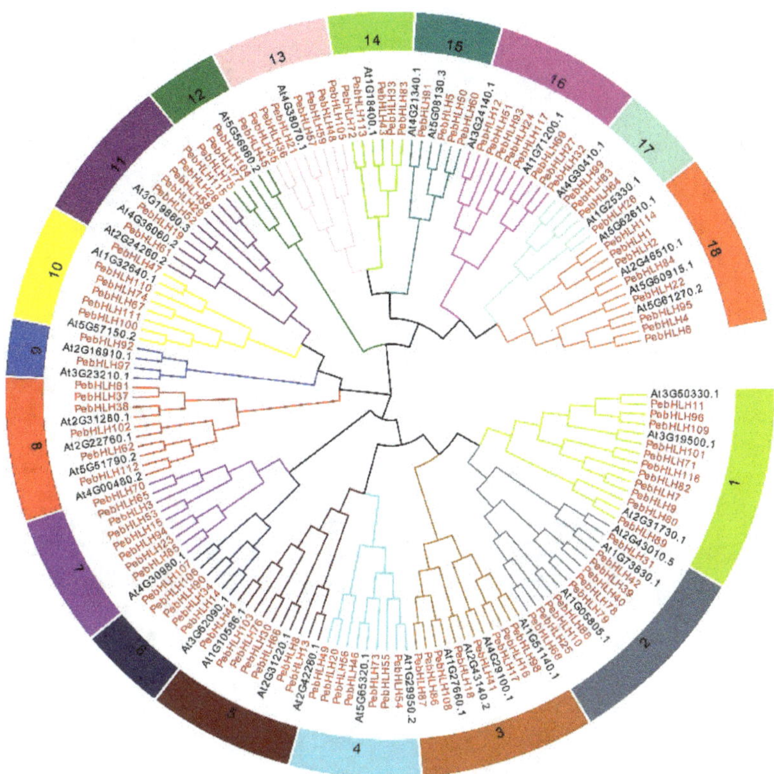

Figure 1. Phylogenetic relationship of bHLHs between passion fruit and *Arabidopsis*. The different colors of the outer circles represented groups 1–18. Pe and At are abbreviations of passion fruit and *Arabidopsis*, respectively.

3.3. Analysis of Gene Structure and Conserved Motifs

A total of 10 conserved motifs in *PebHLHs* were predicted (Figure 2). Although most members have two to four motifs, few members have only one motif; for example, *PebHLH26/48/91* only has motif one. *PebHLH30/38/41/43/52* only has motif two. Specifically, most members contain motifs 1 and 2, such as 111 members contain motif 1, and 114 members contain motif 2. In addition, 30 members contain motif 3, 38 members contain motif 4, 14 members contain motif 5, 17 members contain motif 6, and 4 members contain motif 7.

The exon and intron organization of *PebHLH* gene DNAs were analyzed. The bulk of members had the structure of 5′ and 3′UTR (untranslated region), and 18 only had exons and introns. Among them, nine members (*PebHLH14/27/31/48/55/81/95/98/111*) only had 5′UTR, and 23 members only had 3′UTR. The number of exons was between 2–11(Figure 2).

3.4. Chromosome Distribution of the PebHLHs

The chromosome distribution of each *PebHLH* was obtained, as shown in Figure 3. The total 117 *PebHLHs* were located on nine chromosomes. Seven members were located at unknown chromosomal positions. Among them, Chr1 contained the largest distribution of 41 *PebHLH* genes (~37.27%), followed by Chr6 (24 genes, ~21.82%). Meanwhile, Chr9 distributed the least number of *PebHLH* genes (3 genes, ~2.73%). Chr2, Chr3, Chr4, Chr5, Chr7, and Chr8 presented 7, 6, 10, 5, 4, and 10 *PebHLH* genes, respectively.

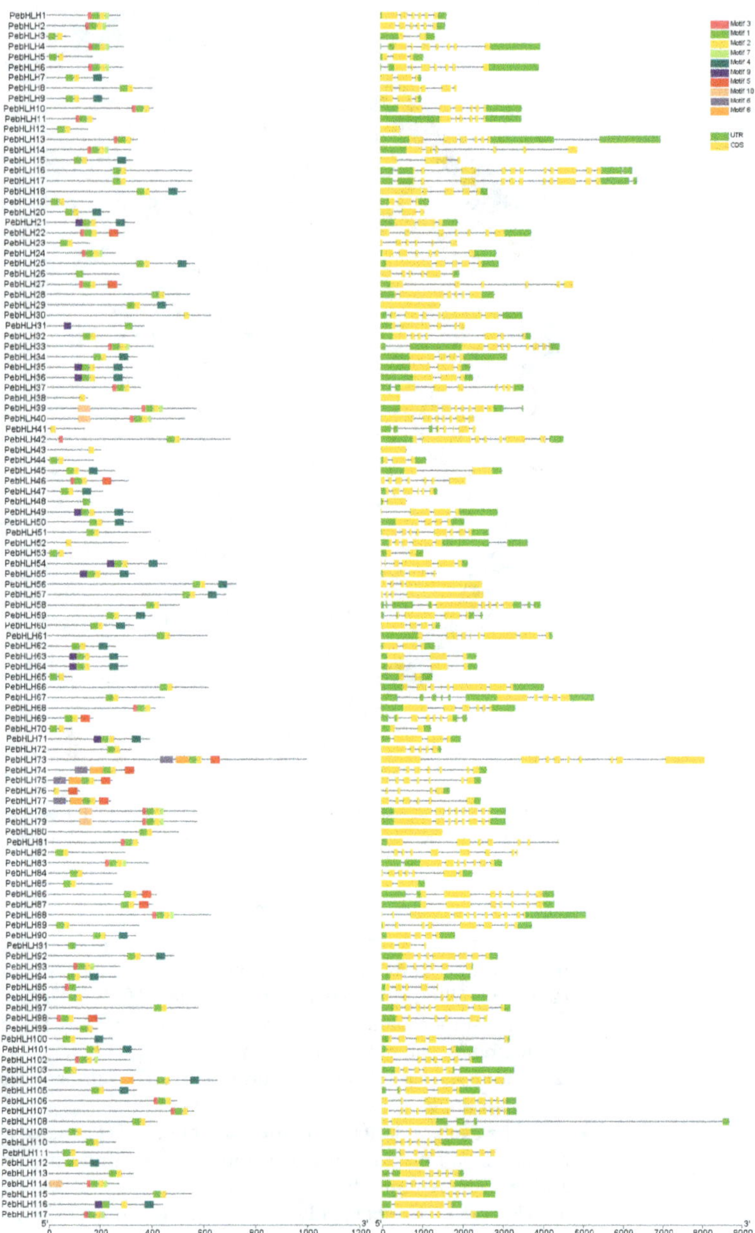

Figure 2. Phylogenetic clustering, conserved motifs, and exon/intron organization of *PebHLHs*. Ten color boxes indicate different motifs, green color represents 5′ and 3′UTR, and yellow color and black lines represent exons and introns, respectively.

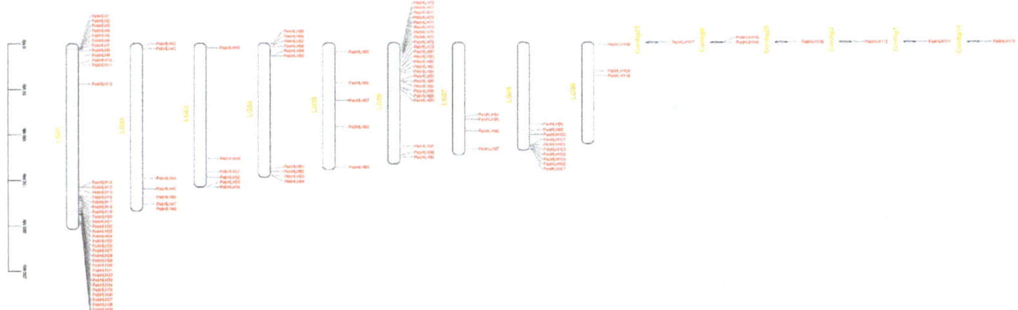

Figure 3. Locations of the 117 identified *PebHLH*s in 9 chromosomes of Passion fruit. The *PebHLH*s were located on No. 1, 2, 3, 4, 5, 6, 7, 8, and 9 chromosomes.

3.5. Promoter Analysis of PebHLH Genes

The TATA and CAAT boxes were the key cis-regulatory elements, which were found in all 117 *PebHLH*s (Figure 4). Additionally, the promoter region of the *PebHLH* family contains a large number of action elements related to abiotic stress, such as the cold-responsive element (CCGAAA) and the salicylic acid-responsive element (CCATCTTTTT and TCAGAA-GAGG). Some cis-acting elements are associated with stress response, such as the wound-responsive element (AAATTTCCT). In addition, there are some hormone-related elements, such as the abscisic acid responsiveness element (ABRE, GACACGTGGC, and ACGTG), the MeJA-responsive motifs (TGACG and CGTCA), the gibberellin-responsive motifs (CCTTTTG, TATCCCA, and TCTGTTG), and the auxin-responsive element (AACGAC).

3.6. Interaction Networks Analysis of PebHLHs

The protein interactions of PebHLHs were predicted using an orthogonal approach to further understand their biological functions. The results showed that 45 PebHLHs had an orthologous relationship with *Arabidopsis*, and 24 interacting proteins had been found. Most of the proteins that interacted with *PebHLH*s were WAKY, JAZ (Jasmonate ZIM-domain), COl (CONSTANS-Like), NINJA (Novel interactor of JA ZIM-domain), EPF (Early Pregnancy Factor), AOS (Allene Oxide Synthase), PFT (Pore-forming Toxin-like), CRY (Crystal protein gene), RBR (Retinoblastoma-related gene), ARF (Auxin response factor) TRY (Tryptophan) and so on (Figure 5). The function of the interacting proteins was related to the stress resistance and development of plants, which indicated that the function of PebHLH was also related to these aspects.

3.7. Collinearity Analysis of PebHLHs

All *PebHLH* genes were found to be displayed on chromosomes except for *PebHLH111-117*. There are 68 collinearity pairs that can be found. Most *PebHLH*s were found with one paralogous gene (Figure 6A).

The collinearity analysis among species was performed, and respectively, there were 12,406, 3118, and 25,224 genes in *Arabidopsis*, rice, and poplar, having a collinearity relationship with passion fruit. Among them, members of the PebHLH family had 91, 31, and 170 connections with *Arabidopsis*, rice, and poplar, respectively (Figure 6B). This result indicates that the homologous relationship between passion fruit and poplar is the closest, followed by *Arabidopsis* and rice.

Out of the 117 members, 27 and 22 genes had two and one homologous genes with the other three species, respectively. Other members have three to eight homologous genes. Among them, *PebHLH56* has 9 homologous genes (At1G01260.3, At1G63650.3, At2G46510.1, At4G00870.1, PNT03691, PNT03984, PNT45271, PNT50177, PNT53769) in collinear connections with *Arabidopsis* and poplar. In addition, 18 genes (*PebHLH27/28/33/35/36/40/46/58/61/67/78/91/97/98/99/109/110/115*) had one orthologous gene in rice.

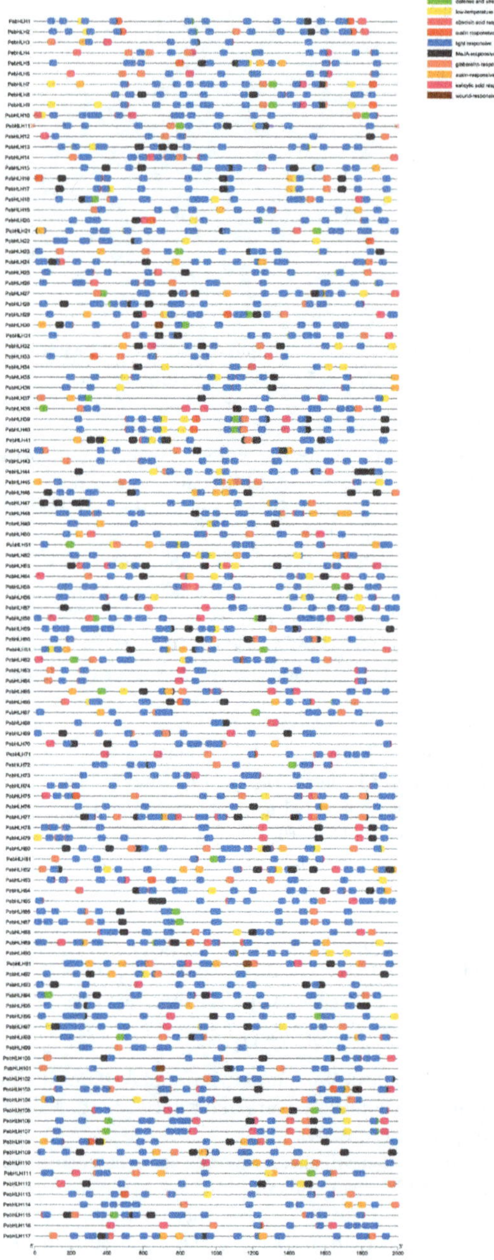

Figure 4. The cis-acting element analysis of 2000 bp promoter upstream of *PebHLH* genes. Functional descriptions of cis-acting elements were labeled by different colors.

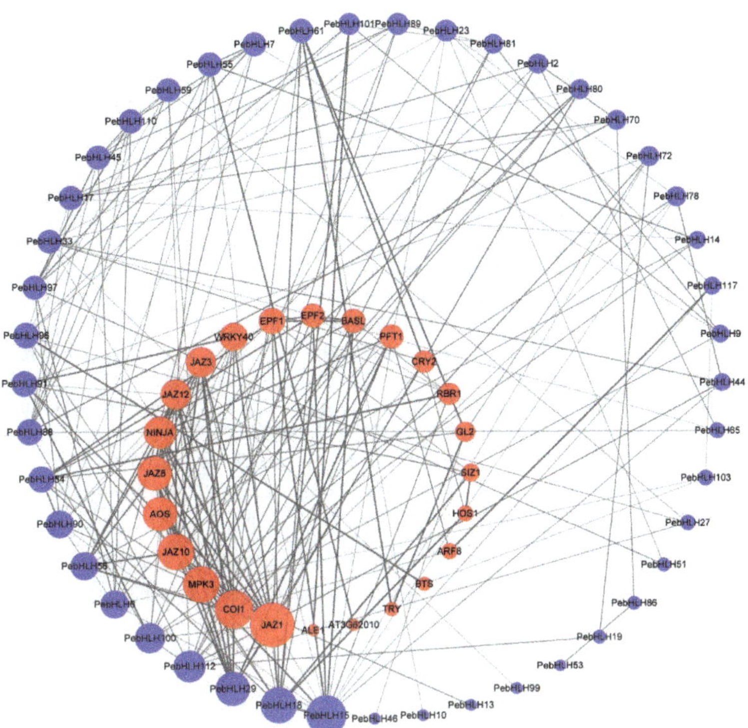

Figure 5. Protein interaction networks prediction of PebHLH. The outer blue circles indicate PebHLH proteins, and the inner red circles indicate other proteins in the reciprocal relationship. Interacting relationships are indicated by straight gray lines.

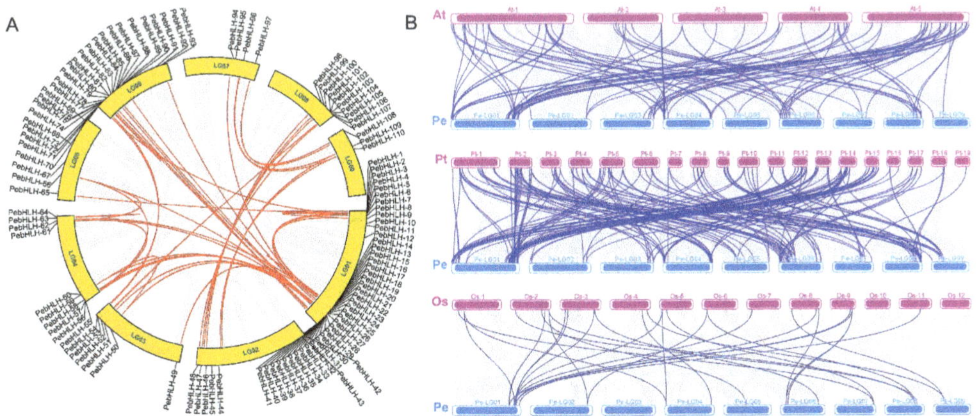

Figure 6. The synteny analysis of bHLHs of passion fruit genomes (**A**) and with other species (**B**). The red line represents the *PebHLH* gene pair, and the blue lines represent homologous gene pairs. Pe, Os, At, and Pt represent passion fruit, rice, *Arabidopsis*, and poplar, respectively.

3.8. Expression Analysis of PebHLHs

Transcriptome analyses showed that the PebHLH genes were differentially expressed under four abiotic stresses (Figure 7). *PebHLH10/11/25/35/36/43/46/68/69/80/117* were induced by drought stress, while *PebHLH6/8/55/63/64/70/83* were suppressed. For the salt treatment, the expression of most genes was highest when treated with salt stress for 3 days. Especially, *PebHLH4* and *PebHLH56* were significantly upregulated under salt stress (NaCl 10d). However, the expression of *PebHLH8/55/63/64/70/83* was suppressed. Under high-temperature stress, some *PebHLHs* were upregulated, such as *PebHLH1/2/5/73/74/92/104*, and some genes were suppressed, such as *PebHLH8/55/63/64/70/83*. Under cold stress, the expression of the most gene had been induced, such as *PebHLH28/29/34/56/57/65/67/85/91/96/107/106/112*. We performed qRT-PCR validation analysis on some of the *PebHLH* genes (Figure 8). The result showed that the gene expression patterns were the same as heatmaps. The *PebHLHs* could respond to various abiotic stresses.

The expression levels of *PebHLHs* were obtained based on transcriptional sequencing results [51] at three different fruit ripening stages (T1, T2, and T3) (Figure 9). Most genes had the highest expression levels in the first period (T1) and regularly decreased in T2 and T3. And the expression of some genes was highest in T2, such as *PebHLH5/6/28/34/41/42/65/68/81/91/95/100/105/106/107*. A few reached the maximum expression in T3, such as *PebHLH13/17/52/62/69/73/74/76/85/101/103/104/109*. This result indicated that the expression of most *PebHLH* genes was negatively correlated with fruit ripening. Some genes were chosen to do the qRT-PCR verification (Figure 10), which showed that the expression trends of *PebHLHs* were consistent with the transcriptome.

3.9. Response of Transgenic Arabidopsis to Cold Stress

In A, GUS staining of the pCAMBIA1304-*PebHLH56p* transgenic seedlings was mainly observed in the leaf, and under the cold stress, GUS staining was enhanced and distributed throughout the transgenic plants. With the increase of cold time, GUS staining gradually deepened, and the GUS enzyme activity also increased (Figure 11).

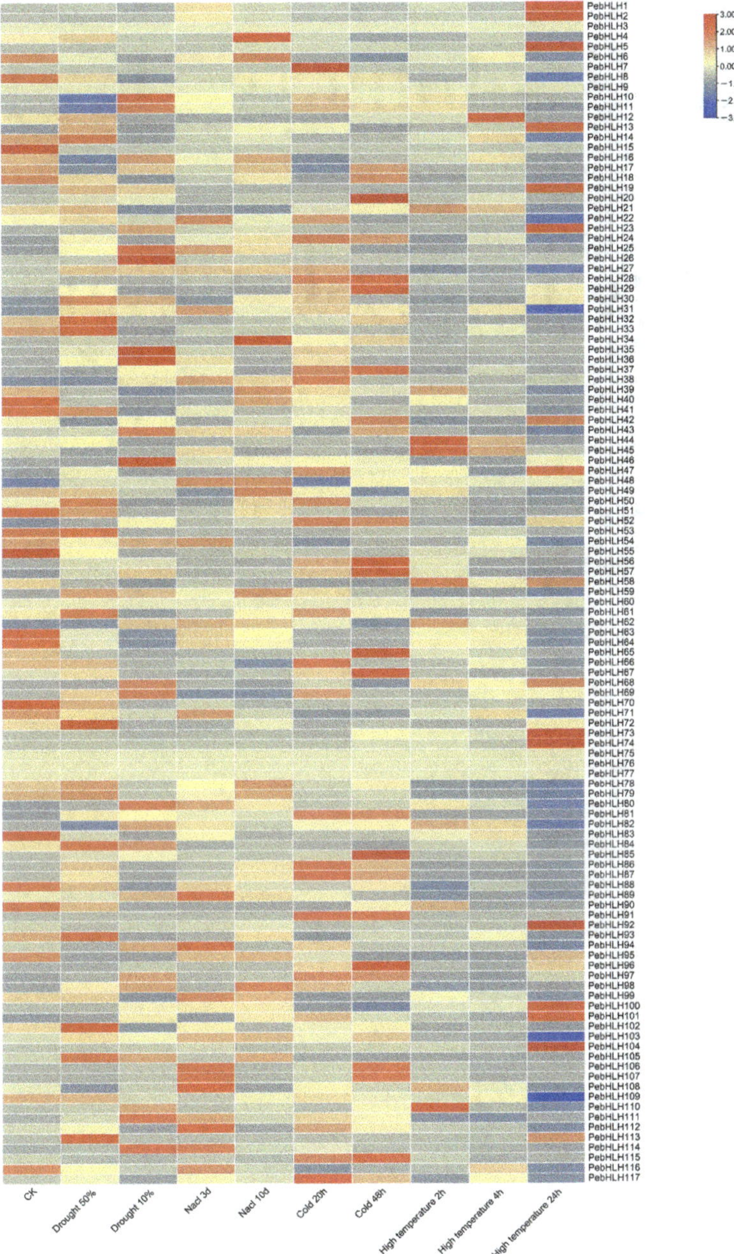

Figure 7. Differential expression of *PebHLHs* responding to different levels of drought, high salt, cold, and high temperature (Table S2). CK is normal growth condition, drought 50% and 10% are soil water content indexes, NaCl 3 d and 10 d are the treatment time under 300 mM concentration, cold 20 h, and 48 h are the treatment under 8 °C, and high temperature 2 h, 4 h, and 24 h are the treatment time under 42 °C, respectively.

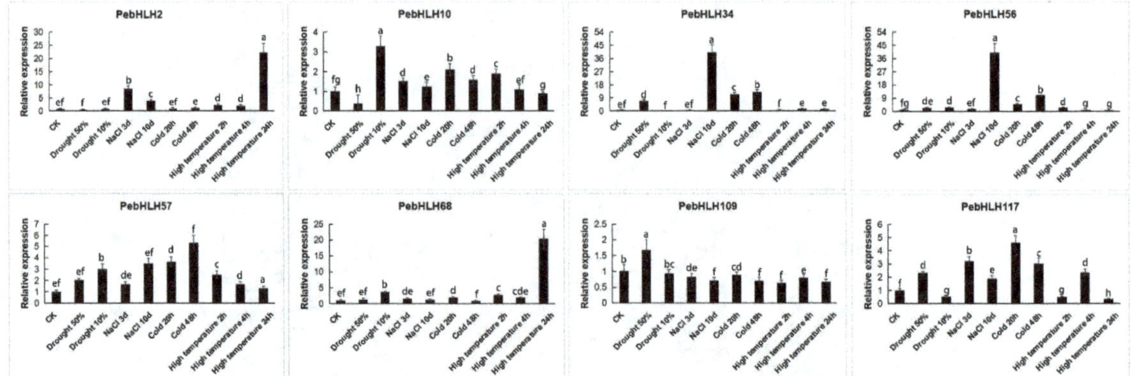

Figure 8. qRT-PCR analysis of 8 *PebHLHs* under 4 different abiotic stresses (Table S3). Horizontal coordinates indicate different stress treatments (the different processing methods are described above), and vertical coordinates represent relative expression values. The different letters mean significance, which was examined by Duncan's range test ($p < 0.05$).

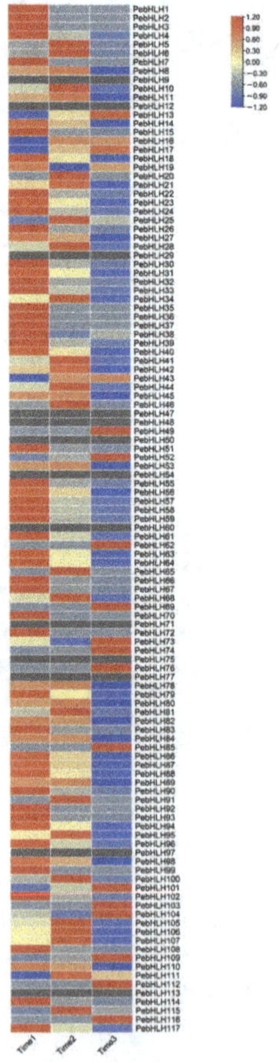

Figure 9. Differential expression of *PebHLHs* during 3 fruit ripening periods (Table S4).

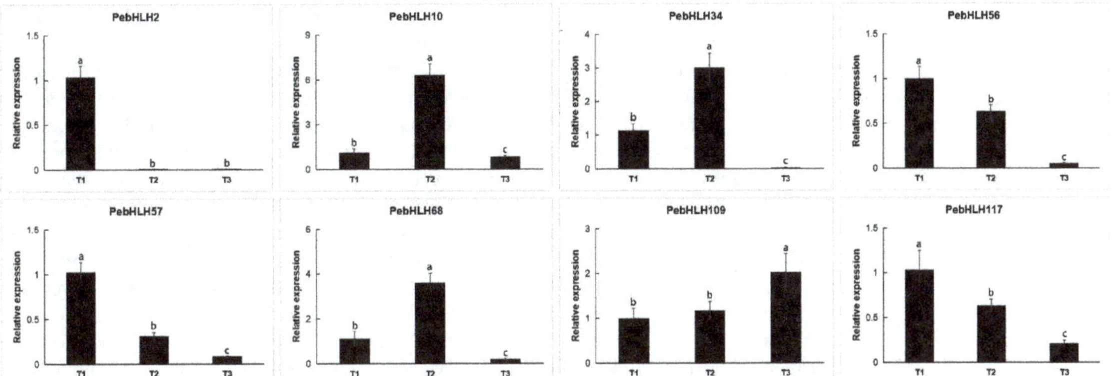

Figure 10. qRT-PCR analysis of 8 *PebHLHs* during 3 fruit ripening periods (Table S5). Horizontal coordinates indicate different stress treatments, and vertical coordinates represent relative expression values. Biological replicates, tests, and *p* values are described above. The different letters mean significance, which was examined by Duncan's range test ($p < 0.05$).

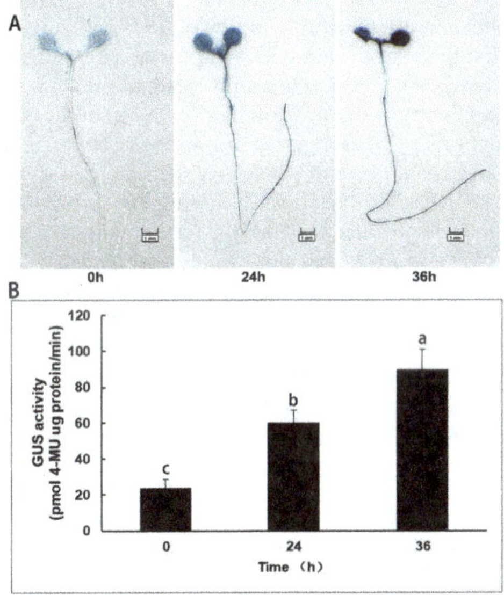

Figure 11. Induction and expression pattern of *PebHLH56* under cold stress. (**A**) GUS staining of overexpressing *Arabidopsis* strains. (**B**) GUS activity quantitative analysis of overexpressing *Arabidopsis*. Horizontal coordinates indicate different stress treatments, and vertical coordinates represent relative expression values. The different letters mean significance, which was examined by Duncan's range test ($p < 0.05$).

4. Discussion and Analysis

Numerous studies have shown that bHLH transcription factors are involved in diverse biological processes and the whole growth cycle [2]. At present, however, the systematic characterization of the bHLH genes in passion fruit is lacking, although bHLHs have been identified in many plants. The first plant bHLH gene was identified in maize (*Zea mays* L.) [52]. Furthermore, in this study, 117 bHLH genes were identified and characterized in passion fruit. This number is similar to some reported species, for example, the pineapple

(121) [53], the pepper (122) [28], the potato (124) [54], the Jilin ginseng (137) [55], and the tomato (159) [56]. The numbers were quite different from some species. For example, the MOSO bamboo was 448 [57], and the wheat was 571 [18]. This shows that the bHLH gene family is diverse in different species. Based on phylogenetic analysis, we classified 117 *PebHLH* genes into 18 subfamilies, which is the same number reported in maize [26]. Previous studies in other species reported six subfamilies in the animal genomes [55], while in plants, the bHLH family genes were divided into 21 in *Arabidopsis*, tomato, and pear [21,23]; 22 in rice [24] and 31 subfamilies in B. napus [53]. Compared with the classification of other plant species, our results show similarities and differences, which also indicate that the classification of bHLH transcription factors in plants is more complex than that in animals, so the classification of plant bHLH families is still needed for more research [58].

The gene structure of the *PebHLH* family was further analyzed. Most family genes contain two or more introns. Among them, *PebHLH12/29/99* have no intron, while seven members (*PebHLH3/45/48/53/65/70/94*) have only one intron. It is generally believed that genes with few or no introns in plants show lower expression levels [59]. The compact gene structure may facilitate the induction of gene expression in response to exogenous stress [60]. For example, intron-less genes, *PebHLH57*, were upregulated under drought, salt, cold, and heat stresses. Genome duplication events have occurred during plant evolution [61]. The evolution of the genome and the expansion of gene families depend primarily on gene duplication events [62]. We have performed the collinearity analysis on passion fruit, rice, and *Populus trichocarpa*, and the results showed that the *PebHLH* gene had the most tandem duplication relationship with the bHLH gene in *Populus trichocarpa*. Therefore, it is speculated that the relationship between passion fruit and *Populus trichocarpa* is the closest. In the same way, *PebHLH31* was found to be an ortholog of *AtPIF4* (AT2G43010.5). The AtPIF gene is a central signaling hub regulating plant growth and development [63].

The functions of bHLH genes in plants are diverse, including plant perception of the growth and development processes such as fruit development [24,64]. *AtCIB1*, together with *AtCRY*, promotes flower opening by stimulating the expression of flowering genes [65,66]. In our results, *PebHLH109* has the highest expression level in fruit, and it is predicted to play a key role in fruit development. It also shows the highest expression level at the T3 stage of fruit ripening. In pepper, *CabHLH33*, a homolog of *AtbHLH31*, was highly expressed in flower buds and petals. Previous studies suggest that *AtbHLH31* regulates petal growth by controlling cell expansion [67].

Recent studies have increasingly focused on the relationship between bHLH genes and abiotic stress. Under stress conditions, certain bHLH TFs are activated, and they combine with the promoters of key genes to regulate the transcription level of the target gene. Several studies have found that *OsbHLH068* of rice and *AtbHLH112* of *Arabidopsis* play an active role in response to salt stress [68]. *MfbHLH38* of the *Myrothamnus flabellifolia* was transformed into *Arabidopsis*, and the drought tolerance of transgenic lines was enhanced by the increase of gene expression [69]. *ZmbHLH55* of Maize can increase salt stress tolerance by regulating the expression of ascorbic acid biosynthesis-related genes [70]. Overexpression of *Ntbhlh123* and *IbbHLH79* can improve the cold tolerance of tobacco [71] and sweet potato, respectively [72]. *ZjbHLH076/ZjICE1* of *Zoysia japonica* can enhance the tolerance of transgenic lines to cold stress [73]. The activity of gene expression is quantitatively efficiently regulated by specific or functional promoters that contain multiple cis-acting elements. These acting elements can respond to a variety of stress responses [52,74]. Plant bHLH responds to various stress responses through the activation of long terminal repeat (LTR) reverse transcription transposons [75,76], with activation factors including drought [77], heat [78], and salt [79]. In this work, the cis-element analysis indicated that *PebHLHs* contained elements (such as cold-responsive element, salicylic acid-responsive element, ABRE) that could be responsive to various stresses, which was consistent with previous studies on potato [18], lotus [80], and Pepper [56] bHLHs.

In this experiment, we focused on the function of the bHLH gene under cold stress. Here, we identified one of the *PebHLHs* (*PebHLH56*), which can respond to cold stress in transgenic *Arabidopsis*. This study provides experimental evidence that bHLH family genes in passion fruit respond to cold stress, and we will further study the resistance mechanism of bHLH genes.

5. Conclusions

The bHLH gene family plays an important role in improving plant response to stress and plant development. In this study, we identified 117 *PebHLHs* from the genome of the passion fruit, and the genomic information about PebHLH family members was analyzed. *PebHLHs* could respond to abiotic stresses, including drought, high salinity, cold and high-temperature stresses, and they were differentially expressed in different fruit developmental stages. We were concerned about a gene *PebHLH56* that was induced under cold stress. The pCAMBIA1304-*PebHLH56p* transgenic *Arabidopsis* had a significantly deeper GUS staining than the control under cold stress, and the GUS enzyme activity also increased. The RT-qPCR result showed that the expression of *PebHLH56* was induced by cold stress. This study provides new insights into the regulatory functions of *PebHLHs* during abiotic stress and fruit ripening and screens a stress-resistance gene as a candidate member to improve the understanding of *PebHLH* gene family characteristics and evolution.

Supplementary Materials: The following supporting information can be downloaded at: https://www.mdpi.com/article/10.3390/horticulturae9020272/s1, Table S1. Basic information of bHLH genes identified in passion fruit; Table S2. The transcriptome data of passion fruit bHLHs under the drought, salt, cold and high-temperature treatment; Table S3. qPCR data of passion fruit bHLHs under the drought, salt, cold and high-temperature treatment; Table S4. The transcriptome data of passion fruit in the three stages of fruit ripening; Table S5. qPCR data of passion fruit bHLHs in the three stages of fruit ripening; Table S6. A list of the oligo primers of PebHLHs used for qRT-PCR.

Author Contributions: Y.X., W.Z., F.M., B.W., D.H., B.X. and W.X. performed experiments, D.C., S.S. and P.S. analyzed the data; Y.X. and S.S. drafted the manuscript. All authors have read and agreed to the published version of the manuscript.

Funding: The work was sponsored by Hainan Provincial Natural Science Foundation (321RC1088), Project of Sanya Yazhou Bay Science and Technology City (SCKJ-JYRC-2022-84, SCKJ-JYRC-2022-93), and National Natural Science Foundation of China (32260737).

Data Availability Statement: Not applicable.

References

1. Pérez-Rodríguez, P.; Riaño-Pachón, D.M.; Corrêa, L.G.; Rensing, S.A.; Kersten, B.; Mueller-Roeber, B. PlnTFDB: Updated content and new features of the plant transcription factor database. *Nucleic Acids Res.* **2010**, *38*, D822–D827. [CrossRef] [PubMed]
2. Zhang, H.; Jin, P.-J.; Tang, L.; Zhao, Y.; Gu, X.-C.; Gao, G.; Luo, J.-C. PlantTFDB 2.0: Update and improvement of the comprehensive plant transcription factor database. *Nucleic Acids Res.* **2011**, *9*, D1114–D1117. [CrossRef] [PubMed]
3. Riechmann, J.L.; Heard, J.; Martin, G.; Reuber, L.; Jiang, C.J.; Keddie, J.; Adam, L.; Pineda, O.; Ratcliffe, O.J.; Samaha, R.R.; et al. *Arabidopsis* transcription factors: Genome-wide comparative analysis among eukaryotes. *Science* **2000**, *290*, 2105–2110. [CrossRef] [PubMed]
4. Wray, G.A.; Hahn, M.W.; Abouheif, E.; Balhoff, J.P.; Pizer, M.; Rockman, M.V.; Romano, L.A. The evolution of transcriptional regulation in eukaryotes. *Mol. Biol. Evol.* **2003**, *20*, 1377–1419. [CrossRef] [PubMed]
5. Schwechheimer, C.; Zourelidou, M.; Bevan, M.W. Plant transcription factor studies. *Annu. Rev. Plant Physiol. Plant Mol. Biol.* **1998**, *49*, 127–150. [CrossRef]
6. Jin, P.J.; Zhang, H.; Kong, L.; Gao, G.; Luo, J.C. PlantTFDB 3.0: A portal for the functional and evolutionary study of plant transcription factors. *Nucleic Acids Res.* **2014**, *42*, D1182–D1187. [CrossRef]
7. Xu, Y.-H.; Liao, Y.-C.; Lu, F.-F.; Zhang, Z.; Sun, P.-W.; Gao, Z.-H.; Hu, K.-P.; Cun, S.; Jin, Y.; Wei, J.-H. Transcription Factor AsMYC2 Controls the Jasmonate-responsive Expression of ASS1 Regulating Sesquiterpene Biosynthesis in *Aquilaria sinensis* (Lour.) Gilg. *Plant Cell Physiol.* **2017**, *58*, 1924–1933. [CrossRef]
8. Feller, A.; Machemer, K.; Braun, E.L.; Grotewold, E. Evolutionary and comparative analysis of MYB and bHLH plant transcription factors. *Plant J.* **2011**, *66*, 94–116. [CrossRef]
9. Yin, J.; Chang, X.; Kasuga, T.; Bui, M.; Reid, M.S.; Jiang, C.Z. A basic helix-loop-helix transcription factor, PhFBH4, regulates flower senescence by modulating ethylene biosynthesis pathway in petunia. *Hortic. Res.* **2015**, *2*, 15059. [CrossRef]

10. Murre, C.; McCaw, P.S.; Baltimore, D. A new DNA binding and dimerization motif in immunoglobulin enhancer binding, daughterless, MyoD, and mycproteins. *Cell* **1989**, *56*, 777–783. [CrossRef]

11. Atchley, W.R.; Terhalle, W.; Dress, A. Positional dependence, cliques, and predictive motifs in the bHLH protein domain. *J. Mol. Evol.* **1999**, *48*, 501–516. [CrossRef] [PubMed]

12. Ellenberger, T.; Fass, D.; Arnaud, M.; Harrison, S.C. Crystal structure of transcription factor E47: E-box recognition by a basic region helix-loop-helix dimer. *Comp. Study* **1994**, *8*, 970–980. [CrossRef] [PubMed]

13. Nesi, N.; Debeaujon, I.; Jond, C.; Pelletier, G.; Caboche, M.; Lepiniec, L. The TT8 gene encodes a basic helix-loop-helix domain protein required for expression of DFR and BAN genes in *Arabidopsis siliques*. *Plant Cell* **2000**, *12*, 1863–1878. [CrossRef] [PubMed]

14. Hong, Y.Q.; Ahmad, N.; Tian, Y.Y.; Liu, J.Y.; Wang, L.Y.; Wang, G.; Liu, X.M.; Dong, Y.Y.F.; Wang, W.; Liu, W.C.; et al. Genome-Wide Identification, Expression Analysis, and Subcellular Localization of *Carthamus tinctorius* bHLH Transcription Factors. *Int. J. Mol. Sci.* **2019**, *20*, 3044. [CrossRef] [PubMed]

15. Li, H.T.; Gao, W.L.; Xue, C.L.; Zhang, Y.; Liu, Z.G.; Zhang, Y.; Zhang, Y.; Meng, X.W.; Liu, M.J.; Zhao, J. Genome-wide analysis of the bHLH gene family in Chinese jujube (*Ziziphus jujuba* Mill.) and wild jujube. *BMC Genom.* **2019**, *20*, 568. [CrossRef]

16. Aslam, M.; Jakada, B.H.; Fakher, B.; Greaves, J.G.; Niu, X.; Su, Z.; Cheng, Y.; Cao, S.J.; Wang, X.M.; Qin, Y. Genome-wide study of pineapple (*Ananas comosus* L.) bHLH transcription factors indicates that cryptochrome-interacting bHLH2 (AcCIB2) participates in flowering time regulation and abiotic stress response. *BMC Genom.* **2020**, *21*, 735. [CrossRef]

17. Zhang, Z.S.; Chen, J.; Liang, C.L.; Liu, F.; Hou, X.L.; Zou, X.X. Genome-Wide Identification and Characterization of the bHLH Transcription Factor Family in Pepper (*Capsicum annuum* L.). *Front. Genet.* **2020**, *11*, 570156. [CrossRef]

18. Wang, R.-Q.; Zhao, P.; Kong, N.; Lu, R.-Z.; Pei, Y.; Huang, C.-X.; Ma, H.-L.; Chen, Q. Genome-Wide Identification and Characterization of the Potato bHLH Transcription Factor Family. *Genes* **2018**, *22*, 54. [CrossRef]

19. Gao, C.; Sun, J.-L.; Wang, C.-Q.; Dong, Y.-M.; Xiao, S.-H.; Wang, X.-J.; Jiao, Z.-G. Genome-wide analysis of basic/helix-loop-helix gene family in peanut and assessment of its rolesin pod development. *PLoS ONE.* **2017**, *27*, e0181843.

20. Zhu, L.; Zhao, M.-Z.; Chen, M.-Y.; Li, L.; Jiang, Y.; Liu, S.-Z.; Jiang, Y.; Wang, K.-Y.; Sun, C.-Y.; Chen, J.; et al. The bHLH gene family and its response to saline stress in *Jilin ginseng, Panax ginseng* C.A. Meyer. *Mol. Genet. Genom.* **2022**, *295*, 877–890. [CrossRef]

21. Niu, X.; Guan, Y.-X.; Chen, S.-K.; Li, H.-F. Genome-wide analysis of basic helix-loop-helix (bHLH) transcription factors in *Brachypodium distachyon*. *BMC Genom.* **2017**, *18*, 619. [CrossRef] [PubMed]

22. Kavas, M.; Baloğlu, M.-C.; Atabay, E.-S.; Ziplar, U.-T.; Daşgan, H.-Y.; Ünver, T. Genome-wide characterization and expression analysis of common bean bHLH transcription factors in response to excess salt concentration. *Mol. Gen. Genom.* **2016**, *291*, 129–143. [CrossRef] [PubMed]

23. Sun, H.; Fan, H.-J.; Ling, H.-Q. Genome-wide identification and characterization of the bHLH gene family in tomato. *BMC Genom.* **2015**, *16*, 9. [CrossRef] [PubMed]

24. Li, X.-X.; Duan, X.-P.; Jiang, H.-X.; Sun, Y.-J.; Tang, Y.-P.; Zhen, Y.; Guo, J.-K.; Liang, W.-Q.; Chen, L.; Yin, J.Y.; et al. Genome-wide analysis of basic/helix-loop-helix transcription factor family in rice and *Arabidopsis*. *Plant Physiol.* **2006**, *141*, 1167–1184. [CrossRef] [PubMed]

25. Mao, K.; Dong, Q.-L.; Li, C.; Liu, C.-H.; Ma, F.-W. Genome Wide Identification and Characterization of Apple bHLH Transcription Factors and Expression Analysis in Response to Drought and Salt Stress. *Front. Plant Sci.* **2017**, *11*, 480. [CrossRef] [PubMed]

26. Zhang, T.-T.; Lv, W.; Zhang, H.-S.; Ma, L.; Li, P.-H.; Ge, L.; Chang, L. Genome-wide analysis of the basic Helix-Loop-Helix (bHLH) transcription factor family in maize. *BMC Plant Biol.* **2018**, *18*, 235. [CrossRef]

27. Song, X.-M.; Huang, Z.-N.; Duan, W.-K.; Ren, J.; Liu, T.-K.; Li, Y.; Hou, X.-L. Genome-wide analysis of the bHLH transcription factor family in Chinese cabbage (*Brassica rapa* ssp. pekinensis). *Mol. Gen. Genom.* **2014**, *289*, 77–91. [CrossRef]

28. Chen, X.-R.; Xiong, R.; Liu, H.-L.; Wu, M.; Chen, F.; Yan, H.-W.; Xiang, Y. Basic helix-loop-helix gene family: Genome wide identification, phylogeny, and expression in Moso bamboo. *Plant Physiol. Biochem.* **2018**, *132*, 104–119. [CrossRef]

29. Wei, K.-F.; Chen, H.-Q. Comparative functional genomics analysis of bHLH gene family in rice, maize and wheat. *BMC Plant Biol.* **2018**, *18*, 309. [CrossRef]

30. Huq, E.; Quail, P.-H. A phytochrome-interacting bHLH factor, functions as a negative regulator of phytochrome B signaling in *Arabidopsis*. *EMBO J.* **2022**, *15*, 2441–2450. [CrossRef]

31. Lee, S.; Lee, S.-h.; Yang, K.-Y.; Kim, Y.-M.; Park, S.-Y.; Kim, S.-Y.; Soh, M.-S. Overexpression of PRE1 and itshomologous genes activates Gibberellin-dependent responses in *Arabidopsis thaliana*. *Plant Cell Physiol.* **2006**, *47*, 591–600. [CrossRef] [PubMed]

32. Hu, D.-G.; Sun, C.-H.; Zhang, Q.-Y.; An, J.-P.; You, C.-X.; Hao, Y.-J. Glucose Sensor MdHXK1 Phosphorylates and Stabilizes MdbHLH3 to Promote Anthocyanin Biosynthesis in Apple. *PLoS Genet.* **2016**, *25*, e1006273.

33. Bernhardt, C.; Lee, M.-M.; Gonzalez, A.; Zhang, F.; Lloyd, A.; Schiefelbein, J. The bHLH genes GLABRA3(GL3) and ENHANCER OF GLABRA3 (EGL3) specify epidermal cell fate in the *Arabidopsis* root. *Development* **2003**, *130*, 6431–6439. [CrossRef] [PubMed]

34. Duek, P.-D.; Fankhauser, C. bHLH class transcription factors take Centre stage in phytochrome signalling. *Trends Plant Sci.* **2005**, *10*, 51–54. [CrossRef]

35. Yao, P.-F.; Li, C.-L.; Zhao, X.-R.; Li, M.-F.; Zhao, H.-X.; Guo, J.-Y.; Cai, Y.; Chen, H.; Wu, Q. Overexpression of a Tartary buckwheat gene, FtbHLH3, Enhances Drought/Oxidative Stress Tolerance in Transgenic *Arabidopsis*. *Front. Plant Sci.* **2017**, *8*, 625. [CrossRef]

36. Yao, P.-F.; Sun, Z.-X.; Chen, Q.-L.; Zhao, X.-Y.; Li, M.-F.; Deng, R.-Y.; Huang, Y.-J.; Zhao, H.-Q.; Chen, H.; Wu, Q. Overexpression of Fagopyrum tataricum FtbHLH2 enhances tolerance to cold stress in transgenic *Arabidopsis*. *Plant Physiol. Biochem.* **2018**, *125*, 85–94. [CrossRef]

37. Zhai, Y.-Q.; Zhang, L.-C.; Xia, C.; Fu, S.-L.; Zhao, G.-Y.; Jia, J.-Z.; Kong, X.Y. The wheat transcription factor, TabHLH39, improves tolerance to multiple abiotic stressors in transgenic plants. *Biochem. Biophys. Res. Commun.* **2016**, *473*, 1321–1327. [CrossRef]

38. Jiang, Y.-Q.; Yang, B.; Deyholos, M.-K. Functional characterization of the *Arabidopsis* bHLH92 transcription factor in abiotic stress. *Mol. Gen. Genom.* **2009**, *282*, 503–516. [CrossRef]

39. Huang, X.-S.; Wang, W.; Zhang, Q.; Liu, J.-H. A basic helix-loop-helix transcription factor, PtrbHLH, of *Poncirus trifoliata* confers cold tolerance and modulates peroxidase-mediated scavenging of hydrogen peroxide. *Plant Physiol.* **2013**, *162*, 1178–1194. [CrossRef]

40. Waseem, M.; Li, N.; Su, D.-D.; Chen, J.-X.; Li, Z.-G. Overexpression of a basic helix–loop–helix transcription factor gene, SlbHLH22, promotes early fowering and accelerates fruit ripening in tomato (*Solanum lycopersicum* L). *Planta* **2019**, *250*, 173–185. [CrossRef]

41. Tan, C.; Qiao, H.-L.; Ma, M.; Wang, X.; Tian, Y.-Y.; Bai, S.; Hasi, A. Genome-Wide Identification and Characterization of Melon bHLH Transcription Factors in Regulation of Fruit Development. *Plants* **2021**, *10*, 2721. [CrossRef] [PubMed]

42. Zhou, D.; Shen, Y.-H.; Zhou, P.; Fatima, M.; Lin, J.-S.; Yue, J.-J.; Zhang, X.-T.; Chen, L.-Y.; Ming, Y. Papaya CpbHLH1/2 regulate carotenoid biosynthesis-related genes during papaya fruit ripening. *Hortic. Res.* **2019**, *6*, 80. [CrossRef] [PubMed]

43. Huang, D.-M.; Xu, Y.; Wu, B.; Ma, F.-N.; Song, S. Comparative analysis of basic quality of passion fruits (*Passiflora edulis sims*) in Guangxi, Guizhou and Fujian, China. *Bangladesh J. Bot.* **2019**, *48*, 901–906.

44. Costa, J.-L.; Jesus, O.-N.-D.; Oliverira, G.-A.-F.; Oliverira, E.-J.-D. Effect of selection on genetic variability in yellow passion fruit. *Crop Breed. Appl. Biotechnol.* **2012**, *12*, 253–260. [CrossRef]

45. Song, S.; Zhang, D.-H.; Ma, F.-N.; Xing, W.-T.; Huang, D.-M.; Wu, B.; Chen, J.; Chen, D.; Xu, B.Q.; Xu, Y. Genome-Wide Identification and Expression Analyses of the Aquaporin Gene Family in Passion Fruit (*Passiflora edulis*), Revealing PeTIP3-2 to Be Involved in Drought Stress. *Int. J. Mol. Sci.* **2022**, *23*, 5720. [CrossRef]

46. Tamura, K.; Dudley, J.; Nei, M.; Kumar, S. MEGA4, molecular evolutionary genetics analysis MEGA software version 4.0. *Mol. Biol. Evol.* **2007**, *24*, 596–1599. [CrossRef]

47. Chen, C.-J.; Chen, H.; Zhang, Y.; Thomas, H.-R.; Frank, M.-H.; He, Y.-H.; Xia, R. TBtools: An integrative toolkit developed for interactive analyses of big biological data. *Mol. Plant* **2020**, *13*, 1199–1202. [CrossRef]

48. Lescot, M.; Dehais, P.; Thijs, G.; Marchal, K.; Moreau, Y.; Peer, Y.-V.-D.; Rouze, P.; Rombauts, S. PlantCARE, a database of plant cis-acting regulatory elements and a portal to tools for in silico analysis of promoter sequences. *Nucleic Acids Res.* **2002**, *30*, 325–327. [CrossRef]

49. Jefferson, R.-A.; Kavanagh, T.-A.; Bevan, M.-W. GUS fusions: Beta-glucuronidase as a sensitive and versatilegene fusion marker in higher plants. *EMBO J.* **1987**, *6*, 3901–3907. [CrossRef]

50. Xu, Y.; Jin, Z.-Q.; Xu, B.-Y.; Li, J.-Y.; Li, Y.-J.; Wang, X.-Y.; Wang, A.-B.; Hu, W.; Huang, D.-M.; Wei, Q.; et al. Identification of transcription factors interacting with a 1274bp promoter of MaPIP1;1 which confers high-level gene expression and drought stress Inducibility in transgenic *Arabidopsis thaliana*. *BMC Plant Biol.* **2020**, *20*, 278. [CrossRef]

51. Xia, Z.-Q.; Huang, D.-M.; Zhang, S.-K.; Wang, W.-Q.; Ma, F.-N.; Wu, B.; Xu, Y.; Xu, B.-Q.; Chen, D.; Zou, M.-L.; et al. Chromosome-scale genome assembly provides insights into the evolution and flflavor synthesis of passion fruit (*Passiflflora edulis* Sims). *Hortic. Res.* **2021**, *8*, 14. [CrossRef] [PubMed]

52. Guo, J.-R.; Sun, B.-X.; He, H.-R.; Zhang, Y.-F.; Tian, H.-Y.; Wang, B.-S. Current Understanding of bHLH Transcription Factors in Plant Abiotic Stress Tolerance. *Int. J. Mol. Sci.* **2021**, *22*, 4921. [CrossRef] [PubMed]

53. Zhou, X.; Liao, Y.-L.; Kim, S.-U.; Chen, Z.-X.; Nie, G.-P.; Cheng, S.-Y.; Ye, J.-B.; Xu, F. Genome-wide identification and characterization of bHLH family genes from *Ginkgo biloba*. *Sci. Rep.* **2020**, *10*, 13723. [CrossRef] [PubMed]

54. Huang, H.-G.; Yu, N.; Wang, L.-J.; Gupta, D.-K.; He, Z.-L.; Wang, K.; Zhu, Z.-Q.; Yan, X.-C.; Li, T.-Q.; Yang, X.-E. The phytoremediation potential of bioenergy crop Ricinus communis for DDTs and cadmium co-contaminated soil. *Bioresour. Technol.* **2011**, *102*, 11034–11038. [CrossRef]

55. Atchley, W.-R.; Fitch, W.-M. A natural classification of the basic helix–loop–helix class of transcription factors. *Proc. Natl. Acad. Sci. USA* **1997**, *94*, 5172–5176. [CrossRef]

56. Kumar, S.; Stecher, G.; Tamura, K. MEGA7: Molecular evolutionary genetics analysis version 7.0 for bigger datasets. *Mol. Biol. Evol.* **2016**, *33*, 1870–1874. [CrossRef]

57. Adeleke, B.-S.; Babalola, O.-O. Oilseed crop sunflower (*Helianthus annuus*) as a source of food: Nutritional and health benefits. *Food Sci. Nutr.* **2020**, *8*, 4666–4684. [CrossRef]

58. Xu, Z.-L.; Liu, X.-Q.; He, X.-L.; Xu, L.; Huang, Y.-H.; Shao, H.-B.; Zhang, D.-Y.; Tang, B.-P.; Ma, H.-X. The soybean basic helix-loop-helix transcription factor ORG3-like enhances cadmium tolerance via increased iron and reduced cadmium uptake and transport from roots to shoots. *Front. Plant Sci.* **2017**, *8*, 1098. [CrossRef]

59. Ren, X.-Y.; Vorst, O.; Fiers, M.-W.-E.-J.; Stiekema, W.-J.; Nap, J.-P. In plants, highly expressed genes are the least compact. *Trends Genet.* **2006**, *22*, 528–532. [CrossRef]

60. Jeffares, D.-C.; Penkett, C.-J.; Bahler, J. Rapidly regulated genes are intron poor. *Trends Genet.* **2008**, *24*, 375–378. [CrossRef]

61. Mehan, M.-R.; Freimer, N.-B.; Ophoff, R.-A. A genome-wide survey of segmental duplications that mediate common human genetic variation of chromosomal architecture. *Hum. Genom.* **2004**, *1*, 335–344. [CrossRef] [PubMed]

62. Vision, T.-J.; Brown, D.-G.; Tanksley, S.-D. The origins of genomic duplications in *Arabidopsis*. *Science* **2000**, *290*, 2114–2117. [CrossRef] [PubMed]

63. Paik, I.; Kathare, P.-K.; Kim, J.-I.; Huq, E. Expanding Roles of PIFs in Signal Integration from Multiple Processes. *Mol. Plant* **2017**, *10*, 1035–1046. [CrossRef] [PubMed]
64. Amoutzias, G.D.; Robertson, D.L.; Oliver, S.G.; Bornberg-Bauer, E. Convergent evolution of gene networks by single-gene duplications in higher eukaryotes. *EMBO Rep.* **2004**, *5*, 274–279. [CrossRef]
65. Liu, H.-T.; Yu, X.-H.; Li, K.-W.; Klejnot, J.; Yang, H.-Y.; Lisiero, D.; Lin, C. Photoexcited CRY2 interacts with CIB1 to regulate transcription and floral initiation in *Arabidopsis*. *Science* **2008**, *322*, 1535–1539. [CrossRef]
66. Liu, H.-T.; Wang, Q.; Liu, Y.-W.; Zhao, X.-Y.T.; Imaizumi, D.-E.; Somers, D.E.; Tobin, E.M.; Lin, C. *Arabidopsis* CRY2 and ZTL mediate blue-light regulation of the transcription factor CIB1 by distinct mechanisms. *Proc. Natl. Acad.Sci. USA* **2013**, *110*, 17582–17587. [CrossRef]
67. Varaud, E.; Brioudes, F.; Szecsi, J.; Leroux, J.; Brown, S.; Perrot-Rechenmann, C. AUXIN RESPONSE FACTOR8 regulates *Arabidopsis* petal growth by interacting with the bHLH transcription factor BIGPETALp. *Plant Cell* **2011**, *23*, 973–983. [CrossRef]
68. Liu, C.; Zhang, Y.; Wang, B.-M.; Ran, Q.-J.; Zhang, J.-R. The bHLH family member ZmPTF1 regulates drought tolerance in maize by promoting root development and abscisic acid synthesis. *J. Exp. Bot.* **2019**, *70*, 5471–5486. [CrossRef]
69. Qiu, J.-R.; Huang, Z.; Xiang, X.-Y.; Xu, W.-X.; Wang, J.-T.; Chen, J.; Song, L.; Xiao, Y.; Li, X.; Ma, J.; et al. MfbHLH38, a *Myrothamnus flabellifolia* bHLH transcription factor, confers tolerance to drought and salinity stresses in *Arabidopsis*. *BMC Plant Biol.* **2020**, *20*, 542. [CrossRef]
70. Yu, C.-M.; Yan, M.; Dong, H.-Z.; Luo, J.; Ke, Y.-C.; Guo, A.-F.; Chen, Y.-H.; Zhang, J.; Huang, X.-S. Maize bHLH55 functions positively in salt tolerance through modulation of AsA biosynthesis by directly regulating GDP-mannose pathway genes. *Plant Sci.* **2021**, *302*, 110676. [CrossRef]
71. Liu, Y.-J.; Ji, X.-Y.; Nie, X.-G.; Qu, M.; Heng, L.-Z.; Tan, Z.-L.; Zhao, H.-M.; Huo, L.; Liu, S.-N.; Zhang, B.; et al. *Arabidopsis* AtbHLH112regulates the expression of genes involved in abiotic stress tolerance bybinding to their E-box and GCG-box motifs. *New Phytol.* **2015**, *207*, 692–709. [CrossRef] [PubMed]
72. Jin, R.; Kim, H.-S.; Yu, T.; Zhang, A.-J.; Yang, Y.-F.; Liu, M.; Yu, W.-H.; Zhao, P.; Zhang, Q.-Q.; Cao, Q.-H.; et al. Identification and function analysis of bHLH genes in response to cold stress in sweetpotato. *Plant Physiol. Biochem.* **2021**, *169*, 224–235. [CrossRef] [PubMed]
73. Zuo, Z.-F.; Sun, H.-J.; Lee, H.-Y.; Kang, H.-G. Identification of bHLH genes through genome-wide association study and antisense expression of ZjbHLH076/ZjICE1 influence tolerance to low temperature and salinity in *Zoysia japonica*. *Plant Sci.* **2021**, *313*, 111088. [CrossRef] [PubMed]
74. Xu, Y.; Liu, J.-H.; Jia, C.-H.; Hu, W.; Song, S.; Xu, B.-Y.; Jin, Z.-Q. Overexpression of a Banana Aquaporin Gene MaPIP1;1 Enhances Tolerance to Multiple Abiotic Stresses in Transgenic Banana and Analysis of Its Interacting Transcription Factors. *Front. Plant Sci.* **2021**, *12*, 699230. [CrossRef] [PubMed]
75. Lin, Y.; Zheng, H.; Zhang, Q.; Liu, C.; Zhang, Z. Functional profiling of EcaICE1 transcription factor gene from Eucalyptus camaldulensis involved in cold response in tobacco plants. *J. Plant Biochem. Biotechnol.* **2014**, *23*, 141–150. [CrossRef]
76. Xu, W.-R.; Jiao, Y.-T.; Li, R.-M.; Zhang, N.-B.; Xiao, D.-M.; Ding, X.-L.; Wang, Z.-P. Chinese wild-growing Vitis amurensis ICE1 and ICE2 encode MYC-type Bhlh transcription activators that regulate cold tolerance in *Arabidopsis*. *PLoS ONE* **2014**, *9*, e102303. [CrossRef]
77. Seo, J.S.; Joo, J.; Kim, M.J.; Kim, Y.K.; Nahm, B.H.; Song, S.I.; Cheong, J.J.; Jong, S.L.; Kim, J.K.; Choi, Y.D. OsbHLH148, a basic helix-loop-helix protein, interacts with OsJAZ proteins in a jasmonate signaling pathway leading to drought tolerance in rice. *Plant J.* **2011**, *65*, 907–921. [CrossRef]
78. Ko, D.K.; Lee, M.O.; Hahn, J.S.; Kim, B.G.; Hong, C.B. Submergence-inducible and circadian rhythmic basic helix-loop-helix protein gene in *Nicotiana tabacum*. *J. Plant Physiol.* **2009**, *166*, 1090–1100. [CrossRef]
79. Liu, W.; Tai, H.-H.; Li, S.-S.; Gao, W.; Zhao, M.; Xie, C.-X.; Li, W.-X. bHLH122 is important for drought and osmotic stress resistance in *Arabidopsis* and in the repression of ABA catabolism. *New Phytol.* **2014**, *201*, 1192–1204. [CrossRef]
80. Mao, T.-Y.; Liu, Y.-Y.; Zhu, H.-H.; Zhang, J.; Yang, J.X.; Fu, Q.; Wang, N.; Wang, Z. Genome-wide a nalyses of the bHLH gene family reveals structural and functional characteristics in the aquatic plant *Nelumbo nucifera*. *PeerJ* **2019**, *7*, e7153. [CrossRef]

Article

Systematic Identification of Long Non-Coding RNAs under Allelopathic Interference of Para-Hydroxybenzoic Acid in *S. lycopersicum*

Guoting Liang [1], Yajie Niu [2] and Jing Guo [2,*]

[1] College of Seed and Facility Agricultural Engineering, Weifang University, Weifang 255430, China
[2] State Forestry and Grassland Administration Key Laboratory of Silviculture in Downstream Areas of the Yellow River, College of Forestry, Shandong Agricultural University, Tai'an 271018, China
* Correspondence: jingguo@sdau.edu.cn

Abstract: The importance of long noncoding RNAs (lncRNAs) in plant development has been established, but a systematic analysis of the lncRNAs expressed during plant allelopathy has not been carried out. We performed RNA-seq experiments on *S. lycopersicum* subjected to different levels of para-hydroxybenzoic acid (PHBA) stress during plant allelopathy and identified 61,729 putative lncRNAs. Of these, 7765 lncRNAs cis-regulated 5314 protein-coding genes (PGs). Among these genes, 1116 lncRNAs and 2239 PGs were involved in a complex web of transcriptome regulation, and we divided these genes into 12 modules. Within these modules, 458 lncRNAs and 975 target genes were found to be highly correlated. Additionally, 989 lncRNAs trans-regulated 1765 PGs, and we classified them into 11 modules, within which 335 lncRNAs were highly correlated with their 633 corresponding target genes. Only 98 lncRNAs in *S. lycopersicum* had homologs in the lncRNA database of *Arabidopsis thaliana*, all of which were affected by the PHBA treatments. MiRNAs that interacted with both mRNAs and lncRNAs were selected on the basis of weighted correlation network analysis (WGCNA) results to make lncRNA-miRNA-mRNA triplets. Our study presents a systematic identification of lncRNAs involved in plant allelopathy in *S. lycopersicum* and provides research references for future studies.

Keywords: *S. lycopersicum*; long noncoding RNAs; allelopathy; cis-regulatory; trans-regulatory; endogenous target mimics; transcriptome

Citation: Liang, G.; Niu, Y.; Guo, J. Systematic Identification of Long Non-Coding RNAs under Allelopathic Interference of Para-Hydroxybenzoic Acid in *S. lycopersicum*. *Horticulturae* **2022**, *8*, 1134. https://doi.org/10.3390/horticulturae8121134

Academic Editor: Sergey V. Dolgov

Received: 30 September 2022
Accepted: 22 November 2022
Published: 2 December 2022

Publisher's Note: MDPI stays neutral with regard to jurisdictional claims in published maps and institutional affiliations.

1. Introduction

Since the proposal of the central dogma of molecular biology in 1961, RNA has been considered an intermediate that translates genetic information from DNA to protein [1,2]. However, with the discovery of noncoding RNAs (ncRNAs), we found that intermediate RNAs represent a fraction of all RNAs [2,3]. As the main products of the eukaryotic transcriptome, ncRNAs function as structural, catalytic, or regulatory RNAs rather than as protein encoders [2,4–6]. Based on RNA length, regulatory ncRNAs can be further classified as small ncRNAs (<200 bp, e.g., miRNAs, siRNAs, and piRNAs) or long ncRNAs (lncRNAs) (>200 bp, e.g., lincRNAs and macroRNAs) [2,7]. In the early 1990s, with the appearance of the X-chromosome-silencing phenomenon, long ncRNAs (lncRNAs) were first discovered, having once been considered the 'dark matter' of transcriptomes [8]. Recently, lncRNAs have been discovered to have strong and universal regulatory effects on gene expression at the post-transcriptional, transcriptional, and epigenetic levels [9–13].

lncRNAs play important roles in many biological processes in plants, especially in developmental regulation and stress responses [14–16]. A total of 1832 lncRNAs in Arabidopsis were changed after cold, drought, high-salt, and abscisic acid (ABA) treatments [17]. Several lncRNAs, such as COLDAIR and COOLAIR, positively respond to cold and function in flowering induction [18,19]. IPS1 and At4, induced by phosphate starvation, regulate the shoot dynamic balance of phosphate through the blockage of the repression of miR399 on its target gene, PHO2 [20–23]. Further exploration of the function of lncRNAs in plant responses to stress would provide approaches to discover the stress response networks [24].

As an important factor affecting agricultural production, allelopathy has become a hot topic in ecology, horticulture, agronomy, and other research areas [25–29]. Allelochemicals have been proven to be responsible for numerous biochemical and physiological changes that cause plant allelopathy [30,31]. The plant growth regulatory system, respiration, photosynthesis, the antioxidant system, and water and nutrient uptake are the key physiological and biochemical systems and processes in plants in which changes are induced by allelochemicals [32,33].

Transcription factors, DNA methylation events, chromatin modifications, and microRNAs have been discovered in the gene regulatory processes described above [34–37]. LncRNAs are critical in plant gene expression regulation in response to stress [38–40], which indicates that lncRNAs might also function in plant response to allelopathy. The involvement of lncRNAs in gene transcription regulation progress have not been explored. The expression of most lncRNAs is tissue-specific, which allows them to be discovered by transcriptome sequencing (RNA-seq). By examining RNA-seq data of leaves under salt stress, long noncoding RNAs were identified and characterized in *Medicago truncatula* [41].

However, plant allelopathy is a complex process. To explore the allelopathy network in plants, it is necessary to study the dynamic changes in gene expression under different degrees of stress. The roles of lncRNAs in plant allelopathy in *S. lycopersicum* were explored in this study. Firstly, lncRNAs expressed in *S. lycopersicum* leaves under different PHBA treatments were identified with RNA-seq. Next, cis- and trans-regulated target PGs (TTPGs) for lncRNAs were predicted, and the coexpressions between lncRNAs and their targets were analyzed. Third, the conservation of lncRNAs in *S. lycopersicum* and *Ahrabidopsis thaliana* was compared. Lastly, we explored the possibility that these lncRNAs might be endogenous pseudotarget mimics (eTMs) of known *S. lycopersicum* miRNAs. Overall, this study helps to understand the role of lncRNAs in the molecular network of plant allelopathy and gives new clues about plant response to stress.

2. Materials and Methods

2.1. Plant Materials and Growth Conditions

The *S. lycopersicum* cultivar 'Diana' was grown at the experimental farm of the Shandong Facility Horticultural Laboratory, Weifang, Shandong Province, China. Tomato seeds were planted in plug plates (Wei Nong Company, Taizhou, China). One month later, we treated the *S. lycopersicum* plants with different concentrations of PHBA (0 mmol/L, 15 mmol/L, and 30 mmol/L) for 2 days. Then, the leaves from the middles of the plants were collected as samples. Specifically, the samples for the CK, T1, and T2 treatments corresponded to the 0 mmol/L, 15 mmol/L, and 30 mmol/L concentrations of PHBA, respectively (Figure 1A). We collected leaf samples from 60 individual tomato seedlings and stored them at $-75\ ^\circ$C in an ultra-low temperature freezer for the following experiments.

2.2. RNA Sequencing and Identification of lncRNAs

Total RNA was extracted from the leaf samples for the different PHBA treatments described above using the MagMAX Plant RNA Isolation Kit (Thermo Fisher Scientific, Carlsbad, CA, USA). RNA quality was measured with a Thermo Scientific NanoDrop 2000 system (Thermo Fisher Scientific, USA), and RNA integrity was examined with an Agilent 2100 Bioanalyzer System (Agilent Technologies, Santa Clara, CA, USA) [42]. RNAs with integrity numbers greater than 7 were selected for strand-specific RNA-seq library construction. First, ribosomal RNAs were removed from the total RNAs using the plant Ribo-Zero rRNA Removal Kit (Illumina, San Diego, CA, USA). A strand-specific RNA-seq library was constructed with the Ribo-Zero RNA fraction via the Illumina TruSeq RNA Library Preparation Kit v2 and sequencing on an Illumina HiSeq 4000 instrument.

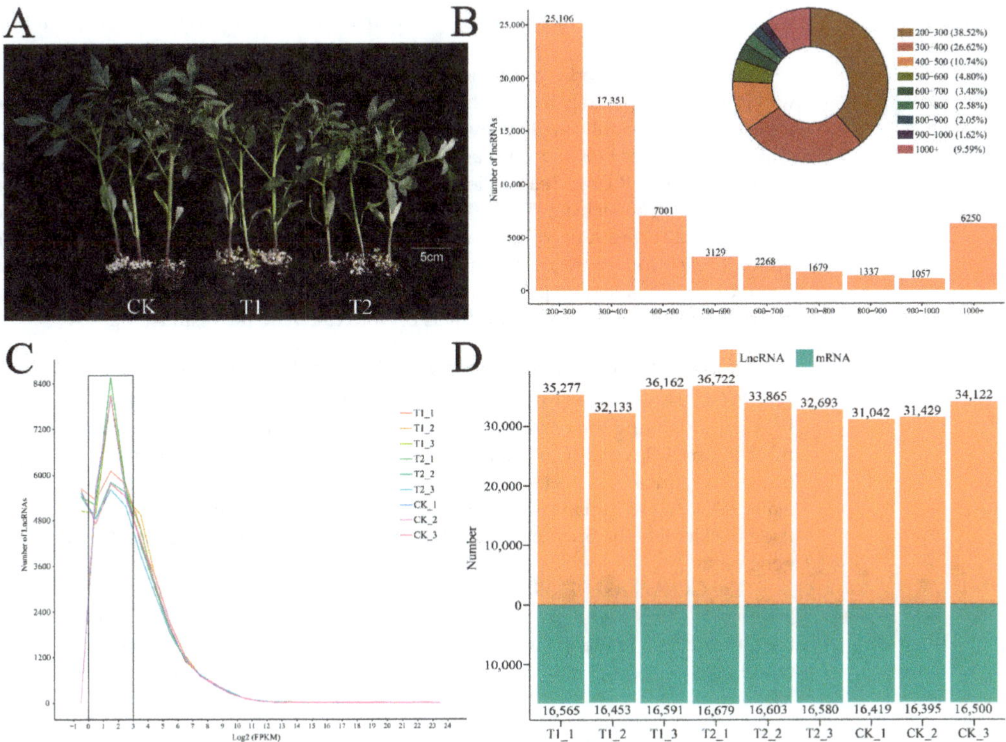

Figure 1. Characteristics of lncRNAs identified in *S. lycopersicum* leaves. (**A**) 'Diana' of *S. lycopersicum* under different treatments of PHBA (scale bar represents 5 cm). (**B**) Length distribution of lncR-NAs. (**C**) LncRNA expression levels. (**D**) The numbers of lncRNAs and PGs expressed in different treatments.

After removing adaptors and low-quality reads from raw reads using Trimmomatic [43], 1,003,711,588 paired-end clean reads were produced (111,523,510 on average for each sample). They were subjected to de novo transcript assembly using Trinity (v2.8.4, Manfred G. Grabherr, Cambridge MA, USA), with strand-specific RNA-Seq read orientation [44]. Duplicates were then removed from the assembled transcripts using CD-HIT (v4.7, Limin Fu, La Jolla CA, USA) with default parameters [45]. Finally, we obtained 82,126 unigenes longer than 200 bp for lncRNA screening. The lncRNA identification process is shown in Figure S1. To improve the accuracy of lncRNA detection, three software programs, RNAplonc (v1.0, Tatianne da Costa Negri, São Paulo, Brazil) [46], PLncPRO (v1.1, Urminder Singh, New Delhi, India) [47], and FEELnc (v0.1.1, Valentin Wucher, Rennes, Cedex, France) [48], with default parameters, were employed to identify lncRNAs. The lncRNAs obtained with these three tools were compared to obtain the shared pre-lncRNA candidates. To further rule out the possibility that the pre-lncRNAs were from coding genes, these pre-lncRNAs were blasted against coding DNA sequences of *S. lycopersicum* (SL.3.0) using BLAT v36 × 2 software [49] with an E-value of 10^{-5}. The filtered lncRNA candidates were then localized to the *S. lycopersicum* reference genome using BLAT to remove repeat sequences, so we could eliminate ones that had more than nine positions and which were considered to be repeats in the genome. Eventually, we identified highly reliable putative lncRNAs in *S. lycopersicum*.

2.3. Expression Analysis of Putative lncRNAs and Quantitative Reverse Transcription PCR (qRT-PCR) Validation

RSEM (v1.3.3, Bo Li, Madison, WI, USA) was used to compare the sequenced reads with the transcripts to calculate the gene expression amounts and carry out a differential gene expression analysis. The levels of gene expression were reflected by FPKM values, and a threshold of FPKM > 0.1 was used for the expressed genes. The DESeq package in R software (v3.6.1, R Core Team, Vienna, Austria) was used for pairwise differential expression analysis. The website http://bioinformatics.psb.ugent.be/webtools/Venn/ (accessed on 14 June 2019) was used for Venn diagram drawing [50]. qRT-PCR was performed to validate the expressions of the putative lncRNAs with the same RNA samples as those used for the RNA-seq [51]. The primers are listed in Table S1. GAPDH was chosen as the reference gene. The DDCt method was used to calculate the results [52].

2.4. Target Gene Prediction and Coexpression Analysis of lncRNAs with PGs

WGCNA was used to explore the cis-regulatory relationships between lncRNAs and PGs by calculating the correlations with chromosomal neighboring genes within 100 kb from the middle of a given lncRNA [50,53]. RNAplex software was used to predict the trans-acting lncRNAs and their target PGs (TPGs) [54]. Their coexpression was analyzed using WGCNA (cis-analysis: soft threshold = 21; trans-analysis: soft threshold = 22). R was used to calculate the Spearman correlations for the expression of lncRNAs and their target gene pairs.

The GO Analysis Toolkit and the Database for the Agricultural Community (agriGO; http://bioinfo.cau.edu.cn/agriGO/analysis.php) (accessed on 19 July 2019) were used for the GO analysis of the TPGs. The hypergeometric test was used as the statistical method, while Yekutieli (FDR under dependency) was employed as the multi-test adjustment method [50].

2.5. Similarity Analysis of lncRNAs from S. lycopersicum and Arabidopsis thaliana

The Plant Long noncoding RNA Database (PLncDB, http://chualab.rockefeller.edu/gbrowse2/homepage.html) (accessed on 16 July 2019) was used to download the lncRNA data for *Arabidopsis thaliana* [50]. BLAST software (E-value set to 10^{-5}) was used to compare the conservation and similarity levels of lncRNAs between *S. lycopersicum* and *Arabidopsis thaliana* [50].

2.6. Prediction of Endogenous Target Mimics for miRNAs

MicroRNA data for *S. lycopersicum* were obtained from miRbase (http://www.mirbase.org/) (accessed on 26 July 2019). WGCNA was used to predict the relationships between mRNAs, lncRNAs, and miRNAs. A threshold weight value of >0.15 was used for the selection of lncRNA-miRNA-mRNA triplets. Cytoscape (https://cytoscape.org/) (accessed on 26 July 2019) was used to plot the networks of lncRNAs, miRNAs, and mRNAs.

2.7. GenBank Accession Numbers

CK_1: GSM4238348; CK_2: GSM4238349; CK_3: GSM4238350; T1_1: GSM4238351; T1_2: GSM4238352; T1_3: GSM4238353; T2_1: GSM4238354; T2_2: GSM4238355; T2_3: GSM4238356. The RNA sequencing read data have been deposited in the Short Read Archive (SRA) data library (accession number: SRP239079).

3. Results

3.1. Identification of lncRNAs in S. lycopersicum

Substantial numbers of novel transcripts could be discovered via the combination of high-throughput RNA-seq and de novo assembly. In this study, the 'Diana' cultivar of *S. lycopersicum* was used for RNA-seq. As a new cultivar of *S. lycopersicum*, 'Diana' is widely cultivated in China. *S. lycopersicum*; an important crop species, it is also affected by continuous cropping obstacles, which are mainly caused by allelopathy. PHBA, a common allelochemical, was studied in depth, and the results showed that photosynthesis, respiratory effects, and gene expression in the leaves were affected by PHBA. Thus, the 'Diana' cultivar of *S. lycopersicum* was chosen to study gene expression in PHBA-induced allelopathy. We built the unigene library with three samples, including leaves undergoing three different treatments (CK: leaves of *S. lycopersicum* under 0 mmol/L PHBA treatment for 2 days; T1: leaves of *S. lycopersicum* under 15 mmol/L PHBA treatment for 2 days; and T2: leaves of *S. lycopersicum* under 30 mmol/L PHBA treatment for 2 days) (Figure 1A).

A total of 82,126 unigenes with an overall length greater than 200 bp and poly(A) tails were identified after de novo assembly. Following the protocol represented in Figure S1, 65,178 unigenes were acquired as credible lncRNAs, among which 61,729 lncRNAs were expressed during the plant allelopathy progression in *S. lycopersicum* in response to PHBA (Table S2). Ten lncRNAs were randomly chosen, and their expression was verified by qRT-PCR; all results were consensual (Figure S2).

The average length of the 61,729 assumed lncRNAs was 565 bp, the minimum length was 201 bp, and the maximum length was 19,161 bp (Figure 1B and Table S2). Of these lncRNAs, 80.68% were shorter than 600 bp, while 9.59% were longer than 1000 bp. About two-thirds of them expressed at low levels, with FPKM values (log2) ranging from 0 to 3 (Figure 1C and Table S2). Interestingly, the trends in the data for the lncRNAs corresponded to those for the PGs across all of the samples (Figure 1D and Table S3).

3.2. Expression of lncRNAs in S. lycopersicum *Leaves under Different PHBA Treatments*

As we knew, 61,729 lncRNAs were expressed in *S. lycopersicum* leaves during the different PHBA treatments, among which 34,000 lncRNAs were expressed throughout all three treatments. Almost half of the lncRNAs were expressed exclusively in two stages, while a large number of lncRNAs were expressed in only one stage (Figure 2A).

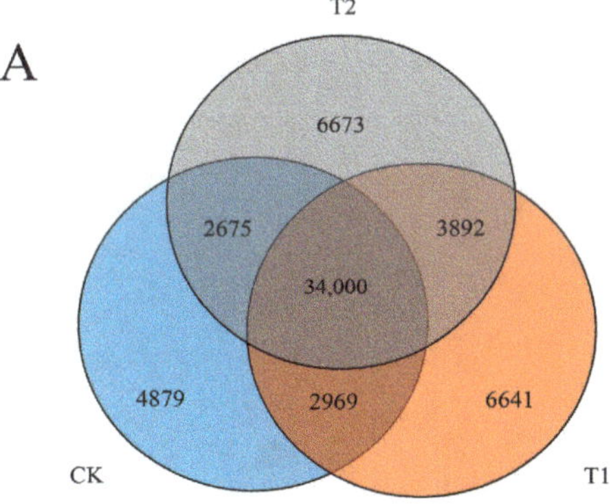

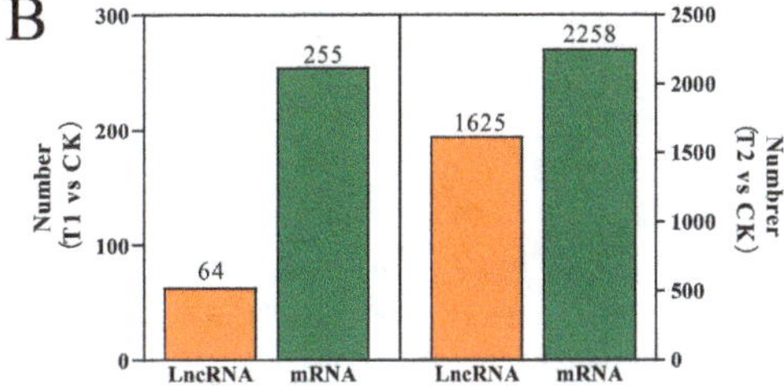

Figure 2. Expression of lncRNAs in *S. lycopersicum* leaves under different PHBA treatments. (**A**) Venn diagram detailing shared and distinct lncRNA expression patterns in *S. lycopersicum* leaves under different treatments of PHBA. (**B**) Differential expressions of lncRNAs and PGs between the PHBA-treated and untreated lines of *S. lycopersicum*.

We compared the differences in expression between the lncRNAs and PGs in *S. lycopersicum* leaves under different PHBA treatments (Table S4). The numbers of differentially expressed PGs and lncRNAs between the treated and untreated *S. lycopersicum* leaves increased gradually with increasing PHBA concentration and reached an apex under the T2 treatment (Figure 2B).

3.3. Prediction of Cis-Regulated Target PGs (CTPGs) of Lncrnas

The expression of proximal and distal PGs has been found to be regulated by lncRNAs through cis- and trans-acting mechanisms [55]. The regulation of genes located on the same chromosome as lncRNAs can be ascribed to cis-regulation [56,57], the investigation of which relies on well-annotated genomes.

The proximal PGs located in a 100 kb genomic window were searched as cis-regulated target genes of lncRNAs [53]. A total of 7765 lncRNAs were found to have potential cis-regulatory effects on 5314 PGs in 38,845 gene pairs altogether (Table S5), among which, 89% of lncRNAs targeted more than one gene, 71% targeted two to seven genes, while only two lncRNAs targeted fourteen genes (Figure 3A). Approximately 5% of the PGs corresponded to only one lncRNA, and only six PGs were cis-regulated by up to nineteen lncRNAs (Figure 3A).

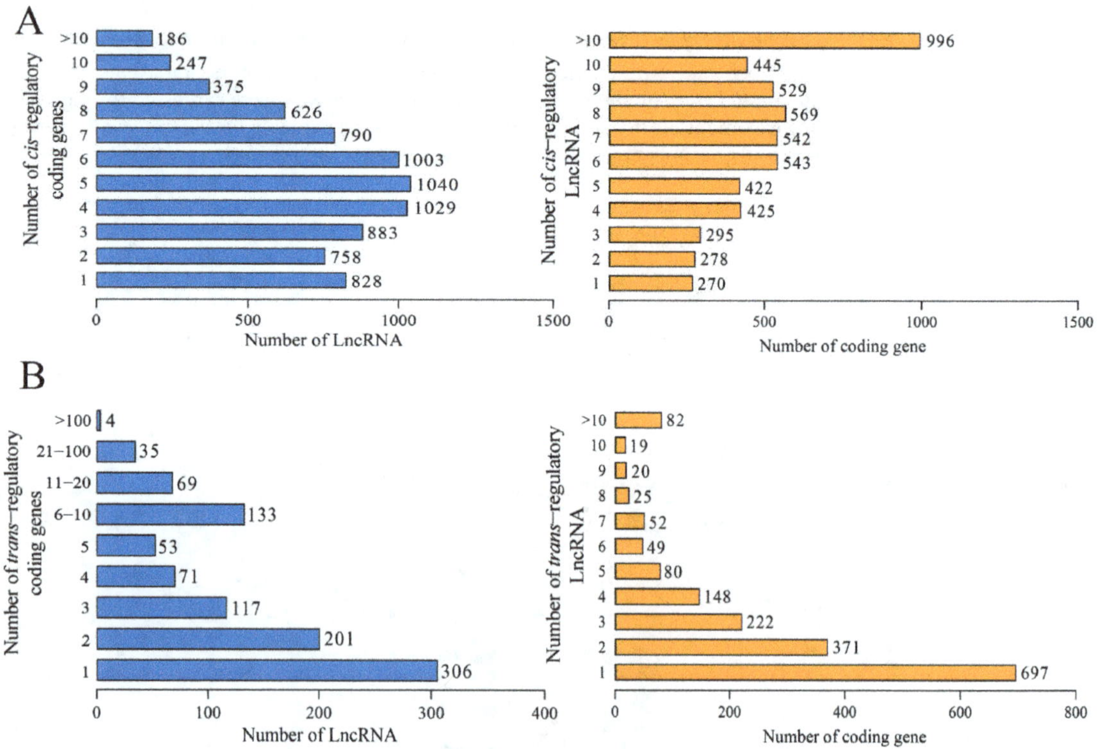

Figure 3. Numbers of target PGs regulated by lncRNAs and numbers of lncRNAs that have potential cis-regulatory and trans-regulatory effects on PGs in *S. lycopersicum* leaves. (**A**) The numbers of TPGs regulated by lncRNAs and the numbers of lncRNAs that have potential cis-regulatory effects on PGs. (**B**) The numbers of TPGs regulated by lncRNAs and the numbers of lncRNAs that have potential trans-regulatory effects on PGs.

Subsequently, the coexpression relationships between lncRNAs and protein-coding gene pairs were studied using WGCNA. A total of 1116 lncRNAs and 2239 PGs were involved in transcriptome regulation, which formed a complex network (Data S1), including 12 modules (Figure 4A and Table S6). Different modules contained different proportions of lncRNAs, ranging from 16.7% in Module 10 (M10) to 57.6% in M6, with an average of 38.2%. Gene expression patterns for each module were distinguished (Figure 5). Significant GO terms were identified in each module, ranging from 11 in M11 to 133 in M1 (Table S7). GO:0032957, which is associated with the inositol trisphosphate metabolic process, was found in M7 and M11, while GO:0050789 (regulation of biological process) was found in M1 and M6. Among these modules, the highly expressed genes in M2 were involved in photosynthesis and cell wall organization and were annotated as being associated with allelopathy (Figure 6A). The genes in M13 were also predicted to be involved in photosynthesis and cell wall biogenesis (Figure 6B).

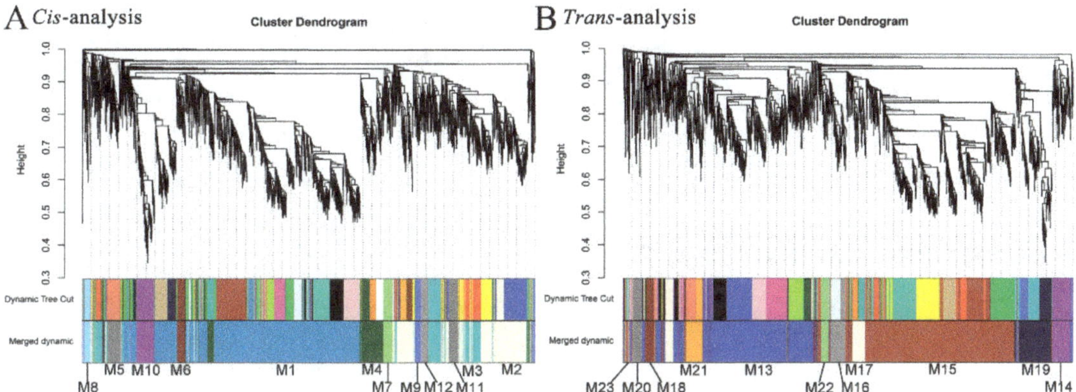

Figure 4. The weighted gene co-expression network analysis of lncRNAs and TPGs in *S. lycopersicum*. (**A**) Modules of lncRNAs and CTPGs. (**B**) Modules of lncRNAs and TTPGs.

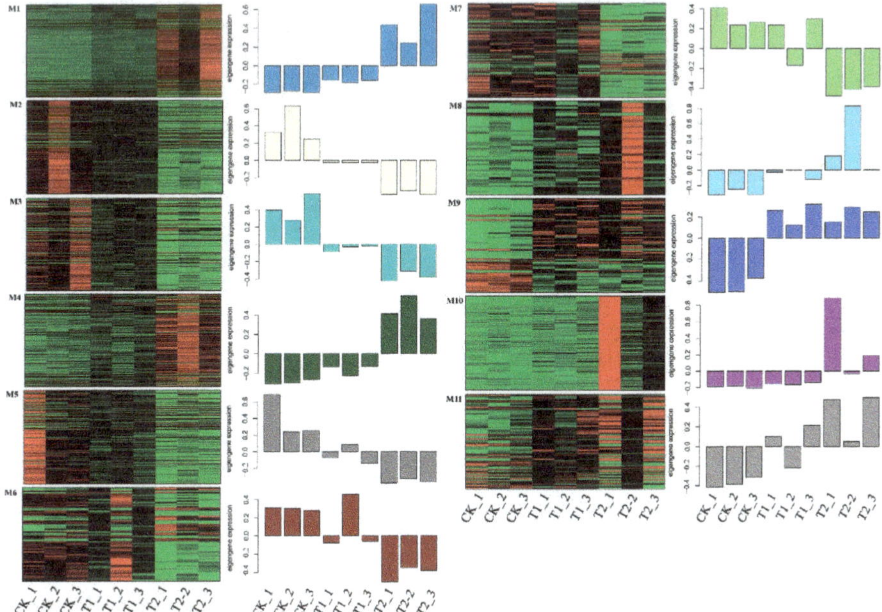

Figure 5. Expression patterns of lncRNAs and CTPGs in different modules.

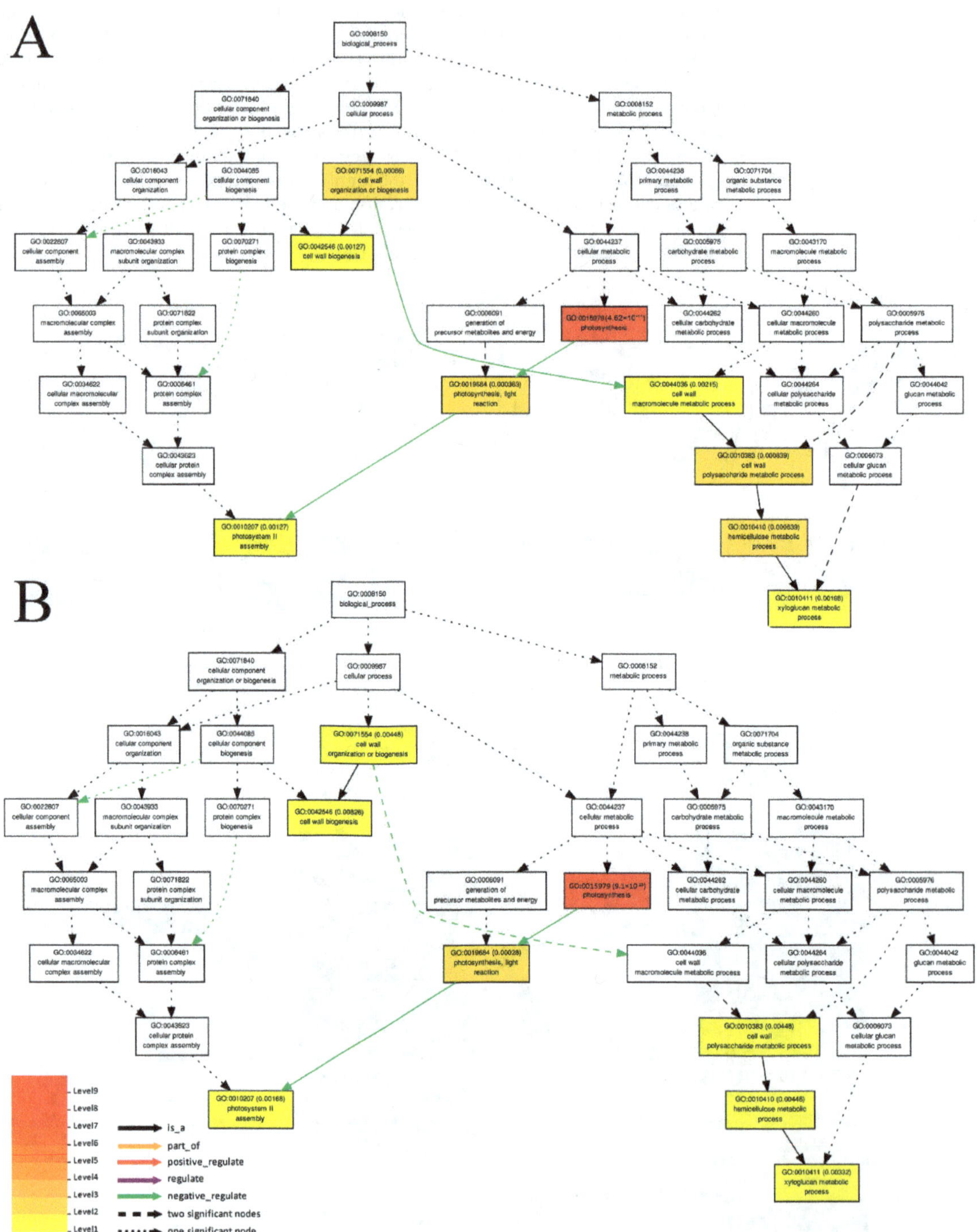

Figure 6. GO annotations of CTPGs in Module 2 (M2) (**A**) and Module 13 (M13) (**B**).

In addition, we calculated the Spearman correlations (correlation coefficient cutoff = 0.9) for the expression of each lncRNA and its CTPG pair. The expressions of 458 lncRNAs and their corresponding 975 CTPGs were found to be strongly correlated (Table S8). A total of 1092 gene pairs were positively correlated, and 963 pairs were negatively correlated. Among these target genes, 1348 were upregulated in *S. lycopersicum* leaves in at least one treatment, and 892 were downregulated (Table S9).

3.4. Prediction of TTPGs of lncRNAs

Using RNAplex software, 989 lncRNAs and 1765 TPGs in 5659 gene pairs were predicted to be trans-regulated (Table S10). Among the 989 lncRNAs, 30.9% of them targeted only one gene, 0.4% of which targeted more than 100 genes (Figure 3B). Among the 1765 targets, 697 (39.5%) were trans-regulated by one lncRNA; in contrast, 82 targets (4.6%) were targeted by more than 10 lncRNAs (Figure 3B).

Using WGCNA, 989 lncRNAs and their 1765 TTPGs were divided into 11 modules (Figure 4B). The proportions of lncRNAs ranged from 11.1% in M14 to 67.1% in M15, with an average of 35.9% in these modules (Table S6). Except for M15, M19, M20, M21, and M23, remarkable GO terms were verified in most modules, with a range from 106 in M15 to 6 in M23 (Table S11). Distinguishable expression patterns were found in all of the modules (Figure 7). For example, genes in M16 and M19 were extremely highly expressed in plants undergoing the T2 treatment (Figure 7) and were annotated as being associated with the following terms: reactive oxygen species metabolic process, antibiotic catabolic process, and hydrogen peroxide catabolic process (Figure 8). Regulatory networks that were widely interconnected between mRNAs and lncRNAs were found in all modules of trans-regulatory activity (Data S1).

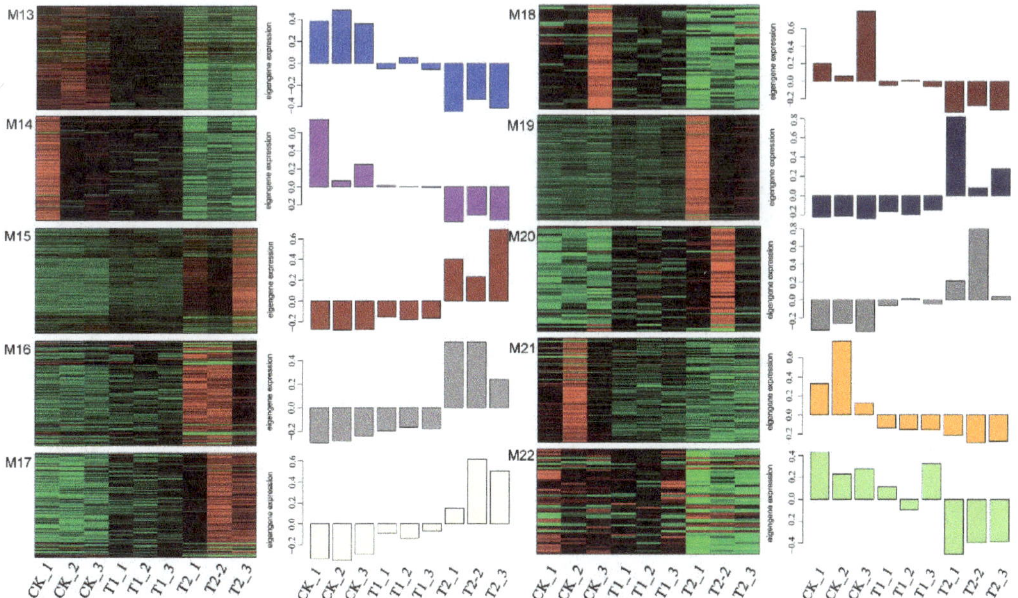

Figure 7. Expression patterns of lncRNAs and TTPGs in different modules.

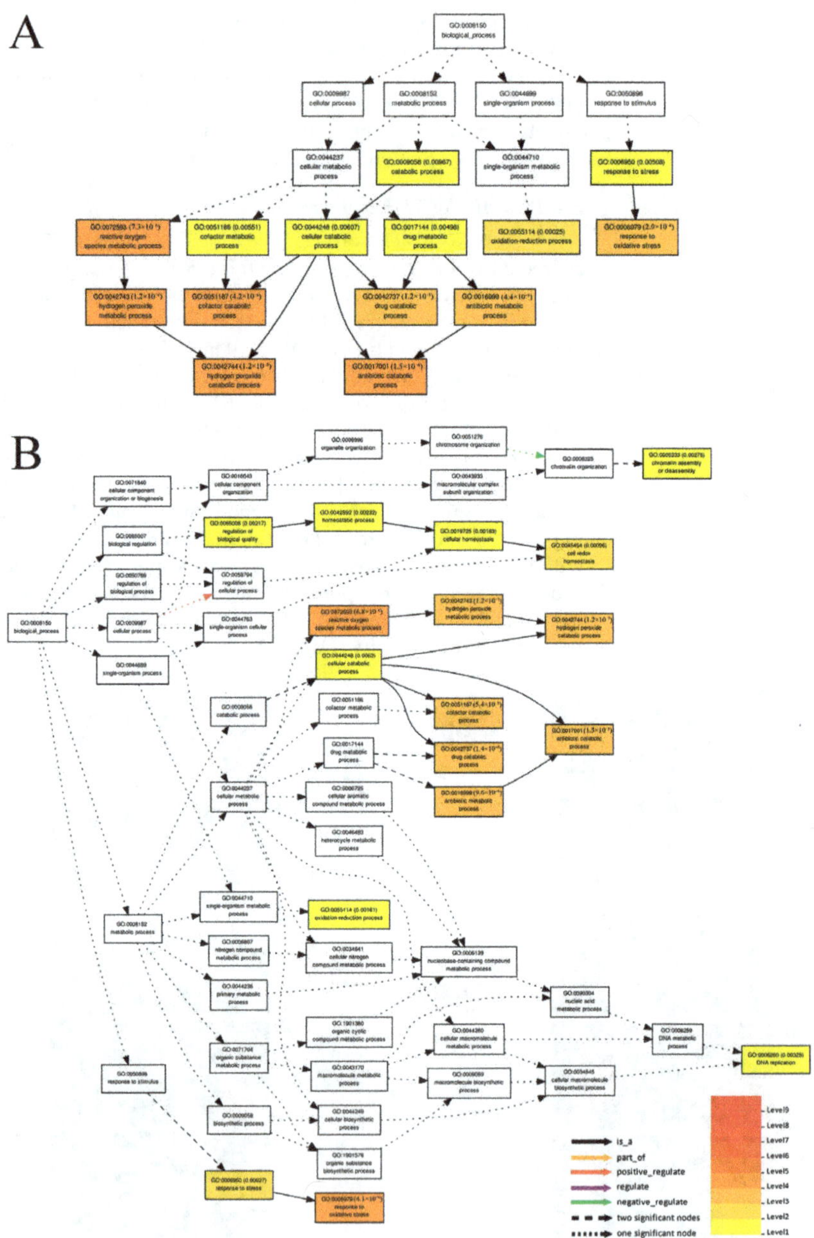

Figure 8. GO annotations of lncRNAs and TTPGs in Module 16 (**A**) and Module 19 (**B**).

A total of 335 trans-acting lncRNAs were highly correlated with their 633 targets (Spearman coefficient rho (rs) > 0.9) (Table S12); 681 gene pairs were positively correlated, while 554 pairs were negatively. The expression patterns of the targets indicated that 1035 were upregulated and that 730 were downregulated in at least one treatment (Table S13).

3.5. Similarity Alignment and Conservation Analysis of lncRNAs in S. lycopersicum and Arabidopsis thaliana

As a model plant and close species of *S. lycopersicum*, *Arabidopsis thaliana* was used to analyze the lncRNA similarities between them. Although there were 65,178 lncRNAs in *S. lycopersicum*, only 98 had homologs in *Arabidopsis thaliana* (Table S14). Expression analysis revealed that these 98 lncRNAs were affected by the PHBA treatments (Figure 9).

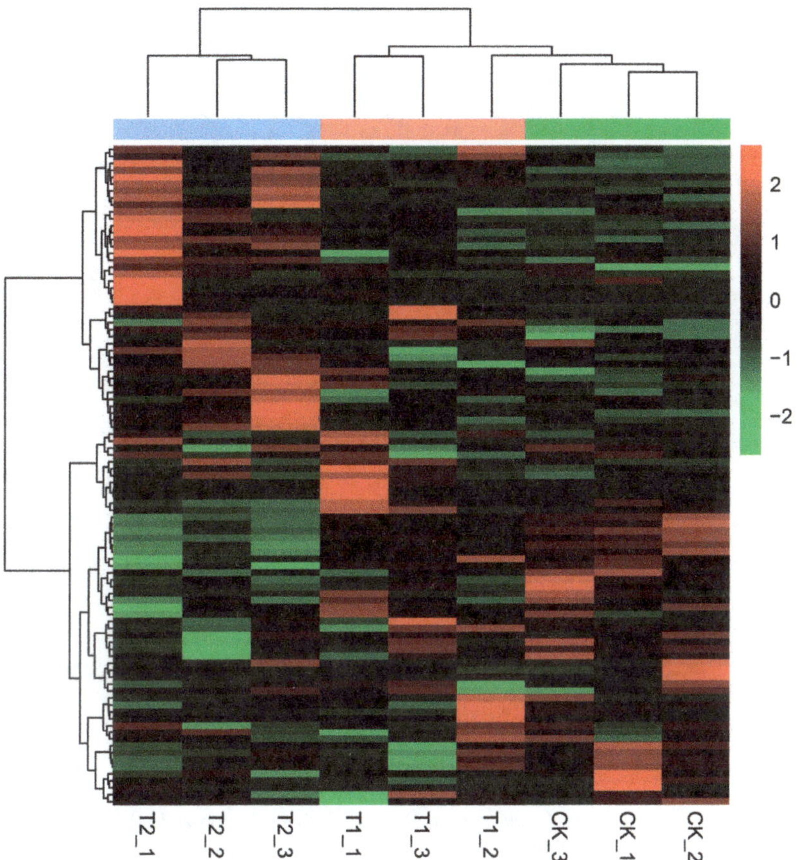

Figure 9. Expression patterns of *S. lycopersicum* lncRNAs with homologs in *Arabidopsis thaliana*.

3.6. Prediction of lncRNAs as Endogenous Target Mimics of miRNAs

Recent research has indicated that lncRNAs interact with miRNAs as competitive endogenous RNAs (ceRNAs) and participate in expression regulation of targets [58,59]. The potential lncRNAs were predicted to be eTM miRNAs using *S. lycopersicum* miRNAs which were previously obtained from miRbase (http://www.mirbase.org/, accessed on 26 July 2019).

The miRNAs that interacted with both mRNAs and lncRNAs were selected from the WGCNA results to make lncRNA-miRNA-mRNA triplets (Table S15). A weight value greater than 0.15 was used to select the lncRNA-miRNA-mRNA triplets (Table S16). Cytoscape was used to plot the interactions between lncRNAs, miRNAs, and mRNAs (Figure 10).

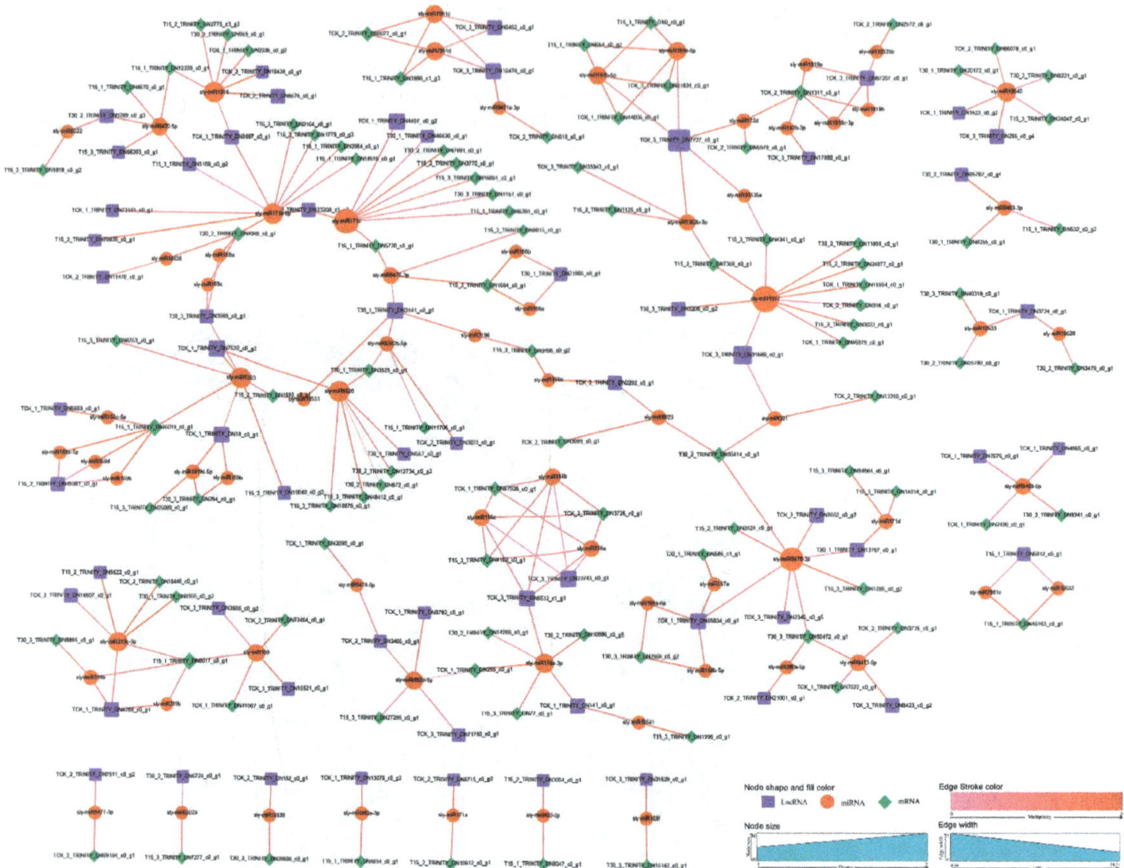

Figure 10. The interactions between lncRNAs, miRNAs, and mRNAs.

4. Discussion

Although genome-wide lncRNA searches and related studies have been carried out in *Arabidopsis thaliana* [60], rice [61], maize [62], wheat [63], and *Populus trichocarpa*, identification of lncRNA functions in plant response to allelopathy is lacking. In this study, an RNA-seq experiment was carried out on tomato plants treated with different concentrations of PHBA, which systematically identified plant allelopathy-related lncRNAs. The tomato material we used in this study is a new variety that is widely planted in North China and thus has an important reference value for plant allelopathy in vegetable production. In addition, the differentially expressed lncRNAs under different PHBA treatments were valuable for studying plant allelopathy in *S. lycopersicum*. Moreover, as an important vegetable crop grown worldwide, *S. lycopersicum* is considered a reference species for *Solanum* genomic studies [64]. The genome-wide characterization of lncRNAs in *S. lycopersicum* will provide a unique annotation resource for the *S. lycopersicum* genome.

Gene expression can be cis- or trans-regulated by lncRNAs [65]. Cis-acting lncRNAs control gene expression around their transcription sites, while trans-acting lncRNAs regulate the expression of genes at independent loci [66–68]. Additionally, the overlap of some cis- and trans-acting lncRNAs increases the complexity of the molecular roles of lncRNAs [50,69]. Since lncRNAs are often coexpressed with their TPGs, a method for inferring the putative biological functions of lncRNAs was developed based on the relationships between the expressions of lncRNAs and TPGs [50,70]. The CTPGs and TTPGs of lncRNAs were predicted, and a coexpression analysis was performed [50]. Strong coexpression between cis- or trans-acting lncRNAs and their TPGs was found in this study. To date, no lncRNA has been characterized in plant allelopathy processes. Further investigation of these lncRNAs that are highly coexpressed with TPGs with known functions may provide some clues that will enable discovery of the functions of lncRNAs in gene expression regulation of allelopathy.

Compared with the study of functionality, the study of evolution needs to be further explored. For plant lncRNAs, a recent BLAST search of *B. rapa* lncRNAs against the lncRNAs of *Arabidopsis thaliana* yielded few homologs (approximately 5%) [50], indicating that evolutionary conservation is limited in plant lncRNAs [50]. In this study, we found that only approximately 0.15% (98 out of 65,178) of lncRNAs in *S. lycopersicum* had homologs in *Arabidopsis thaliana*. This result also suggested limited evolutionary conservation in plant lncRNAs.

Recent studies have found lncRNAs involved in miRNA regulation mechanisms as ceRNAs, which were initially studied in *Arabidopsis thaliana* and tobacco [71,72]. In this study, we predicted the relationships between lncRNAs, miRNAs, and mRNAs using the miRNAs that were previously identified in *S. lycopersicum*. We propose that certain interactions between these lncRNAs, mRNAs, and miRNAs play fundamental roles in plant allelopathy.

5. Conclusions

In conclusion, screening and functional analysis of the genome of *S. lycopersicum* enabled the identification of a set of lncRNAs. The analysis of their functional links with mRNAs and miRNAs and the similarity alignment and conservation analyses of lncRNAs in *S. lycopersicum* and *Arabidopsis thaliana* may provide some molecular clues about plant allelopathy due to PHBA in *S. lycopersicum*.

Supplementary Materials: The following supporting information can be downloaded at: https://www.mdpi.com/article/10.3390/horticulturae8121134/s1, Data S1: The coexpression relationships between lncRNAs and protein-coding gene pairs, Figure S1: Informatics pipeline for the identification of lncRNAs in *S. lycopersicum*, Figure S2: Experimental validation of the lncRNA expression by qRT-PCR. (The horizontal axis represents different treatments of PHBA, the vertical axis represents the relative expression level of lncRNAs.), Table S1: The primers used in this study, Table S2: LncRNAs identified in *S. lycopersicum*, Table S3: Numbers of lncRNAs and PGs detected in *S. lycopersicum*, Table S4: Differential expressions of lncRNAs and mRNAs, Table S5: LncRNAs and their CTPGs, Table S6: The weighted gene co-expression network analysis modules of lncRNAs and corresponding TPGs, Table S7: GO terms and annotations of cis-regulated modules, Table S8: LncRNAs and cis-regulated target mRNAs that have correlation coefficients greater than 0.9 and *p*-values less than 1.0×10^{-5}, Table S9: Cis-regulated target mRNAs that are differentially expressed, Table S10: LncRNAs and their trans-regulated corresponding target mRNAs analyzed using LncTar, Table S11: GO terms and annotations of trans-regulated modules, Table S12: LncRNAs and trans-regulated target mRNAs that have correlation coefficients greater than 0.9 and *p*-values less than 1.0×10^{-5}, Table S13: Trans-regulated target mRNAs that are differentially expressed, Table S14: Similarity alignment of lncRNAs in *S. lycopersicum* and *Arabidopsis thaliana*, Table S15: MiRNA target prediction by psRNATarget, Table S16: The lncRNA-miRNA-mRNA pairs with weight values greater than 0.15.

Author Contributions: Conceptualization, G.L.; Formal analysis, G.L.; Funding acquisition, J.G.; Software, G.L.; Validation, Y.N.; Writing—original draft, G.L.; Writing—review & editing, J.G. All authors have read and agreed to the published version of the manuscript.

Funding: This research was founded by the National Natural Science Foundation of China, grant numbers: 31770706 and 32071789.

Institutional Review Board Statement: Not applicable.

Informed Consent Statement: Not applicable.

Data Availability Statement: The RNA sequencing read data have been deposited in the Short Read Archive (SRA) data library (accession number: SRP239079).

Conflicts of Interest: The authors declare no conflict of interest.

References

1. Crick, F.H.C.; Barnett, L.; Brenner, S.; Watts-Tobin, R.J. General Nature of the Genetic Code for Proteins. *Nature* **1961**, *192*, 1227–1232. [CrossRef] [PubMed]
2. Yang, G.; Lu, X.; Yuan, L. LncRNA: A link between RNA and cancer. *Biochim. Et Biophys. Acta (BBA)-Gene Regul. Mech.* **2014**, *1839*, 1097–1109. [CrossRef] [PubMed]
3. Eddy, S.R. Non–coding RNA genes and the modern RNA world. *Nat. Rev. Genet.* **2001**, *2*, 919–929. [CrossRef] [PubMed]
4. Carninci, P.; Hayashizaki, Y. Noncoding RNA transcription beyond annotated genes. *Curr. Opin. Genet. Dev.* **2007**, *17*, 139–144. [CrossRef]
5. Eddy, S.R. Computational genomics of noncoding RNA genes. *Cell* **2002**, *109*, 137–140. [CrossRef]
6. Griffiths-Jones, S. Annotating noncoding RNA genes. *Annu. Rev. Genomics Hum. Genet.* **2007**, *8*, 279–298. [CrossRef]
7. Zhu, Q.-H.; Wang, M.-B. Molecular functions of long non-coding RNAs in plants. *Genes* **2012**, *3*, 176–190. [CrossRef]
8. Yamada, K.; Lim, J.; Dale, J.M.; Chen, H.; Shinn, P.; Palm, C.J.; Southwick, A.M.; Wu, H.C.; Kim, C.; Nguyen, M. Empirical analysis of transcriptional activity in the *Arabidopsis* genome. *Science* **2003**, *302*, 842–846. [CrossRef]
9. Morris, K.V.; Mattick, J.S. The rise of regulatory RNA. *Nat. Rev. Genet.* **2014**, *15*, 423. [CrossRef]
10. Kong, R.; Zhang, E.-b.; Yin, D.-d.; You, L.-h.; Xu, T.-p.; Chen, W.-m.; Xia, R.; Wan, L.; Sun, M.; Wang, Z.-x. Long noncoding RNA PVT1 indicates a poor prognosis of gastric cancer and promotes cell proliferation through epigenetically regulating p15 and p16. *Mol. Cancer* **2015**, *14*, 82. [CrossRef]
11. Wierzbicki, A.T. The role of long non-coding RNA in transcriptional gene silencing. *Curr. Opin. Plant Biol.* **2012**, *15*, 517–522. [CrossRef] [PubMed]
12. Sacco, L.D.; Baldassarre, A.; Masotti, A. Bioinformatics tools and novel challenges in long non-coding RNAs (lncRNAs) functional analysis. *Int. J. Mol. Sci.* **2012**, *13*, 97–114. [CrossRef] [PubMed]
13. Zhang, L.; Ge, X.; Du, J.; Hu, J. Genome-wide Identification of Long Non-coding RNAs and Their Potential Functions in Poplar Growth and Phenylalanine Biosynthesis. *Front. Genet.* **2021**, *12*, 2113. [CrossRef] [PubMed]
14. Kim, E.-D.; Sung, S. Long noncoding RNA: Unveiling hidden layer of gene regulatory networks. *Trends Plant Sci.* **2012**, *17*, 16–21. [CrossRef]
15. Shafiq, S.; Li, J.; Sun, Q. Functions of plants long non-coding RNAs. *Biochim. Et Biophys. Acta (BBA)-Gene Regul. Mech.* **2016**, *1859*, 155–162. [CrossRef]
16. Chekanova, J.A. Long non-coding RNAs and their functions in plants. *Curr. Opin. Plant Biol.* **2015**, *27*, 207–216. [CrossRef]
17. Liu, J.; Jung, C.; Xu, J.; Wang, H.; Deng, S.; Bernad, L.; Arenas-Huertero, C.; Chua, N.-H. Genome-wide analysis uncovers regulation of long intergenic noncoding RNAs in *Arabidopsis*. *Plant Cell* **2012**, *24*, 4333–4345. [CrossRef]
18. Swiezewski, S.; Liu, F.; Magusin, A.; Dean, C. Cold-induced silencing by long antisense transcripts of an *Arabidopsis* Polycomb target. *Nature* **2009**, *462*, 799. [CrossRef]
19. Heo, J.B.; Sung, S. Vernalization-mediated epigenetic silencing by a long intronic noncoding RNA. *Science* **2011**, *331*, 76–79. [CrossRef]
20. Shin, H.; Shin, H.S.; Chen, R.; Harrison, M.J. Loss of At4 function impacts phosphate distribution between the roots and the shoots during phosphate starvation. *Plant J.* **2006**, *45*, 712–726. [CrossRef]
21. Liu, C.; Muchhal, U.S.; Raghothama, K. Differential expression of TPS11, a phosphate starvation-induced gene in tomato. *Plant Mol. Biol.* **1997**, *33*, 867–874. [CrossRef] [PubMed]
22. Martín, A.C.; Del Pozo, J.C.; Iglesias, J.; Rubio, V.; Solano, R.; De La Peña, A.; Leyva, A.; Paz-Ares, J. Influence of cytokinins on the expression of phosphate starvation responsive genes in *Arabidopsis*. *Plant J.* **2000**, *24*, 559–567. [CrossRef] [PubMed]
23. Franco-Zorrilla, J.M.; Valli, A.; Todesco, M.; Mateos, I.; Puga, M.I.; Rubio-Somoza, I.; Leyva, A.; Weigel, D.; García, J.A.; Paz-Ares, J. Target mimicry provides a new mechanism for regulation of microRNA activity. *Nat. Genet.* **2007**, *39*, 1033. [CrossRef]
24. Zhang, Y.C.; Chen, Y.Q. Long noncoding RNAs: New regulators in plant development. *Biochem. Biophys. Res. Commun.* **2013**, *436*, 111–114. [CrossRef] [PubMed]
25. Chou, C.-H. Roles of allelopathy in plant biodiversity and sustainable agriculture. *Critical Reviews in Plant Sciences* **1999**, *18*, 609–636. [CrossRef]
26. EL-GAWAD, A.M.A. Ecology and allelopathic control of Brassica tournefortii in reclaimed areas of the Nile Delta, Egypt. *Turk. J. Bot.* **2014**, *38*, 347–357. [CrossRef]
27. de Albuquerque, M.B.; dos Santos, R.C.; Lima, L.M.; de Albuquerque Melo Filho, P.; Nogueira, R.J.M.C.; Da Câmara, C.A.G.; de Rezende Ramos, A. Allelopathy, an alternative tool to improve cropping systems. A review. *Agron. Sustain. Dev.* **2011**, *31*, 379–395. [CrossRef]
28. Cheng, Z.-h.; Xu, P. Lily (*Lilium* spp.) root exudates exhibit different allelopathies on four vegetable crops. *Acta Agric. Scand. Sect. B–Soil Plant Sci.* **2013**, *63*, 169–175.

29. Hortal, S.; Bastida, F.; Moreno, J.L.; Armas, C.; García, C.; Pugnaire, F.I. Benefactor and allelopathic shrub species have different effects on the soil microbial community along an environmental severity gradient. *Soil Biol. Biochem.* **2015**, *88*, 48–57. [CrossRef]
30. Gniazdowska, A.; Bogatek, R. Allelopathic interactions between plants. Multi site action of allelochemicals. *Acta Physiol. Plant.* **2005**, *27*, 395–407. [CrossRef]
31. Nasrollahi, P.; Razavi, S.M.; Ghasemian, A.; Zahri, S. Physiological and Biochemical Responses of Lettuce to Thymol, as Allelochemical. *Russ. J. Plant Physiol.* **2018**, *65*, 598–603. [CrossRef]
32. Cheng, F.; Cheng, Z. Research progress on the use of plant allelopathy in agriculture and the physiological and ecological mechanisms of allelopathy. *Front. Plant Sci.* **2015**, *6*, 1020. [CrossRef] [PubMed]
33. Latif, S.; Chiapusio, G.; Weston, L.A. Allelopathy and the role of allelochemicals in plant defence. *Adv. Bot. Res.* **2017**, *82*, 19–54.
34. Zhang, Q.; Zheng, X.-Y.; Lin, S.-X.; Gu, C.-Z.; Li, L.; Li, J.-Y.; Fang, C.-X.; He, H.-B. Transcriptome analysis reveals that barnyard grass exudates increase the allelopathic potential of allelopathic and non-allelopathic rice (*Oryza sativa*) accessions. *Rice* **2019**, *12*, 30. [CrossRef]
35. Liang, X.; Hou, X.; Li, J.; Han, Y.; Zhang, Y.; Feng, N.; Du, J.; Zhang, W.; Zheng, D.; Fang, S. High-resolution DNA methylome reveals that demethylation enhances adaptability to continuous cropping comprehensive stress in soybean. *BMC Plant Biol.* **2019**, *19*, 79. [CrossRef]
36. Fang, C.; Li, Y.; Li, C.; Li, B.; Ren, Y.; Zheng, H.; Zeng, X.; Shen, L.; Lin, W. Identification and comparative analysis of micro RNAs in barnyardgrass (E chinochloa crus-galli) in response to rice allelopathy. *Plant Cell Environ.* **2015**, *38*, 1368–1381. [CrossRef]
37. Wang, J.; Meng, X.; Yuan, C.; Harrison, A.P.; Chen, M. The roles of cross-talk epigenetic patterns in *Arabidopsis thaliana*. *Brief. Funct. Genom.* **2015**, *15*, 278–287. [CrossRef]
38. Xu, X.-W.; Zhou, X.-H.; Wang, R.-R.; Peng, W.-L.; An, Y.; Chen, L.-L. Functional analysis of long intergenic non-coding RNAs in phosphate-starved rice using competing endogenous RNA network. *Sci. Rep.* **2016**, *6*, 20715. [CrossRef]
39. Lv, Y.; Liang, Z.; Ge, M.; Qi, W.; Zhang, T.; Lin, F.; Peng, Z.; Zhao, H. Genome-wide identification and functional prediction of nitrogen-responsive intergenic and intronic long non-coding RNAs in maize (*Zea mays* L.). *BMC Genom.* **2016**, *17*, 350. [CrossRef]
40. Chen, M.; Wang, C.; Bao, H.; Chen, H.; Wang, Y. Genome-wide identification and characterization of novel lncRNAs in *Populus* under nitrogen deficiency. *Mol. Genet. Genom.* **2016**, *291*, 1663–1680. [CrossRef]
41. Wang, T.-Z.; Liu, M.; Zhao, M.-G.; Chen, R.; Zhang, W.-H. Identification and characterization of long non-coding RNAs involved in osmotic and salt stress in Medicago truncatula using genome-wide high-throughput sequencing. *BMC Plant Biol.* **2015**, *15*, 131. [CrossRef] [PubMed]
42. Xin, J.W.; Chai, Z.X.; Zhang, C.F.; Yang, Y.M.; Ji, Q.M. Transcriptome analysis identified long non-coding RNAs involved in the adaption of yak to high-altitude environments. *R. Soc. Open Sci.* **2020**, *7*, 200625. [CrossRef] [PubMed]
43. Bolger, A.M.; Lohse, M.; Usadel, B. Trimmomatic: A flexible trimmer for Illumina sequence data. *Bioinformatics* **2014**, *30*, 2114–2120. [CrossRef] [PubMed]
44. Grabherr, M.G.; Haas, B.J.; Yassour, M.; Levin, J.Z.; Thompson, D.A.; Amit, I.; Adiconis, X.; Fan, L.; Raychowdhury, R.; Zeng, Q. Full-length transcriptome assembly from RNA-Seq data without a reference genome. *Nat. Biotechnol.* **2011**, *29*, 644–652. [CrossRef]
45. Limin, F.; Beifang, N.; Zhengwei, Z.; Sitao, W.; Li, W. CD-HIT: Accelerated for clustering the next-generation sequencing data. *Bioinformatics* **2012**, *28*, 3150–3152.
46. Costa, N.T.D.; Luz, A.W.A.; Henrique, B.P.; Maeda, S.P.T.; Silva, D.D.; Rossi, P.A. Pattern recognition analysis on long noncoding RNAs: A tool for prediction in plants. *Brief. Bioinform.* **2018**, *2*, 682–689.
47. Urminder, S.; Niraj, K.; Mohan, S.; Rajkumar, R.; Garg, M. PLncPRO for prediction of long non-coding RNAs (lncRNAs) in plants and its application for discovery of abiotic stress-responsive lncRNAs in rice and chickpea. *Nucleic Acids Res.* **2017**, *45*, e183.
48. Valentin, W.; Fabrice, L.; Benoît, H.; Guillaume, R.; Lætitia, L.; Tosso, L.; Vidhya, J.; Edouard, C.; Audrey, D.; Hannes, L. FEELnc: A tool for long non-coding RNA annotation and its application to the dog transcriptome. *Nuclc Acids Res.* **2017**, *45*, e57.
49. Kent, W.J. Blat-the BLAST-like alignment tool. *Genome Res.* **2001**, *12*, 656–664.
50. Huang, L.; Dong, H.; Zhou, D.; Li, M.; Liu, Y.; Zhang, F.; Feng, Y.; Yu, D.; Lin, S.; Cao, J. Systematic identification of long non-coding RNAs during pollen development and fertilization in Brassica rapa. *Plant J. Cell Mol. Biol.* **2018**, *96*, 203–222. [CrossRef]
51. Wan, S.; Zhang, Y.; Duan, M.; Huang, L.; Yu, Y. Integrated Analysis of Long Non-coding RNAs (lncRNAs) and mRNAs Reveals the Regulatory Role of lncRNAs Associated With Salt Resistance in *Camellia sinensis*. *Front. Plant Sci.* **2020**, *11*, 218. [CrossRef] [PubMed]
52. Livak, K.J.; Schmittgen, T.D. Analysis of Relative Gene Expression Data using Real-Time Quantitative PCR. *Method* **2002**, *25*, 402–408. [CrossRef] [PubMed]
53. Casero, D.; Sandoval, S.; Seet, C.S.; Scholes, J.; Zhu, Y.; Ha, V.L.; Luong, A.; Parekh, C.; Crooks, G.M. Long non-coding RNA profiling of human lymphoid progenitor cells reveals transcriptional divergence of B cell and T cell lineages. *Nat. Immunol.* **2015**, *16*, 1282–1291. [CrossRef] [PubMed]
54. Hakim, T.; Ivo, H. RNAplex: A fast tool for RNA-RNA interaction search. *Bioinformatics* **2008**, *24*, 2657–2663. [CrossRef] [PubMed]
55. Zhu, L.; Zhu, J.; Liu, Y.; Chen, Y.; Li, Y.; Huang, L.; Chen, S.; Li, T.; Dang, Y.; Chen, T. Methamphetamine induces alterations in the long non-coding RNAs expression profile in the nucleus accumbens of the mouse. *BMC Neuroence* **2015**, *16*, 1–13. [CrossRef] [PubMed]

56. Kornienko, A.E.; Guenzl, P.M.; Barlow, D.P.; Pauler, F.M. Gene regulation by the act of long non-coding RNA transcription. *BMC Biol.* **2013**, *11*, 1–14. [CrossRef]
57. Lv, J.; Liu, Z.; Yang, B.; Deng, M.; Zou, X. Systematic identification and characterization of long non-coding RNAs involved in cytoplasmic male sterility in pepper (*Capsicum annuum* L.). *Plant Growth Regul.* **2020**, *91*, 277–288. [CrossRef]
58. Guo, L.L. Competing endogenous RNA networks and gastric cancer. *World J. Gastroenterol.* **2015**, *21*, 11680. [CrossRef]
59. Xia, T.; Liao, Q.; Jiang, X.; Shao, Y.; Guo, J. Long noncoding RNA associated-competing endogenous RNAs in gastric cancer. *Sci. Rep.* **2014**, *4*, 6088. [CrossRef]
60. Edouard, S.; Luigi, F.; Suraj, J.; Marco, B.; Yang, K.Z.; Francesca, B.; Jacqueline, B.L.; Virginia, F.; Angenent, G.C.; Immink, R.G.H. *Arabidopsis thaliana* ambient temperature responsive lncRNAs. *BMC Plant Biol.* **2018**, *18*, 145.
61. Zhang, Y.C.; Liao, J.Y.; Li, Z.Y.; Yu, Y.; Zhang, J.P.; Li, Q.F.; Qu, L.H.; Shu, W.S.; Chen, Y.Q. Genome-wide screening and functional analysis identify a large number of long noncoding RNAs involved in the sexual reproduction of rice. *Genome Biol.* **2014**, *15*, 512. [CrossRef]
62. Li, L.; Eichten, S.R.; Shimizu, R.; Petsch, K.; Yeh, C.T.; Wu, W.; Chettoor, A.M.; Givan, S.A.; Cole, R.A.; Fowler, J.E.; et al. Genome-wide discovery and characterization of maize long non-coding RNAs. *Genome Biol.* **2014**, *8*, e9585. [CrossRef] [PubMed]
63. Keshi, M.; Wenshuo, S.; Mengyue, X.; Jiaxi, L.; Feixiong, Z. Genome-Wide Identification and Characterization of Long Non-Coding RNA in Wheat Roots in Response to Ca2+ Channel Blocker. *Front. Plant Ence* **2018**, *9*, 244.
64. Aflitos, S.; Schijlen, E.; de Jong, H.; de Ridder, D.; Smit, S.; Finkers, R.; Wang, J.; Zhang, G.; Li, N.; 100 Tomato Genome Sequencing Consortium; et al. Exploring genetic variation in the tomato (Solanum section Lycopersicon) clade by whole-genome sequencing. *Plant J. Cell Mol. Biol.* **2015**, *80*, 136–148.
65. Yan, P.; Luo, S.; Lu, J.Y.; Shen, X. Cis- and trans-acting lncRNAs in pluripotency and reprogramming. *Curr. Opin. Genet. Dev.* **2017**, *46*, 170–178. [CrossRef] [PubMed]
66. Fatica, A.; Bozzoni, I. Long non-coding RNAs: New players in cell differentiation and development. *Nat. Rev. Genet.* **2014**, *15*, 7–21. [CrossRef] [PubMed]
67. Vance, K.W.; Ponting, C.P. Transcriptional regulatory functions of nuclear long noncoding RNAs. *Trends Genet.* **2014**, *30*, 348–355. [CrossRef]
68. Wang, J.; Shen, Y.C.; Chen, Z.N.; Yuan, Z.C.; Wang, H.; Li, D.J.; Liu, K.; Wen, F.Q. Microarray profiling of lung long non-coding RNAs and mRNAs in lipopolysaccharide-induced acute lung injury mouse model. *Biosci. Rep.* **2019**, *39*, BSR20181634. [CrossRef]
69. Guil, S.; Esteller, M. Cis-acting noncoding RNAs: Friends and foes. *Nat. Struct. Mol. Biol.* **2012**, *19*, 1068. [CrossRef]
70. Huarte, M.; Guttman, M.; Feldser, D.; Garber, M.; Rinn, J.L. A Large Intergenic Noncoding RNA Induced by p53 Mediates Global Gene Repression in the p53 Response. *Cell* **2010**, *142*, 409–419. [CrossRef]
71. Li, F.; Wang, W.; Zhao, N.; Xiao, B.; Cao, P.; Wu, X.; Ye, C.; Shen, E.; Qiu, J.; Zhu, Q.H. Regulation of Nicotine Biosynthesis by an Endogenous Target Mimicry of MicroRNA in Tobacco. *Plant Physiol.* **2015**, *169*, 1062–1071. [CrossRef] [PubMed]
72. Rubio-Somoza, I.; Weigel, D.; Franco-Zorilla, J.M.; García, J.A.; Paz-Ares, J. ceRNAs: miRNA target mimic mimics. *Cell* **2011**, *147*, 1431–1432. [CrossRef] [PubMed]

horticulturae

Article

Mechanism by Which High Foliar Calcium Contents Inhibit Sugar Accumulation in Feizixiao Lychee Pulp

Xian Shui [1,2], Wenjing Wang [1,2], Wuqiang Ma [1,2], Chengkun Yang [1,2] and Kaibing Zhou [1,2,*]

[1]   Sanya Nanfan Research Institute, Hainan University, Sanya 572025, China
[2]   College of Horticulture, Hainan University, Haikou 570228, China
*   Correspondence: zkb@hainanu.edu.cn

Abstract: The problem of Feizixiao lychee fruit cracking is typically solved by the application of calcium to the leaves. However, lychee trees are sensitive to excessive amounts of calcium, and in practice, it is easy to spray excessive amounts that result in fertilizer burns. This paper intends to explore the effects of excessive calcium fertilizer application on lychee fruit pulp quality and the underlying molecular physiological mechanism. Adult Feizixiao lychee trees were used as test materials; concerning treatment, a 54 μM anhydrous $CaCl_2$ aqueous solution was sprayed onto the leaves, and water was used as a control (CK). The levels of pulp sugar and the activities of key enzymes involved in glucose metabolism were observed, and transcriptome analysis and genetic screening were performed on the pulp. Spraying excessive amounts of calcium onto the leaf surfaces caused the downregulation of trehalase-encoding genes and SUS-encoding genes, thus inhibiting the activities of trehalase, SS-I and SS-II, and further inhibiting the accumulation of glucose, fructose, and sucrose. Moreover, upregulation of VIN gene expression enhanced AI activity and inhibited sucrose accumulation, thus inhibiting upregulation of NI gene expression during fruit growth and expansion; in turn, this inhibited the increase in NI activity in the fruit pulp, which then decreased the glucose and fructose accumulation in the pulp in the high-calcium treatment group compared with the CK group. The downregulation and expression of CHS family genes may lead to a decrease in chalcone accumulation, which may lead to damage caused by active oxygen production in the fruit pulp, thus inhibiting the accumulation of soluble sugars in that tissue.

Keywords: calcium; fertilizer; Feizixiao lychee; sugar accumulation

Citation: Shui, X.; Wang, W.; Ma, W.; Yang, C.; Zhou, K. Mechanism by Which High Foliar Calcium Contents Inhibit Sugar Accumulation in Feizixiao Lychee Pulp. *Horticulturae* **2022**, *8*, 1044. https://doi.org/10.3390/horticulturae8111044

Academic Editor: Rosario Paolo Mauro

Received: 6 October 2022
Accepted: 3 November 2022
Published: 7 November 2022

Publisher's Note: MDPI stays neutral with regard to jurisdictional claims in published maps and institutional affiliations.

1. Introduction

Litchi (Litchi chinensis *Sonn.*) is a widely cultivated specialty fruit tree in tropical and subtropical areas of China and has high economic value [1]. The tender, juicy, and delicious fruit of "Concubine Xiao" litchi makes it one of the most popular litchi varieties in China, occupying a large share in the litchi market. Therefore, good appearance and inner quality are fundamental guarantees for "smile of princess" litchi fruit's market competitiveness; however, "smile of princess" litchi fruit experience dehiscent problems during the process of growth and development, which seriously influence yield and quality [2], weaken market competitiveness, and damage growers' economic benefit. Many technical measures have been explored to solve the problem of litchi fruit cracking, among which the simplest one is foliar spraying of Ca fertilizer [3].

Foliar calcium fertilizer has a good effect that prevents the fruit cracking, not only on a variety of fruits, but it also improves the intrinsic quality at the same time. Foliar application of Ca fertilizer effectively reduces grape berry cracking [4], increases the content of sugar in pulp, and promotes the degradation of citric acid in ripe grape pulp [5]. Foliar application of Ca fertilizer reduces cracking [6] and sunburn damage of pomegranate fruits, increases the content of total sugar and chlorophyll, and effectively accelerates the vegetative growth process of the tree [5]. Foliar application of Ca fertilizer can increase the hardness of apple meat by acting on the cell wall [5]. Foliar application of Ca fertilizer improved fruit quality and reduced cracking, increased fruit Ca levels and total phenolic content, and reduced the percentage of dehydrated and rotten berries after storage [5]. Therefore, foliar spraying Ca to prevent fruit cracking has become an effective measure for many fruits.

In actual production, owing to the difficulty in accurate control of operation, the problem of spraying Ca fertilizer solution when the concentration is too high leads to the fruit being damaged by Ca fertilizer, resulting in the deterioration of fruit quality, ion toxicity, and salt stress reaction [7]. At the same time, we found that after spraying 0.6% calcium chloride to the canopy, the branches were unchanged but leaves and only a few fruits were covered with some necrotic spots, and the common abnormal performance was a decline in the soluble sugar content of the fruit. However, the physiological, biochemical, and molecular mechanisms of excessive application of Ca fertilizer to fruit trees are rarely studied. The excessive use of Ca fertilizer in pineapple causes a decrease in glucose, fructose, and β-carotene content [8]. High Ca treatment of grapes results in a decrease in sugar–acid ratio, an increase in tannin content, and a decrease in fruit taste quality [9]. In other plants, excessive calcium also affects the normal growth and development of plants: MICU gene inhibits root growth and reduces mRNA transcription levels in land plants under high Ca conditions [10]. Under high Ca stress, the root volume of wheat plants decreased, and the buds showed greening symptoms and shortened length [11]. Under the influence of high Ca, the germination rate of herbage seeds decreased [12]. High Ca is absorbed by cells and accumulated in mitochondria, which may cause cell death [10].

At present, in litchi production, foliar calcium injection is often used to prevent the fruit from cracking [13], but at the same time, it is also found that the litchi tree body is sensitive to calcin; even a little excessive use causes leaf necrosis and a decline in fruit sugar content, similar to the harm to the fruits and other plants described above. Therefore, it is necessary to explore the fertilizer damage caused by excessive Ca in litchi and its mechanism. This paper intends to start from the mechanism of high leaf calcium inhibiting sugar accumulation in "Feizixiao" litchi pulp, so as to provide a theoretical foundation for the development of comprehensive and effective measures for foliar calcium spraying to prevent litchi fruit from cracking.

2. Materials and Methods

2.1. Experimental Setup and Materials

The lychee orchard at the Yongfa Scientific Research Demonstration Station of the Tropical Fruit Tree Research Institute of Hainan Academy of Agricultural Sciences served as the experimental site (latitude and longitude of 19.9° and 109.8°, respectively). The site is located in the town of Yongfa in northern Chengmai County, Hainan Province. The terrain is a plateau and lies in the tropical monsoon climate area. The annual average temperature is 23.8 °C, the annual average number of sunshine hours is 2059 h, the annual average rainfall is 1786.1 mm, and there is no frost year-round. Rain and heat occur in the same season, and the soil is fertile and lateritic. At the time of this study, the soil organic matter content was 23.37 ± 0.87 g/kg, the content of available K was 125.63 ± 2.38 mg/kg, the content of available Ca was 600.51 ± 12.35 mg/kg, and the content of available Mg was 150.47 ± 3.86 mg/kg. Ten 20-year-old grafted Feizixiao lychee trees displaying the same growth trend with no signs of poor growth were selected for sampling. Feizixiao lychee plants mature quickly when grown in this area—nearly 3–4 weeks earlier than those grown in the Guangdong production area. The main phenological periods of lychee in this region are as follows: the fruit dropping period is in early April, the fruit expansion period is in late April, maturity and harvest occur in early May, and the autumn shoot growth period is from July to September.

2.2. Field Experimental Design and Sampling Methods

The treatments involved spraying a 54 μM anhydrous $CaCl_2$ aqueous solution onto the leaf surfaces (high-Ca treatment hereafter; the $CaCl_2$ used was chemically pure), and a water treatment served as the control (CK). Five replications of single-tree plots were established. The field experiment was initiated at the beginning of fruit expansion (the aril completely covered the seeds). The foliar fertilizer was sprayed once every 7 d for a total of 3 times. The high-Ca treatment was applied between 8 and 9 a.m. on 10, 18, and 25 April 2019 (corresponding to 35, 43, and 50 d after anthesis). The first sampling occurred on 18 April 2019, and subsequent sampling was conducted every 7 d, i.e., at 43, 50, 57, 64, 71, and 77 d after anthesis. As references, five fruits of moderate size and shape were taken from four sides of the tree around the middle of the crown. Thirty fruits similar to the reference fruits were collected from each tree at each collection time. The fruits were quickly frozen in liquid nitrogen and taken back to the laboratory. Then, they were put into a $-80\ °C$ ultralow-temperature freezer for storage.

2.3. Measurement Methods of Observed Indicators

The soluble sugar content was determined by anthrone colorimetry. Then, 0.1 g of the sample was ground in a boiling water bath, centrifuged, and placed in a second boiling water bath. After cooling, the anthrone reacted with free hexose or hexosyl, pentosyl, and hexuronic acid in the polysaccharide. After the reaction, the solution was blue-green and had maximum absorption at 620 nm, and the specific content was obtained by comparison with the standard curve [14]. The contents of sugars such as sucrose, fructose, and glucose were determined by the use of a high-performance liquid chromatograph. Here, 0.5 g of pulp was weighed, and 5 mL of 90% ethanol was added to fully grind the sample. Then, the sample was centrifuged, the supernatant collected, and ethanol used for secondary extraction. The two supernatants were combined, placed in a water bath, evaporated to dryness, and brought up to 10 mL with deionized water. The sample was filtered through a 0.45 μm filter membrane prior to testing. The sugar content was determined using a Waters 2695 high-performance liquid chromatograph manufactured in the United States and equipped with an evaporative light-scattering detector [15]. The fresh sample was weighed and then dried to an ashed sample. The ashed sample was allowed to cool and 10 drops of deionized water followed by 3–4 mL of nitric acid were added to the sample. Excess nitric acid was evaporated by placing the sample on a hot plate set at 100–120 °C. The sample was returned to the furnace and ashed for an additional 1 h and after being cooled, the ash was dissolved in 10 mL hydrochloric acid and transferred quantitatively to 50 mL volumetric acid. Analysis for calcium was carried out using atomic absorption spectroscopy set at different wavelengths for the optimum working conditions of the minerals.

Extraction of the acid invertase (AI), sucrose phosphate synthase (SPS), sucrose synthase II (SS-II), neutral invertase (NI), and sucrose synthase I (SS-I) enzymes was performed according to Nielsen's method [16], and the enzyme activity was determined according to the avidin biotin peroxidase complex enzyme-linked immunosorbent assay and ELISA reagent test kit (Catalogue No.: KT8013-A, KT50452-A, KT5045-A, KT8107-A, KT5044-A; Meimian, Yancheng, Jiangsu, China). For example: Take 2 g of pulp, add 2 mL of acetic acid buffer, grind into a paste with a mortar in an ice bath, 12 000 r/min for 10 min, and retain the supernatant for enzyme activity determination. Using 2 brace-plug graduated tubes, add 0.8 mL of acetic acid buffer, 0.2 mL of 0.5 mmol/L sucrose solution, and 1 mL of appropriately diluted enzyme solution to each tube, and use the same treatment but no enzyme solution as a blank control, and place it at room temperature for 10 min. Then add 1 mL of Nelson reagent to each tube and place the tubes in a boiling water bath for 20 min. Cool to room temperature, add 1 mL of arsenic–molybdic acid reagent to each tube, and after 5 min, add 7 mL of distilled water to each tube, colorimetric at 510 nm, and determine the optical density (OD) 510nm.

Several key time points were selected: 35 d after anthesis was the starting point and 64 or 71 d after anthesis was the end point. The samples were sent to Wuhan Metwell

Biotechnology Co., Ltd., for transcriptome sequencing. Four unigene databases, namely, NCBI nonredundant (NR) protein, SwissProt, Kyoto Encyclopedia of Genes and Genomes (KEGG), and Clusters of Orthologous Groups of proteins, were used in conjunction with alignment via BLASTX software to obtain protein sequences with high similarity; thus, we obtained the protein functional annotation information corresponding to each unigene. To analyze the differentially expressed genes and their clustering, RSEM software was used for quantitative analysis of gene expression based on fragments per kilobase of exon per million mapped fragment (FPKM) values [17], and DESeq2 was used for screening differentially expressed genes [18] and for analyzing subsequent Gene Ontology (GO) and KEGG functional enrichment. Eight differentially expressed genes were randomly selected from the transcriptome results, and their expression was measured via real-time quantitative PCR (qRT–PCR) [19]; the $2^{-\Delta\Delta CT}$ value was calculated and used to evaluate whether the transcriptome sequencing results were reliable.

2.4. Statistical Analysis

The online version of SAS software was used for statistical analysis of the data. ANOVA was used for variance analysis, and the Duncan method was used for multiple comparison analysis. T tests were used to analyze the significance of the differences between the high-Ca treatment and the CK treatment.

3. Data Analysis and Results

3.1. Effects of High-Ca Treatment on Sugar Content

As shown in Figure 1a,b, the glucose and fructose content in the fruit pulp exhibited the same increasing trend; after 71 d, the contents in the high-Ca treatment group were significantly lower than those in the CK group, so the accumulation of reducing sugars was inhibited by the high-Ca treatment.

As shown in Figure 1c, the dynamic changes in the sucrose content showed a single-peak curve, but the peak in the high-Ca treatment group occurred 57 d after anthesis, and the peak in the CK occurred 50–64 d after anthesis. There was no significant difference in the sucrose content between the high-Ca treatment and CK before 50 d after anthesis, yet at 57 d after anthesis, the content in the high-Ca treatment group was significantly higher than that in the CK group. Additionally, at 57 d after anthesis, the content in the high-Ca treatment was significantly lower than that in the CK, so the inhibition of sucrose accumulation in response to the high-Ca treatment occurred at this time.

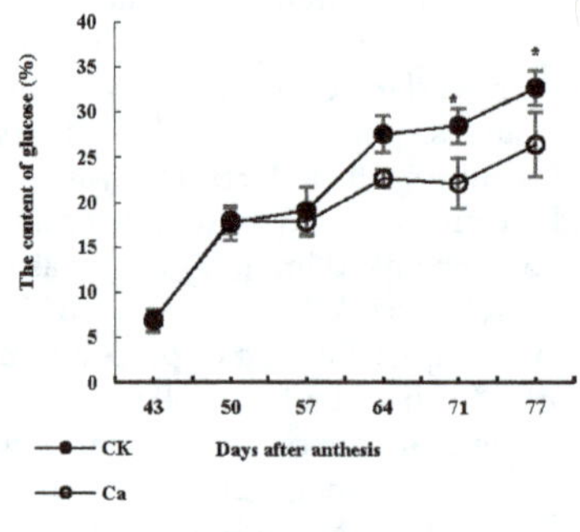

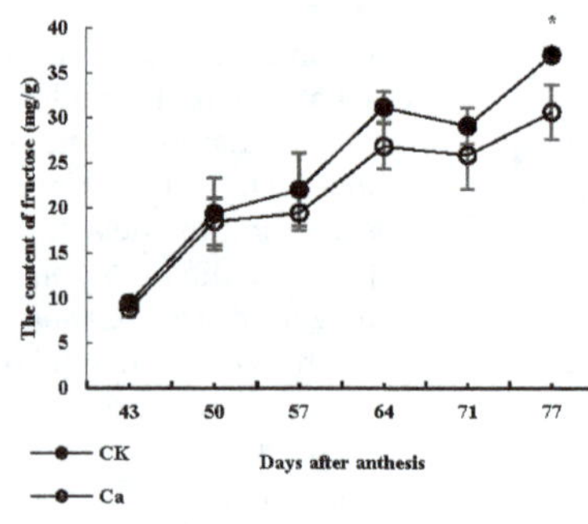

Figure 1. *Cont.*

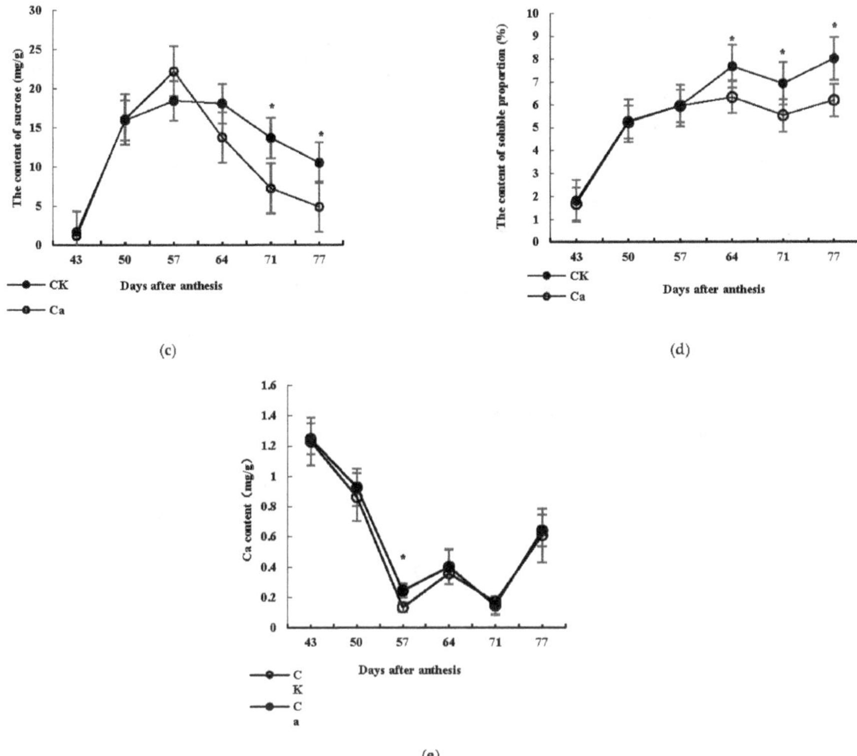

Figure 1. Effects of high-Ca treatment on sugar accumulation and Ca content: (**a**) content of glucose; (**b**) content of fructose; (**c**) content of sucrose; (**d**) content of soluble proportion; (**e**) content of Ca. Note: The symbols with * in the figure indicate significant differences between the CK and high-Ca treatments for the same data type; otherwise, there was no significant difference. The same scheme applies to the figures below.

As shown in Figure 1d, the dynamic change trend of the soluble sugar content of the high-Ca treatment and CK was different, and the phenomenon of "sugar withdrawal" appeared in the flesh at the mature stage. The difference was that there was no significant change from 43 to 57 d after anthesis, and at 64 d after anthesis, the value in the high-Ca treatment was significantly lower than that in the CK. As a result, the accumulation of soluble sugar was significantly inhibited by high-Ca treatment.

3.2. Ca Content

It can be seen from Figure 1e that the Ca content of treatment and CK shows a "W" shaped trend of decreasing, rising, decreasing, and rising. There were two troughs, 57 and 71 d after flowering. Treatment was significantly higher than CK at 57 days after flowering; there was no significant difference between treatment and CK on other dates. It shows that pretreatment can increase the Ca content.

3.3. Effects of High-Ca Treatment on the Activities of Key Enzymes Involved in Glucose Metabolism
3.3.1. AI

The changes in AI activity are shown in Figure 2a. The AI activity of the CK group first increased, then decreased and increased again. Finally, there was no significant difference between the CK group and the high-Ca treatment group. The AI activity of the high-Ca treatment group showed a single-peak curve, and the highest peak occurred at 64 d after anthesis; this peak was significantly lower and occurred earlier than that of the CK group. There was no significant difference in AI activity between the high-Ca treatment group and CK group at 64 d after anthesis. These results indicated that high-Ca treatment inhibited AI activity in the early stage of sugar accumulation, thus inhibiting the accumulation of reducing sugars.

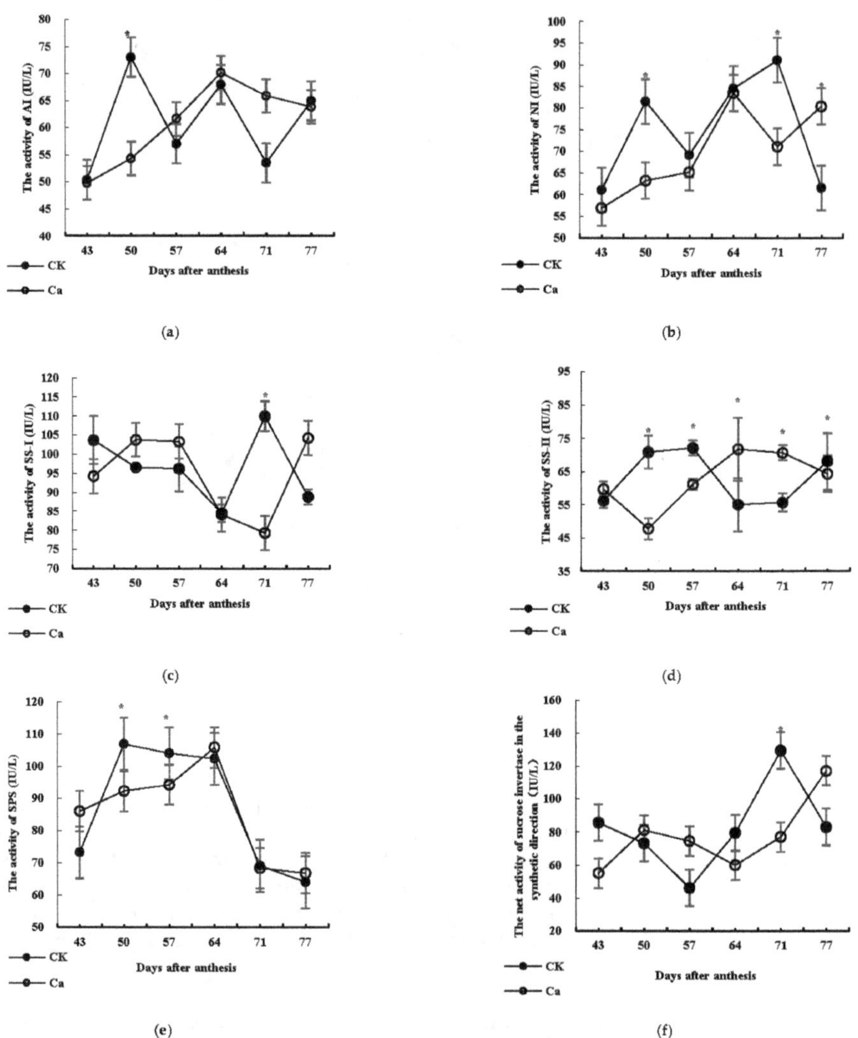

Figure 2. Effects of high-Ca fertilizer on the activities of key enzymes involved in fructose metabolism: (**a**) AI; (**b**) NI; (**c**) SS-I; (**d**) SS-II; (**e**) SPS. (**f**) The net activity of sucrose invertase in the synthetic direction. Note: The symbols with * in the figure indicate significant differences between the CK and high-Ca treatments for the same data type; otherwise, there was no significant difference.

3.3.2. NI

The changes in NI activity are shown in Figure 2b. The change trend of the enzyme activity in the high-Ca treatment group decreased only at 71 d after anthesis, and the enzyme activity in the CK group experienced two increases and decreases. However, the NI activity in the high-Ca treatment group was significantly higher than that in the CK group at 77 d after anthesis and was significantly lower than that in the CK group overall, indicating that the high-Ca treatment had an obvious inhibitory effect on the accumulation of reducing sugars in the fruit pulp.

3.3.3. SS-I

The changes in SS-I activity are shown in Figure 2c. The dynamic changes in enzyme activity in the high-Ca treatment group and CK group were different. The values in the high-Ca treatment group were significantly lower than those in the CK group at 71 d after anthesis. There was no significant difference at the other time points. Therefore, combined with the results of the sugar content determination, the results here indicate that high-Ca treatment may inhibit the conversion of sucrose to reducing sugars by reducing SS-I activity and subsequently inhibiting the accumulation of reducing sugars in the fruit pulp.

3.3.4. SS-II

The changes in SS-II activity are shown in Figure 2d. The SS-II activity in the high-Ca treatment group was significantly lower than that in the CK group before 64 d after anthesis and vice versa after 64 d after anthesis, but there was no significant difference at 43 and 77 d after anthesis. These results indicated that high-Ca treatment inhibited SS-II activity before 64 d after anthesis, resulting in the inhibition of sucrose accumulation.

3.3.5. SPS

The changes in SPS activity are shown in Figure 2e. The dynamic changes in enzyme activity in the high-Ca treatment group and CK group showed a single-peak curve; the peak value of the activity in the high-Ca treatment group occurred at 64 d after anthesis, and the peak value in the CK group was at 50 d after anthesis. There was no significant difference in SPS activity between the high-Ca treatment group and the CK group at 64 d after anthesis, but the activity was significantly lower in the high-Ca treatment group than in the CK group at 64-50 d after anthesis. High-Ca treatment inhibited SPS enzyme activity, resulting in the inhibition of sucrose accumulation.

3.3.6. Synthesis Direction and Net Activity of the Sucrose Invertase System

The characteristics of net activity change in the synthesis direction of the sucrose invertase system are shown in Figure 2f. The change trend of the net activity in the high-Ca treatment group was "up, down, and up", while that in the CK showed the opposite trend. At 71 d after anthesis, the net activity of the sucrose invertase system in the high-Ca treatment group was significantly higher than that in the CK treatment group. At this time, a large amount of sucrose in the CK group was transformed into glucose and fructose, while less sucrose in the high-Ca treatment group was converted. Overall, the sucrose in the CK group decomposed into fructose and glucose, while the high-Ca treatment inhibited the conversion of sucrose to reducing sugars, and the accumulation of reducing sugars was inhibited relative to that in the CK group.

3.4. Transcriptome Analysis of Fruit Pulp
3.4.1. Assembly and Analysis of Transcriptome Sequencing Data

The sequencing yield of the lychee pulp transcriptome is shown in Table 1. The GC content of each sample is greater than 44.86%, the percentage of Q30 bases is greater than 91.43%, and the error rate is only 0.02%, which indicates that the quality of the bases is high and that the sequencing results are ideal, which ensures the accuracy and reliability of subsequent analysis.

Figure 3a shows that the number of transcripts and unigenes with a length of more than 200 to 2000 bp exceeds 1000, and the number of transcripts and unigenes with a length ranging from 200 to 300, 300 to 400, and 400 to 500 bp exceeds 10,000. Table 2 shows that a total of 306,396 transcripts with an N50 of 1913 bp, an N90 of 456 bp, and an average length of 1119 bp were assembled. A total of 115,413 unigenes were obtained. The average length of these sequences was 894 bp, the N50 was 1581 bp, and the N90 was 335. Taken together, these findings indicate that the sequencing results are ideal and that the analysis results are reliable.

Table 1. Statistics of transcriptome sequencing data from the pulp of Feizixiao lychee fruit in the high-Ca treatment and CK treatment groups.

Sample	Raw Reads	Clean Reads	Clean Bases (Gb)	Error Rate (%)	Q30 (%)	GC Content (%)
35d-CK	48,281,265	47,115,740.7	7.0	0.02	92.5	44.9
64d-CK	48,142,673	47,328,952	7.0	0.02	91.4	44.9
64d-Ca	49,977,736	49,345,295.3	7.4	0.02	92.3	44.8
71d-CK	47,102,587	46,452,051.3	6.9	0.02	91.6	44.9
71d-Ca	47,311,827	46,672,307.3	7.0	0.02	91.8	44.8

Notes: 35 d-CK, 64 d-CK, and 71 d-CK represent transcripts in the control Feizixiao lychee fruit pulp at 35, 64, and 71 d after anthesis, respectively; 35d-CK, 64d-Ca, and 71d-Ca represent transcripts of Ca-treated Feizixiao lychee fruit pulp at 35, 64, and 71 d after anthesis, respectively.

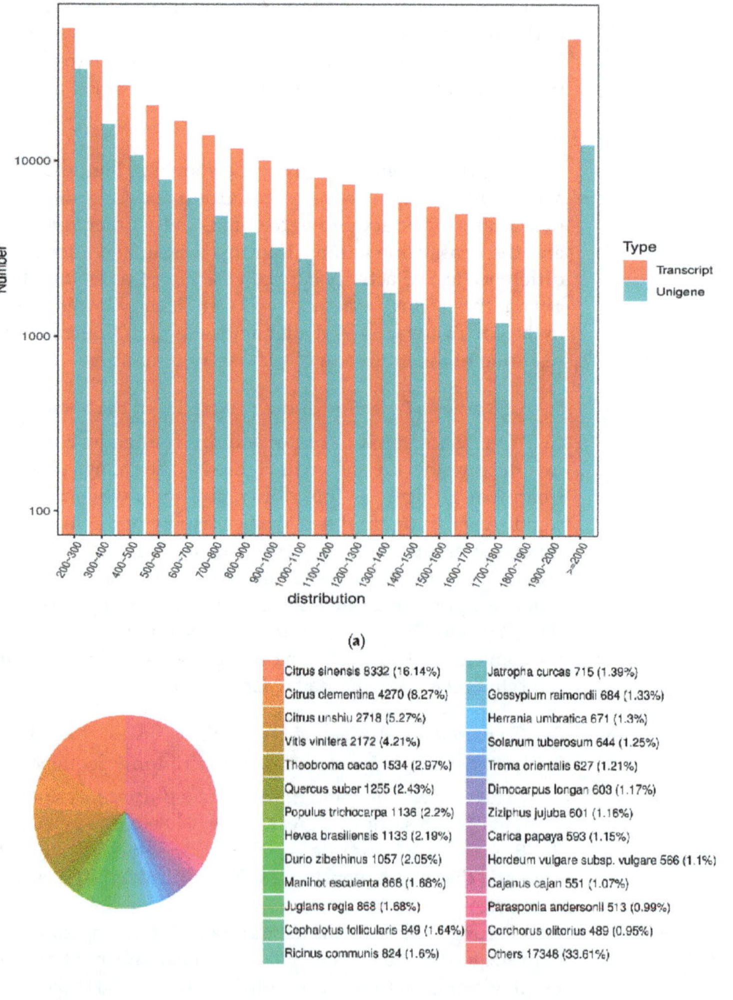

Figure 3. *Cont.*

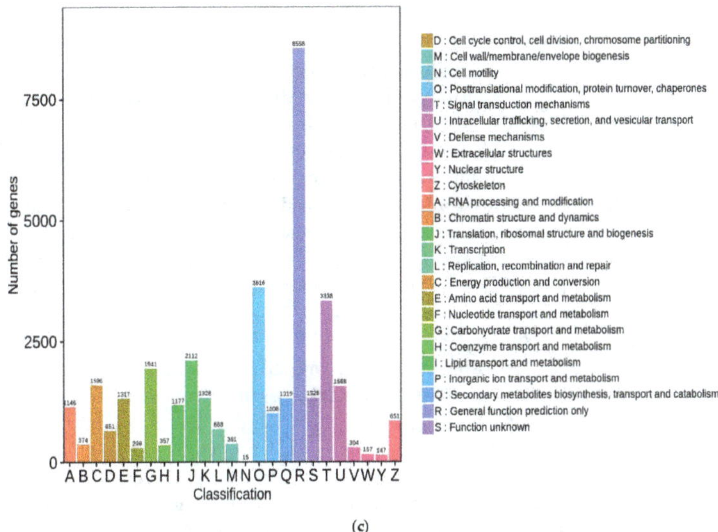

(c)

Figure 3. Assembly and annotation of the transcriptome from the pulp of Feizixiao lychee: (**a**) sequence length distribution; (**b**) statistical chart of species with gene homology; (**c**) statistical chart of KOG function classification.

Table 2. The statistics of assembly results.

Type	Number	Mean Length	N50	N90	Total Bases
Transcript	306,396	1119	1913	456	342,893,342
Unigene	115,413	894	1581	335	103,163,097

3.4.2. Functional Annotation and Analysis of Unigenes

The unigenes were subjected to functional enrichment analysis, and the results are shown in Table 3. By comparing the unigene sequences with sequences in the KEGG, NR, GO, and other databases, we found that 55,352 unigenes were annotated in at least one database, accounting for 47.96% of the total. Among these unigenes, the number of unigene annotations obtained from the TrEMBL database was the largest at 51,397, accounting for 44.53%; this was followed by the annotations obtained from the NR and GO databases. In addition, 36,242 unigenes were annotated in the SwissProt database, and 41,174 and 34,652 unigenes were annotated in the KEGG pathway and Pfam databases, respectively.

Table 3. Unigene annotation results.

Database	Number of Genes	Percentage (%)
KEGG	41,174	35.6
NR	51,132	44.3
SwissProt	36,242	31.4
TrEMBL	51,397	44.5
KOG	32,067	27.7
GO	43,328	37.5
Pfam	34,652	30.0
Annotated in at least one database	55,352	47.9

The results of our evolutionary analysis are shown in Figure 3b. The number of homologous genes is ranked from large to small. The top ten species with homologous genes were *Citrus sinensis*, *Citrus clementina*, *Citrus unshiu*, *Vitis vinifera*, *Theobroma cacao*, *Quercus suber*, *Populus trichocarpa*, *Hevea brasiliensis*, *Durio zibethinus*, and *Manihot regia*.

With 603 homologous genes, Dimocarpus longan, which is closest to lychee, ranks 19th. Based on an in-depth study of citrus fruit growth and development, this study provides a reference method and results.

As shown in Figure 3c, the obtained unigenes were compared with sequences in the EuKaryotic Orthologous Groups (KOG) database to analyze their possible functions. The results showed that these gene sequences were related to 25 biological processes.

Among these biological processes (in terms of general function prediction only), the most abundant terms involved posttranslational modification, protein turnover, chaperones, signal transduction mechanisms, translation, ribosomal structure, and biogenesis. Carbon transport and metabolism; energy production and conversion; and secondary metabolite biosynthesis, transport, and catabolism were the next most abundant terms.

3.4.3. Functional Annotation and Enrichment Analysis of Differentially Expressed Genes

Additionally, we examined the positive and negative regulatory processes related to biological processes and the regulation of enzyme activity related to molecular function. As shown in Figure 4a, there were differences in gene expression during the growth and development of the pulp of the fruit in the high-Ca treatment group and the CK group. With respect to sugar metabolism in the pulp, there were different degrees of gene upregulation and downregulation in the CK group compared with the high-Ca treatment group. In total, 48 genes were upregulated in the high-Ca treatment group compared with the CK group at 50 d after anthesis, and 83 genes were downregulated; at 64 d after anthesis, 92 genes were upregulated, and 19 genes were downregulated. Moreover, at 71 d after anthesis, 49 genes were upregulated, and 38 genes were downregulated, which indicated that high-Ca treatment significantly altered the expression of genes during fructose metabolism.

Figure 4b shows that at 64 d after anthesis, in the biological process category, the differentially expressed genes in the high-Ca treatment and CK groups were involved mostly in metabolic processes, cellular processes, and cellular component organization or biogenesis. In the broad cellular components category, the genes in the high-Ca treatment and CK groups were involved mostly in extracellular regions, cells, and cell parts. In the broad molecular function category, the differentially expressed genes in the high-Ca treatment and CK groups were mostly involved in catalytic activity.

Figure 4c shows that there are many subclasses of genes differentially expressed at 71 d after anthesis compared with those at 64 d after anthesis, but the number is similar to that at 64 d after anthesis. At 71 d after anthesis, there were also a greater number of genes involved in response to stimuli and a smaller number involved in cell components, tissues, or biogenesis. In the cellular component category, the number of genes annotated as extracellular regions was relatively small. In the molecular function category, binding was also notably annotated.

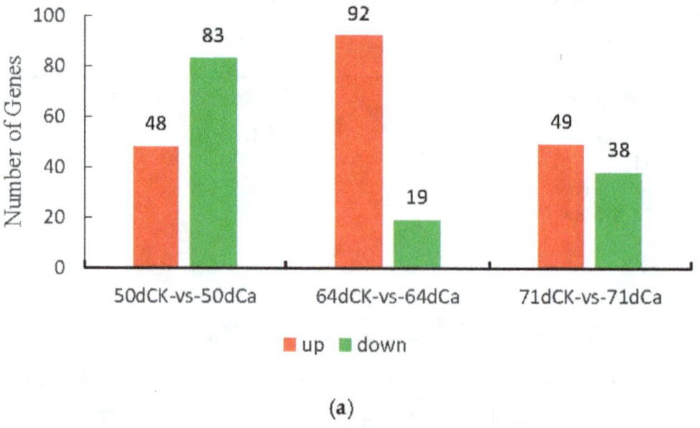

(a)

Figure 4. *Cont.*

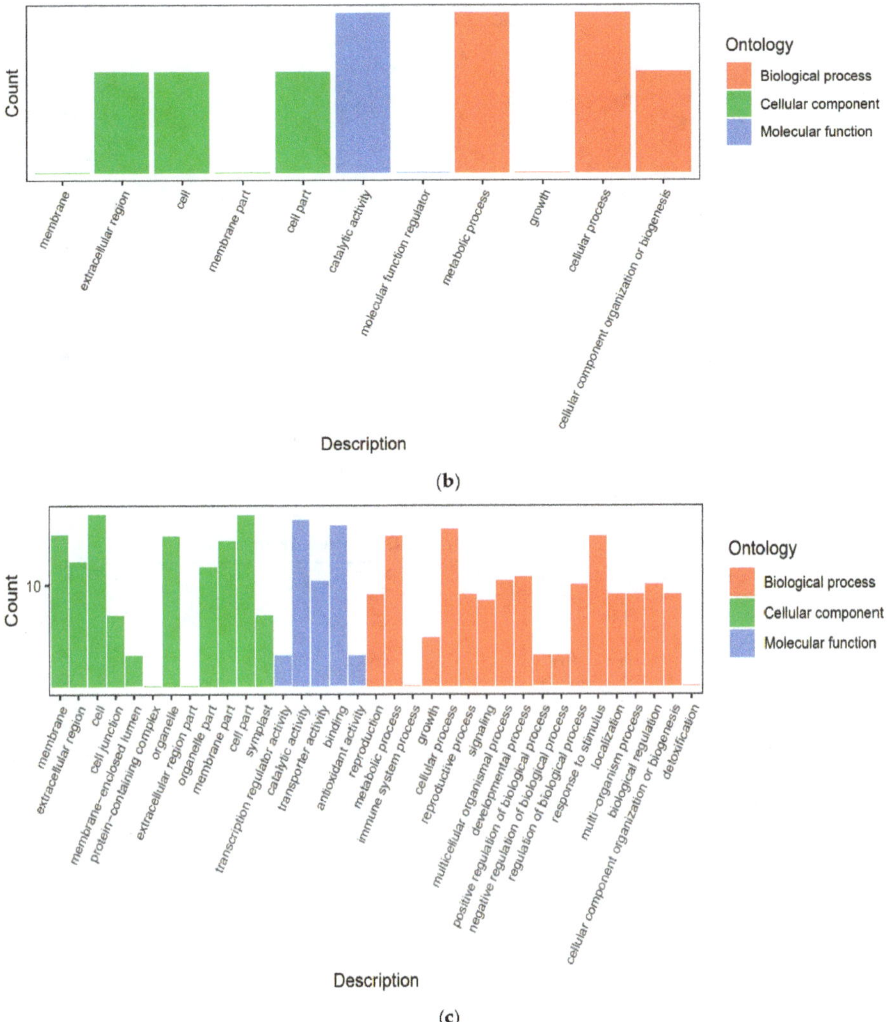

Figure 4. Statistics and functional classification of differentially expressed genes: (**a**) statistical chart of gene difference between treatment and control; (**b**) histogram of GO classification of DEGs in CK and treatment at 64 days after flowering; (**c**) histogram of GO classification of DEGs in CK and treatment at 71 days after flowering.

Pathway enrichment analysis was carried out on the genes differentially expressed between the CK group and the high-Ca treatment group. The pathways that were significantly enriched are shown in Figure 5a,b. The fructose content of the high-Ca treatment and CK groups reached a maximum at 64 d after anthesis, and the fructose content of the high-Ca treatment group was significantly lower than that of the CK group at that time. The pathways most significantly enriched were pentose and glucuronic acid interconversion, flavonoid biosynthesis, and others. At 71 d after anthesis, the sugar withdrawal phenomenon was detected in both the high-Ca treatment group and the CK group, and the enriched pathways causing this difference were mainly involved in plant hormone signal transduction and ABC transporters. This indicates that the inhibitory effect of high-Ca treatment on sugar accumulation is mainly reflected in the effects on the above pathways.

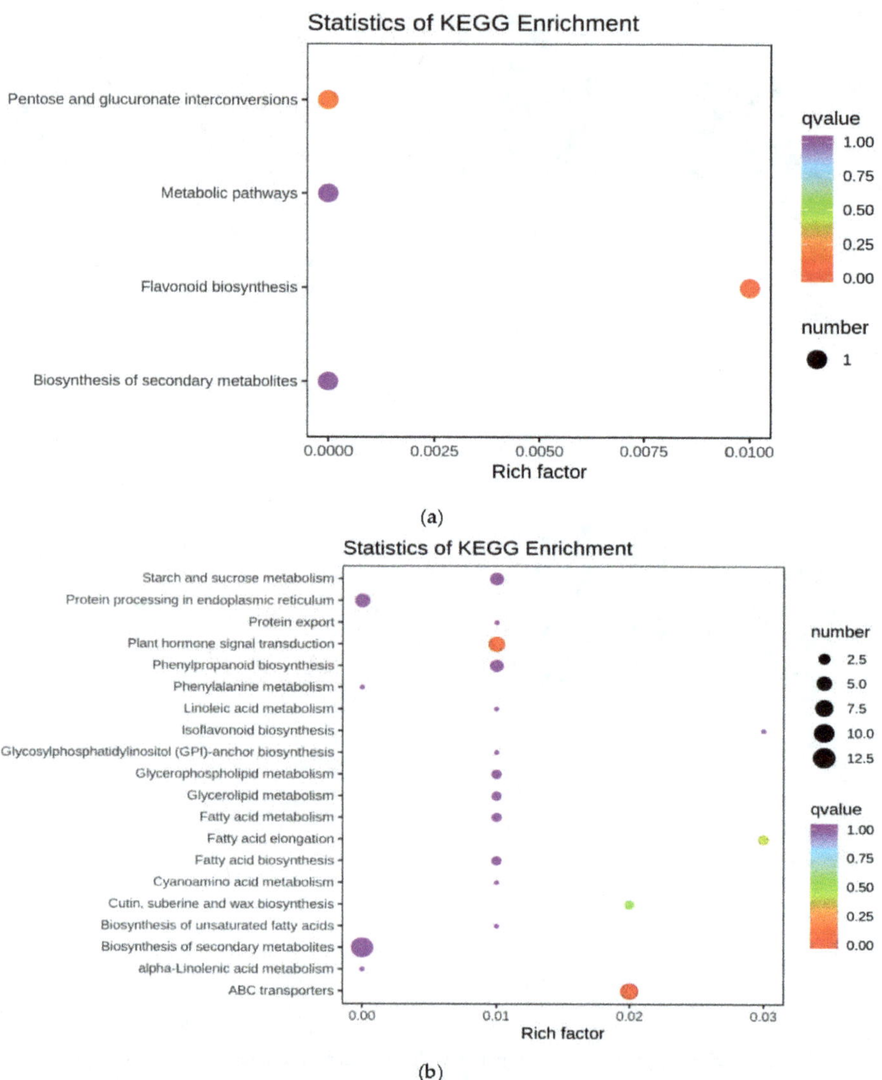

Figure 5. Enrichment statistics of metabolic pathways involving differentially expressed genes: (a) KEGG pathway enrichment of DEGs in CK and treatment at 64 days after flowering; (b) KEGG pathway enrichment of DEGs in CK and treatment at 71days after flowering.

Hierarchical cluster analysis was performed on the expression (FPKM values) after analysis and standardization of the differentially expressed genes, and a cluster heatmap of the groups of genes differentially expressed between the high-Ca treatment group and the CK group at 64 and 71 d after anthesis was constructed, as shown in Figure 6a,b. The results corresponding to the two time points and the CK are grouped into one category. The differentially expressed genes could be divided into two categories: those with high expression and those with low expression.

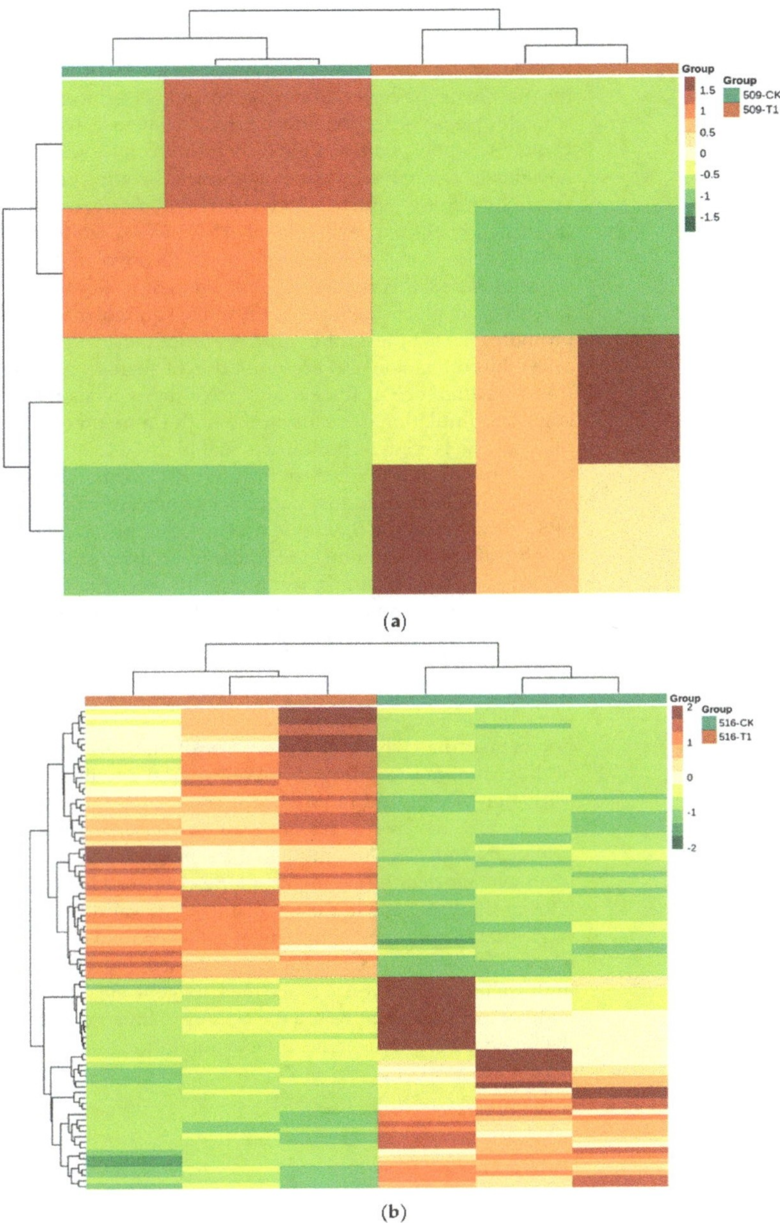

Figure 6. Hierarchical clustering heatmap of differentially expressed genes: (**a**) hierarchical clustering heatmap of differential genes between 509−CK and 509−T1; (**b**) hierarchical clustering heatmap of differential genes between 516−CK and 516−T1. Note: The abscissa represents the sample name and hierarchical clustering results, and the ordinate represents the differentially expressed gene and hierarchical clustering results shown in Figure 6. Red indicates high expression, and green indicates low expression; 509-CK and 509-T1 correspond to the control and high-Ca treatment groups, respectively, at 64 d after anthesis; 516−CK and 516−T1 correspond to the control and high-Ca treatment groups, respectively, at 71 d after anthesis.

Figure 7a shows the metabolic pathway of starch and sucrose metabolism, which mainly controls carbohydrate metabolism. In sucrose metabolism, there are three pathways labeled 3.2.1.21 involving β-D-glucoside glucose hydrolase, which involves an enzyme that promotes the conversion of other sugars to glucose. Pathway number 5.3.1.9 involves α-D-glucose-6-phosphate aldose ketone isomerase, which is involved in the catalysis of α-D-glucose 6-phosphate and the β-reversible conversion between D-fructose and furanose 6-phosphate. The pathway number of the SUS-coding gene is 2.4.1.13; this pathway involves sucrose synthase, which is an enzyme that catalyzes the reversible synthesis of sucrose. Interestingly, the expression level of SUS was downregulated. In starch metabolism, the gene expression in pathway number 3.2.1.28 (involving α-trehalose glucose hydrolase) was downregulated; this enzyme is involved in the catalysis of α,α-trehalose. The hydrolysis of the o-bond of glucoside releases the initial equimolar amount of α- and β-D-glucose, which promotes the synthesis of D-glucose. Since the enzymes encoded by these genes directly affect the conversion and accumulation of sucrose, glucose, and fructose, it is suggested that the difference in the expression of these genes may be related to the inhibition of sugar accumulation detected in the high-Ca treatment group. The enzyme encoded by the sus gene is SS, which indicates that high-Ca treatment leads to downregulation of the SS enzyme-encoding gene. In the analysis of differentially expressed genes involved in glucose metabolism in this paper, no significant change was found in the expression of SPS genes, indicating that there may be other physiological and biochemical mechanisms involved in the regulation of SPS activity in the high-Ca treatment group.

Figure 7b shows the biosynthetic pathway of flavonoids, which compose a main category of plant secondary metabolites. Flavonoids are synthesized from phenylpropanoid derivatives by condensation with malonyl-CoA. The genes in pathway numbers 2.3.1.74 (involving 4-β-D-glucan glucose hydrolase) and 2.3.1.170 (involving malonyl-CoA) were downregulated. Both belong to the chalcone synthase (CHS) superfamily. CHS is the first key structural enzyme in flavonoid synthesis and is responsible for catalyzing the formation of chalcone, which, as a precursor of many flavonoids, participates in the formation of downstream secondary metabolites. The downregulation of these genes inhibits the formation of chalcone and may ultimately inhibit the accumulation of flavonoids.

Figure 7c shows a cluster heatmap of the expression levels of the invertase-related genes CIN and VIN. According to the expression level, it can be concluded that the high-Ca treatment affected the expression of the gene-encoding invertase. Moreover, the expression level of the VIN gene in the CK group significantly decreased during sugar withdrawal, while that in the high-Ca treatment group did not significantly decrease. The expression of the CIN gene was significantly increased, but there was no significant change in the high-Ca treatment group. Since the CIN gene encodes NI and the VIN gene encodes AI, high-Ca treatment suppressed the activity of invertase by downregulating the expression of the invertase-encoding gene, thereby inhibiting the accumulation of glucose and fructose.

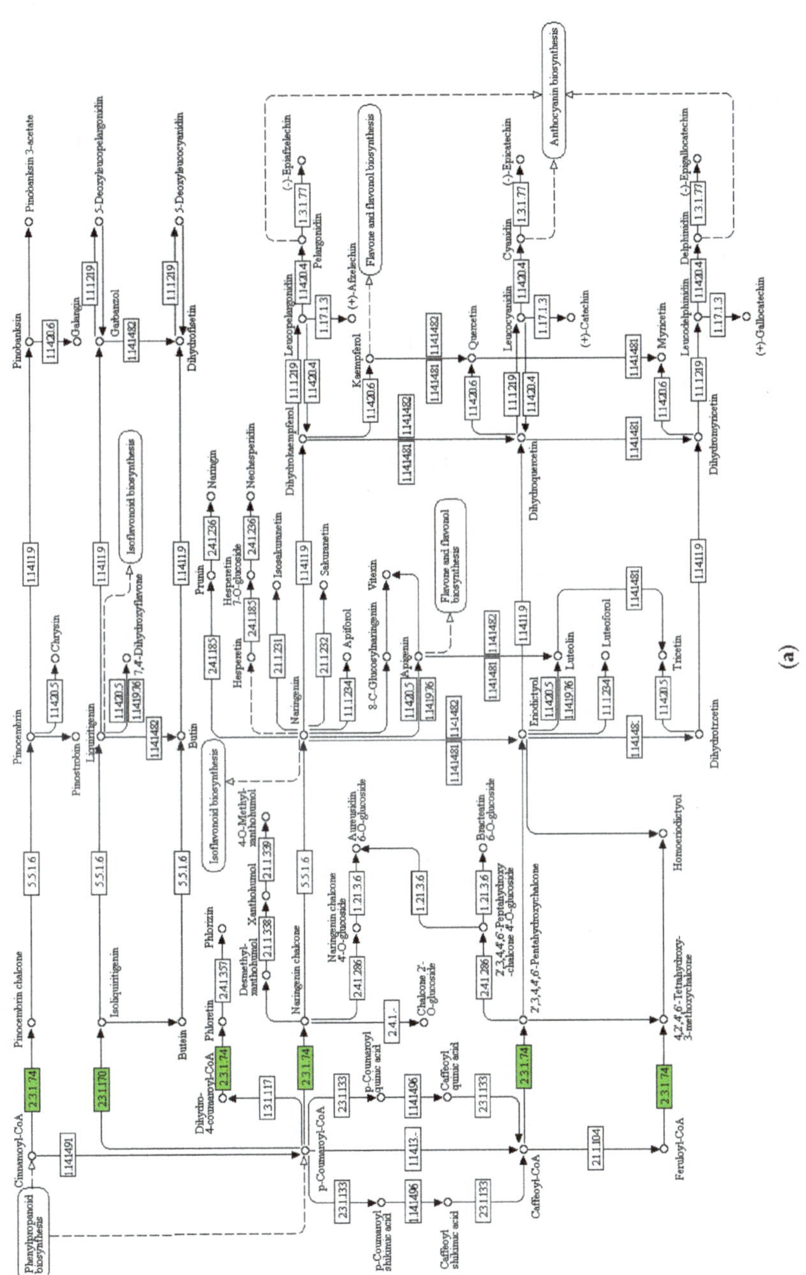

Figure 7. *Cont.*

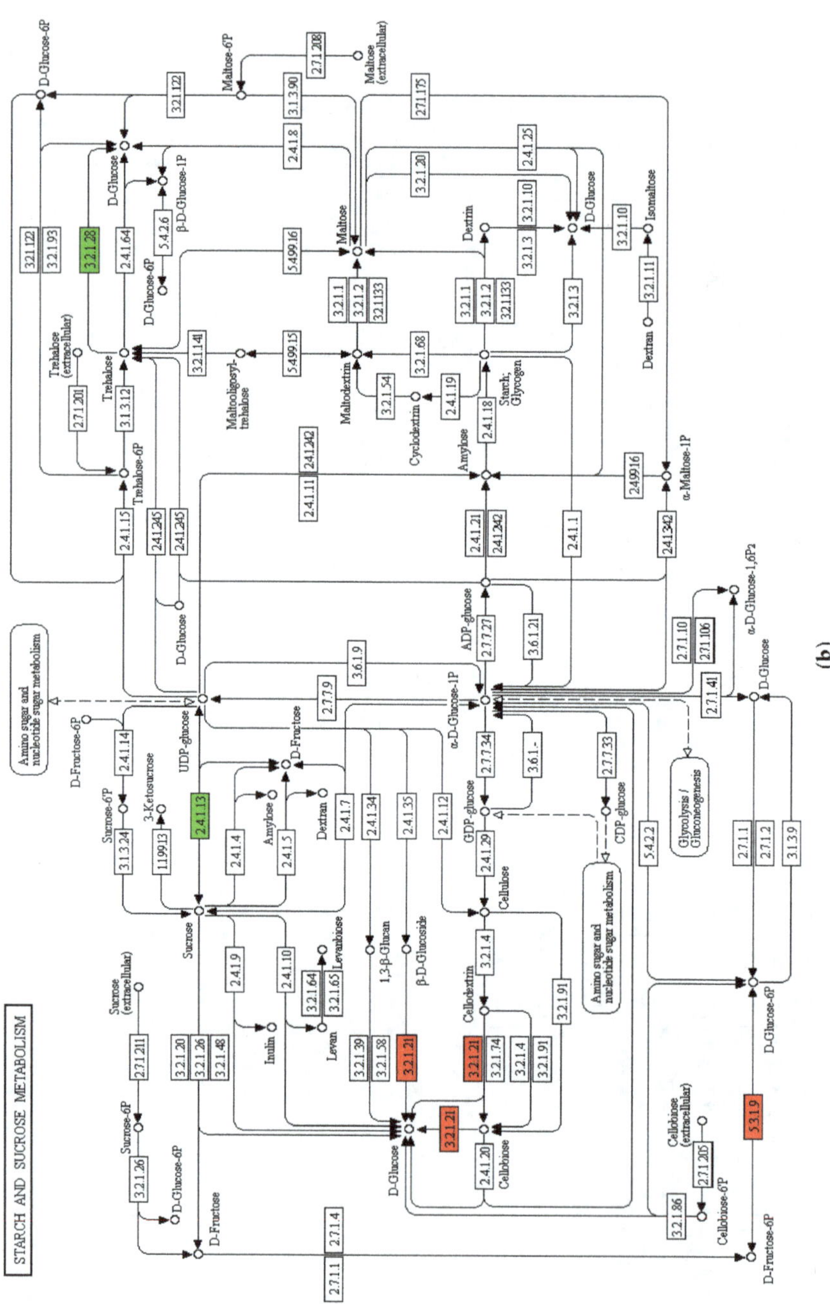

Figure 7. *Cont.*

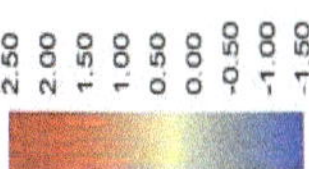

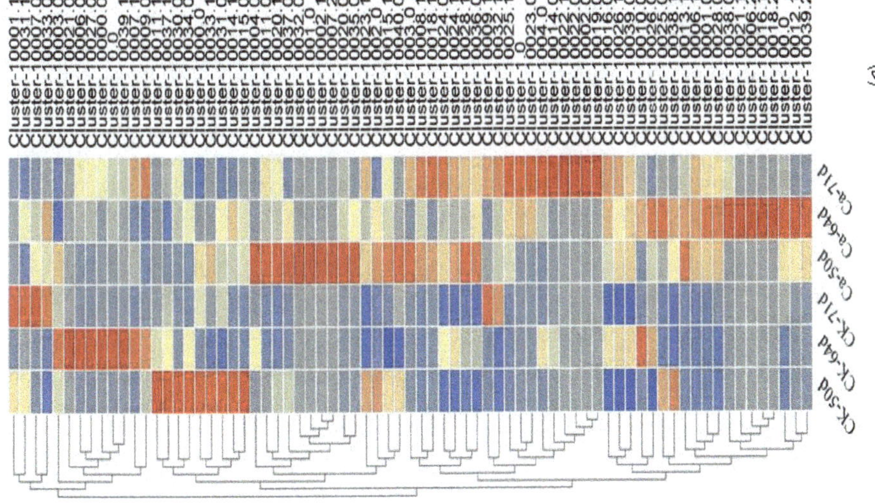

(c)

Figure 7. Statistics and cluster heatmap of differentially expressed genes involved in specific metabolic pathways: (**a**) metabolic pathway of starch and sucrose metabolism; (**b**) flavonoid biosynthesis pathway; (**c**) starch and sucrose metabolism pathway.

3.5. qRT–PCR Validation Analysis

qRT–PCR analysis was performed on eight unigenes. As shown in Figure 8, the linear relationship between the RNA sequencing (RNA-seq) data and qRT–PCR data was significantly positive, indicating that the results of the transcriptome analysis were accurate and reliable.

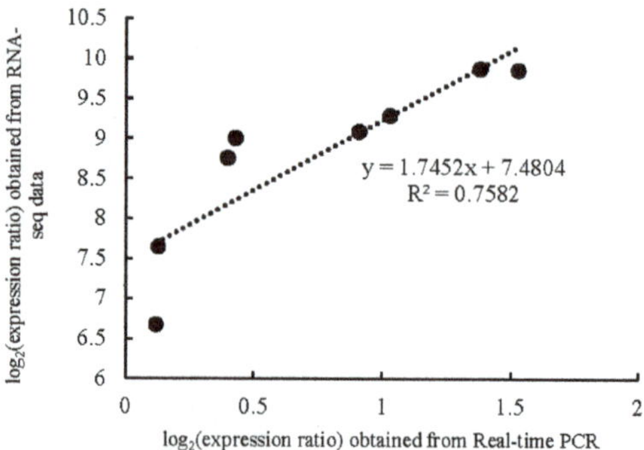

Figure 8. Correlation analysis of gene expression levels obtained from RNA-seq data and qRT–PCR data. Note: The y values represent $\log_2$(FPKM) from RNA-seq, and the x values represent $-(\Delta\Delta Ct)$ from qRT–PCR.

4. Discussion and Analysis

Soluble sugars are important quality indicators of fruits [20]. In lychee, the main sugars are glucose, fructose, and sucrose [21]. Different kinds of lychees have different types of sugars that accumulate as a result of sugar metabolism, and Feizixiao lychees accumulate reducing sugars [15]. In addition, the sugar withdrawal phenomenon can occur [13], which needs further attention. The results of the present study showed that the soluble sugars in lychee pulp were mainly glucose and fructose, and the sucrose content was significantly lower than that of the two monosaccharides, the results of which were consistent with our previous results. The results also showed that the high-Ca treatment inhibited the accumulation of soluble sugars, which was consistent with data from pineapple treated with high Ca [8].

AI, NI, SS-I, SS-II, and SPS [22–25] play an important role in sugar metabolism. The results showed that the high-Ca treatment obviously inhibited the activities of AI and NI during the sugar accumulation period, inhibited the conversion of sucrose to glucose and fructose, and reduced the accumulation of reducing sugars in the fruit pulp. The high-Ca treatment increased SS-I activity and decreased SS-II activity in the early stage of sugar accumulation, and it also decreased SS-I activity and enhanced SS-II activity after sugar withdrawal; that is, at the early stage of sugar accumulation, SS mainly catalyzed the decomposition of sucrose into fructose, and at the time of sugar withdrawal, SS mainly catalyzed the synthesis of sucrose from fructose. At the same time, SPS activity was inhibited, resulting in a decrease in overall sugar accumulation in the pulp. The Feizixiao lychee fruit pulp industry is mainly controlled by the use of these key enzymes. This result is the same as those in previous reports that sugar metabolism of Feizixiao lychee fruit is mainly regulated by AI and SS [26].

Transcriptome analysis showed that high-Ca treatment did not cause changes in the expression of SPS-encoding genes, indicating that the inhibition of the activities of these two enzymes may occur posttranscriptionally, during translation and posttranslation; nonetheless, the underlying molecular mechanism needs further study.

Among the differentially expressed genes encoding α- and β-D-glucose-6-phosphate aldose ketone isomerase [27], the upregulation of D-glucoside glucohydrolase gene expression [28] suggests that the content of reducing sugars in the fruit pulp under high-Ca treatment remains relatively stable. α-Trehalose and glucose hydrolase can catalyze the decomposition of trehalose to produce glucose, and the downregulation of related genes directly causes a reduction in glucose synthesis. SUS-encoding genes control SS enzyme synthesis, which catalyzes the NDP-α-D-glucose + D-fructose to ribonucleoside 5′- diphosphate + H^+ + sucrose (reversible) conversion reaction [29]. The expression of the SUS gene was downregulated in the high-Ca treatment group, which was consistent with the change trend of SS-I and SS-II activities in the high-Ca treatment group, the findings of which were opposite those of the CK group. In the forward direction, SS-II led to a decrease in the synthesis of sucrose in the pulp, and in the reverse direction, SS-I led to a decrease in the amount of sucrose converted to fructose; thus, the accumulation of soluble sugars in the pulp was inhibited.

High-Ca treatment upregulated VIN gene expression and downregulated CIN gene expression. Combined with the measurements of the sugar components, the sugar withdrawal phenomenon occurred after the sugar content peaked. At that time, the VIN expression in the CK group was significantly decreased, while the VIN expression in the high-Ca treatment group was not significantly decreased. VIN gene expression positively regulated AI activity [30], which was consistent with the AI activity trend. Moreover, AI promoted sucrose decomposition to provide a carbon source for growth; that is, by inhibiting the downregulation of VIN expression, high-Ca treatment may cause AI activity to remain at a high level, thus reducing sucrose accumulation. The expression of the CIN gene in the CK group significantly increased when sugar withdrawal occurred; the CIN gene positively regulated NI activity, but there was no change in CIN activity in the high-Ca treatment group. Among them, the upregulation of the CIN gene promoted the accumulation of reducing sugars [25], while the high-Ca treatment in this experiment inhibited the upregulation of NI expression, thus inhibiting the accumulation of reducing sugars. The results of this paper also confirm that invertase plays an important role in fruit sugar metabolism [31].

By analyzing the transcriptome results at the critical stage of fructose accumulation, we found that the genes of the CHS family involved in the flavonoid biosynthesis pathway were downregulated [32]. CHS is mainly responsible for catalyzing the conversion of coumaroyl-CoA to synthesize naringenin chalcone, which, as a precursor, participates in the formation of flavonoids. According to the results of a test analysis, the downregulation of related genes caused a reduction in chalcone synthesis, which indirectly inhibited the accumulation of flavonoids. High-Ca treatment inhibited the accumulation of glucose, fructose, sucrose, and soluble sugars. It can thus be inferred that the content of flavonoids was reduced [32], and the pulp was damaged by reactive oxygen species, thus inhibiting the accumulation of sugars. However, this needs to be further confirmed.

5. Conclusions

The physiological mechanism that occurs in response to excessive Ca fertilizer being sprayed onto leaves to inhibit sugar accumulation is as follows: First, Ca causes the expression of trehalase-encoding genes and SUS genes to be downregulated and inhibits the activities of trehalase, SS-I, and SS-II, thus inhibiting the accumulation of glucose, fructose, and sucrose. It does not cause the expression of SPS-related genes to change but does inhibit the accumulation of this sugar by inhibiting SPS activity. There may be other mechanisms involved in inhibiting the activity of this kind of enzyme, which needs further study (glucose metabolism). High-Ca fertilizer also causes decreased expression of the VIN gene, thus inhibiting AI activity and inhibiting glucose and fructose accumulation. In turn, these phenomena inhibit the upregulation of CIN gene expression in the fruit during the fruit growth and expansion period, thereby inhibiting the increase in NI activity in the pulp and causing a decrease in glucose and fructose accumulation in the treated pulp relative to that in the fruit in the CK group. Inhibition of CIN expression possibly causes a downregulation of CHS family gene expression, causing a decrease in chalcone accumulation, which may lead to damage caused by active oxygen production in the pulp, thus inhibiting the accumulation of fructose. However, these phenomena need further confirmation.

Author Contributions: Writing—original draft, X.S.; Data curation, W.W.; Data curation, W.M.; Data curation, C.Y.; Data curation, K.Z. All authors have read and agreed to the published version of the manuscript.

Funding: National Natural Science Foundation of China (NSFC) (No. 31960570).

Data Availability Statement: Not applicable.

Conflicts of Interest: The authors declare no conflict of interest.

References

1. Liao, H.-Z.; Lin, X.-K.; Du, J.-J.; Peng, J.-J.; Zhou, K.-B. Transcriptomic analysis reveals key genes regulating organic acid synthesis and accumulation in the pulp of *Litchi chinensis* Sonn. cv. Feizixiao. *Sci. Hortic.* **2022**, *303*, 111220.
2. Khadivi-Khub, A. Physiological and genetic factors influencing fruit cracking. *Acta Physiol. Plant.* **2014**, *37*, 1718. [CrossRef]
3. Martínez Bolaños, M.; Martínez Bolaños, L.; Guzmán Deheza, A.; Gómez Jaimes, R.; Reyes Reyes, A.L. Calcio y ácido giberélico en el bretado de frutos de litchi (Litchi chinensis Soon.) cultivar Mauritius. *Rev. Mex. Cienc. Agrícolas* **2017**, *8*, 837–848. [CrossRef]
4. Zhang, C.; Cui, L.; Zhang, P.; Dong, T.; Fang, J. Transcriptome and metabolite profiling reveal that spraying calcium fertilizer reduces grape berry cracking by modulating the flavonoid biosynthetic metabolic pathway. *Food Chem. Mol. Sci.* **2021**, *2*, 100025. [CrossRef] [PubMed]
5. Wang, P.; Huang, Y.; Wang, X.; Wang, J.; Liu, F. Effects of calcium on gluconic acid accumulation in 'Shine Muscat' grape. *IOP Conf. Ser. Earth Environ. Sci.* **2021**, *792*, 012039. [CrossRef]
6. Davarpanah, S.; Tehranifar, A.; Abadía, J.; Val, J.; Davarynejad, G.; Aran, M.; Khorassani, R. Foliar calcium fertilization reduces fruit cracking in pomegranate (*Punica granatum* cv. Ardestani). *Sci. Hortic.* **2018**, *230*, 86–91. [CrossRef]
7. Wang, D.-R.; Yang, K.; Wang, X.; You, C.-X. A C_2H_2-type zinc finger transcription factor, MdZAT17, acts as a positive regulator in response to salt stress. *J. Plant Physiol.* **2022**, *275*, 153737. [CrossRef]
8. Loekito, S.; Afandi Afandi, A.; Nishimura, N.; Koyama, H.; Senge, M. The Effects of Calcium Fertilizer Sprays during Fruit Development Stage on Pineapple Fruit Quality under Humid Tropical Climate. *Int. J. Agron.* **2022**, *2022*, 3207161. [CrossRef]
9. Wang, R.; Qi, Y.; Wu, J.; Shukla, M.K.; Sun, Q. Influence of the application of irrigated water-soluble calcium fertilizer on wine grape properties. *PLoS ONE* **2019**, *14*, e0222104. [CrossRef]
10. Wang, M.; Teng, Y. Genome-wide identification and analysis of MICU genes in land plants and their potential role in calcium stress. *Gene* **2018**, *670*, 174–181. [CrossRef]
11. Aslam, R.; Williams, L.E.; Bhatti, M.F.; Virk, N. Genome-wide analysis of wheat calcium ATPases and potential role of selected ACAs and ECAs in calcium stress. *BMC Plant Biol.* **2017**, *17*, 174. [CrossRef] [PubMed]
12. Wang, Z.; Tian, S.; Wang, J.; Shuai, H.; Zhang, Y.; Wang, Y.; Jin, B.; Zhao, X. Effects of pH and calcium salt stress on the seed germination performance of three herbage species. *Authorea* **2022**.
13. Su, Y.; Zhou, X.; Gao, D.; Zhou, K. Studies on the content change characteristics of K, Ca and Mg in pericarp of Litchi chinensis Sonn. cv. 'Feizixiao' and their relation to the pericarp's coloring. *J. Yunnan Agric. Univ.* **2016**, *31*, 274–280.
14. Tian, S.; Zhou, X.; Gong, H.; Ma, X.; Zhang, F. Orthogonal test design for optimization of the extraction of polysaccharide from Paeonia sinjiangensis K.Y. Pan. *Pharmacogn. Mag.* **2011**, *7*, 4–8. [CrossRef]
15. Wang, H.; Huang, H.; Huang, X.; Hu, Z. Sugar and acid compositions in the arils of Litchi chinensis Sonn.: Cultivar differences and evidence for the absence of succinic acid. *J. Hortic. Sci. Biotechnol.* **2006**, *81*, 57–62. [CrossRef]
16. Nielsen, T.H.; Skjærbæk, H.C.; Karlsen, P. Carbohydrate metabolism during fruit development in sweet pepper (*Capsicum annuum*) plants. *Physiol. Plant.* **1991**, *82*, 311–319. [CrossRef]
17. Li, B.; Dewey, C.N. RSEM: Accurate transcript quantification from RNA-Seq data with or without a reference genome. *BMC Bioinform.* **2011**, *12*, 323. [CrossRef]
18. Love, M.I.; Huber, W.; Anders, S. Moderated estimation of fold change and dispersion for RNA-seq data with DESeq2. *Genome Biol.* **2014**, *15*, 550. [CrossRef] [PubMed]
19. Chen, C.; Chen, H.; Zhang, Y.; Thomas, H.R.; Frank, M.H.; He, Y.; Xia, R. TBtools: An Integrative Toolkit Developed for Interactive Analyses of Big Biological Data. *Mol. Plant* **2020**, *13*, 1194–1202. [CrossRef]
20. Borsani, J.; Budde, C.O.; Porrini, L.; Lauxmann, M.A.; Lombardo, V.A.; Murray, R.; Andreo, C.S.; Drincovich, M.F.; Lara, M.V. Carbon metabolism of peach fruit after harvest: Changes in enzymes involved in organic acid and sugar level modifications. *J. Exp. Bot.* **2009**, *60*, 1823–1837. [CrossRef]
21. Chan, H.T.; Kwok, S.C.M.; Lee, C.W.Q. Sugar composition and invertase activity in lychee. *J. Food Sci.* **1975**, *40*, 772–774. [CrossRef]
22. Kambiranda, D.; Vasanthaiah, H.; Basha, S.M. Relationship between acid invertase activity and sugar content in grape species. *J. Food Biochem.* **2011**, *35*, 1646–1652. [CrossRef]
23. Chen, L.; Hangxian, X.; Xueliang, C.; Yuxue, Z.; Jinbo, Y.; Jinping, S.; Lei, Z. Genome-wide identification and expression pattern of alkaline/neutral invertase gene family in Dendrobium catenatum. *Biotechnol. Biotechnol. Equip.* **2021**, *35*, 527–537.
24. Abdullah, M.; Cao, Y.; Cheng, X.; Meng, D.; Chen, Y.; Shakoor, A.; Gao, J.; Cai, Y. The Sucrose synthase gene family in Chinese Pear (Pyrus bretschneideri Rehd.): Structure, Expression, Evolution. *Molecules* **2018**, *23*, 1144. [CrossRef]

25. Yang, Z.; Wang, T.; Wang, H.; Huang, X.; Qin, Y.; Hu, G. Patterns of enzyme activities and gene expressions in sucrose metabolism in relation to sugar accumulation and composition in the aril of Litchi chinensis Sonn. *J. Plant Physiol.* **2013**, *170*, 731–740. [CrossRef] [PubMed]

26. Huicong, W.; Huibai, H.; Xuming, H. Sugar accumulation and related enzyme activities in the litchi fruit of 'Nuomici' and 'Feizixiao'. *Acta Hortic. Sin.* **2003**, *30*, 1.

27. Gao, F.; Zhang, H.; Zhang, W.; Wang, N.; Zhang, S.; Chu, C.; Liu, C. Engineering of the cytosolic form of phosphoglucose isomerase into chloroplasts improves plant photosynthesis and biomass. *New Phytol.* **2021**, *231*, 315–325. [CrossRef]

28. Srivastava, N.; Rathour, R.; Jha, S.; Pandey, K.; Srivastava, M.; Thakur, V.K.; Sengar, R.S.; Gupta, V.K.; Mazumder, P.B.; Khan, A.F.; et al. Microbial Beta Glucosidase Enzymes: Recent Advances in Biomass Conversation for Biofuels Application. *Biomolecules* **2019**, *9*, 220. [CrossRef]

29. Schmölzer, K.; Gutmann, A.; Diricks, M.; Desmet, T.; Nidetzky, B. Sucrose synthase: A unique glycosyltransferase for biocatalytic glycosylation process development. *Biotechnol Adv.* **2016**, *34*, 88–111. [CrossRef]

30. He, X.; Wei, Y.; Kou, J.; Xu, F.; Chen, Z.; Shao, X. PpVIN2, an acid invertase gene family member, is sensitive to chilling temperature and affects sucrose metabolism in postharvest peach fruit. *Plant Growth Regul.* **2018**, *86*, 169–180. [CrossRef]

31. Russell, C.; Morris, D. Invertase activity, soluble carbohydrates and inflorescence development in the tomato (*Lycopersicon esculentum* Mill.). *Ann. Bot.* **1982**, *49*, 89–98. [CrossRef]

32. Wang, Z.; Yuan, M.; Li, S.; Gao, D.; Zhou, K. Applications of magnesium affect pericarp colour in the Feizixiao lychee. *J. Hortic. Sci. Biotechnol.* **2017**, *92*, 559–567. [CrossRef]

Article

Development of InDel Markers for *Gypsophila paniculata* Based on Genome Resequencing

Chunlian Jin [1,†], Bin Liu [2,†], Jiwei Ruan [1], Chunmei Yang [1] and Fan Li [1,*]

1   Floriculture Research Institute, Yunnan Academy of Agricultural Sciences,
    National Engineering Research Center for Ornamental Horticulture,
    Key Laboratory for Flower Breeding of Yunnan Province, Kunming 650205, China
2   School of Agriculture, Yunnan University, Kunming 650500, China
*   Correspondence: lifan@yaas.org.cn
†   These authors contributed equally to this work.

Abstract: *Gypsophila paniculata* is the only species in the genus *Gypsophila* that has been used as cut flowers, and the sequencing of its genome has just been completed, opening a new chapter in its molecular genetic breeding. The molecular marker system is the basis for genetic molecular research in the era of genomics, whereas it is still a gap for *G. paniculata*. In this study, we constructed a genome-wide InDel marker system of *G. paniculata* after genome resequencing of another wild-type accession with white flowers. Consequently, 407 InDel markers at a distance of ~2 Mb were designed for all 17 chromosomes. Later, the validation of these markers by PCR revealed that 289 markers could distinguish alleles of the two wild-type alleles clearly. The predicted polymorphisms of two wild-type alleles were then transferred to the commercial cultivars, which displayed a rich polymorphism among four commercial cultivars. Our research established the first genome-level genetic map in *G. paniculata*, providing a comprehensive set of marker systems for its molecular research.

Keywords: *Gypsophila paniculata*; InDel marker; genetic map; polymorphism

Citation: Jin, C.; Liu, B.; Ruan, J.; Yang, C.; Li, F. Development of InDel Markers for *Gypsophila paniculata* Based on Genome Resequencing. *Horticulturae* **2022**, *8*, 921. https://doi.org/10.3390/horticulturae8100921

Academic Editor: Jose V. Die

Received: 25 August 2022
Accepted: 5 October 2022
Published: 7 October 2022

Publisher's Note: MDPI stays neutral with regard to jurisdictional claims in published maps and institutional affiliations.

1. Introduction

Gypsophila paniculata is a perennial herbaceous shrub from the genus *Gypsophila*, which comprises about 150 species of annual, biennial and perennial plants [1,2]. It is usually used as a filler in flower arrangements, making it an important cut flower in the global market. Differing from crops, flower type and colour as well as its inner quality and biotic or abiotic stress resistance are the main goals for ornamental breeding [3]. Although breeding efforts have been invested in the creation of new varieties and the improvement of desirable traits, conventional crosses and subsequent phenotypic selection for specific traits remain the dominant breeding methods used in the breeding of *G. paniculata*, which has severely hindered its breeding efficiency [4]. As a consequence, there are currently fewer commercial varieties on the world floricultural market, such as 'Million Stars', 'Perfect', 'Dream Pink' and 'Huixing 1' [5]. In contrast to its important position in the floricultural industry, the molecular genetics research of *G. paniculata* is limited, which hinders the improvement of the cultivars to some extent.

Molecular markers have been widely used in genetic and evolutionary research of various ornamental species including *Rosa*, *Paeonia*, *Dendrobium*, etc., in their germplasm characterization, genetic mapping, diversity analysis and molecular marked-assisted selection in breeding [6–11]. In the development history of molecular markers, DNA-based marker systems such as RFLP (restriction fragment length polymorphism) have been replaced progressively by PCR-based markers such as RAPD (random amplified polymorphic DNA), SSR (simple sequence repeat), SNPs (single-nucleotide polymorphisms) and InDels (insertions/deletions) [12]. A few pieces of research about the assessment of genetic diversity among wild species and commercial hybrids from *Gypsophila* using RAPD markers and chloroplast simple sequence repeat (cpSSR) markers have been reported [13,14], but the genetic map or marker system covering the whole genome of *G. paniculata* lacks. Benefiting from the recent progress in genome sequencing technology, reference genomes with high quality have been accessed in various ornamental plants, bringing floral research to a genome-wide level [15,16]. In addition, the SNP and InDel markers have become the most used molecular markers for plant research due to their abundant polymorphisms, genome-wide distribution and co-dominance [17–19]. For instance, high-density genetic maps based on SNPs facilitate the identification the genetic regulators of key ornamental traits such as flower type as well as the resistance gene in roses, carnation and chrysanthemum [20–23]. Although powerful, the genotyping of SNPs relies on the KASP assay, which requires a special machine and is expensive. In contrast, InDel makers are easy to use, low in cost and efficient, as the InDel marker-based mapping system relies on simple PCR and gel electrophoresis procedures [24].

This study aims to construct a genome-wide InDel marker system of *G. paniculata*, providing a useful tool to facilitate its breeding. Recently, the genome sequence of *G. paniculata* has been assembled and released to the public [25], meeting our goal to develop markers through genome resequencing. Thus, a wild-type accession of *G. paniculata* with white flower (WT-W) was re-sequenced using next-generation sequencing technology, and a series of molecular markers distributed genome-wide were identified, including SNP, InDel, SV (structure variation) and CNV (copy number variations by comparing with the reference genome of a wild-type accession with pink flower (WT-P)). As hypothesized, a set of InDel markers with a high level of polymorphism was developed using the information generated by genome resequencing. Moreover, the InDel markers also displayed polymorphism among four commercial cultivars. Our work provides the first genome-wide genetic map of *G. paniculata*, supporting the further genetic study and molecular breeding of this species.

2. Materials and Methods

2.1. Plant Materials

Two *G. paniculata* wild-type accessions with pink (WT-P) and white flowers (WT-W) were used in this study (Figure 1). The WT-P plant was used for the de novo genome sequenced previously, providing the reference genome data thereby. The WT-W plant was used for genome resequencing to generate InDel markers. Meanwhile, we selected four commercial cultivars of *G. paniculata* ('YX1', 'YX2', 'YX3' and 'YX4') to identify and validate the polymorphic InDel markers. 'YX1', 'YX2' and 'YX4' are three representative commercial varieties of *G. paniculata* with white petals, and the difference is the flower size ('YX1'>'YX4'>'YX2', from large to small'), whereas the flower colour of 'YX3' is pink with a similar size as 'YX2' (Figure 1). All of the above plant materials were provided by Yuxi Yunxing Biological Technology Co., Ltd. (Yuxi, China).

2.2. Variation Detection by Genome Resequencing

The fresh young leaves of *G. paniculata* WT-W were used for genome resequencing. The MGISEQ-2000 PE150 sequencer was applied to conduct genome sequencing, after which the original reads (8.66 Gb) were filtered to generate clean reads (8.05 Gb) for subsequent analysis. Using in-house scripts, we filtered any sequencing reads with the following: reads with adapter sequences, consecutive bases on the ends with base quality < Q20, read length < 50 bp and singletons. The clean reads were then aligned to the *G. paniculata* reference genome using BWA mem (v0.7.17) with default settings [26]. The alignment results were sorted using Samtools (v1.9) [27].

SNP and InDel were called using GATK HaplotypeCaller (v4.1.4.1, Broad Institute, Cambridge, MA, USA) with default settings [28]. We further filtered the calls using GATK VariantFiltration with the following parameters: SNP filtering (QD < 2.0, FS > 60.0, MQ < 40.0, MQRankSum< −12.5, ReadPosRankSum < −8.0); InDel filtering (QD < 2.0, FS > 200.0, ReadPosRankSum < −20.0). CNV were detected using CNVnator (v0.3.2) with default settings [29]. SV were identified using Manta (v1.6.0) [30]. Mutational positions, genomic regions and potential amino acid changes were assessed using ANNOVAR (v2019, Wang Kai, PA, USA) [31]. Circos (v0.69, Martin Krzywinski, Vancouver, BC, Canada) was used to plot the genome-wide distribution of variation [32].

Figure 1. The phenotype of *G. paniculata* wild-type accessions and commercial cultivars used in this study. (**A**). The pink flower wild type of *G. paniculata* (WT-P). (**B**). The white flower wild type of *G. paniculata* (WT-W). (**C**). The flower phenotype of four commercial cultivars ('YX1', 'YX2', 'YX3' and 'YX4', from left to right). Bar = 1 cm.

2.3. Development of InDel Markers

We selected the InDels that were over 10 bp long and distributed ~2 Mb. The positions with excess InDels which might interfere with the PCR verification were excluded. After selecting the suitable InDels, a ~400 bp genome sequence covering each InDel was used as the template for primer design. The primers were designed on NCBI and named after the chromosome number and the physical position (N-XX.XX, Table S1).

2.4. PCR Analyses of InDel Markers

The total DNA of two wild-type plants and four commercial cultivars was extracted from fresh leaves using the CTAB method [33]. Template DNA was amplified with the designed primers in a 10 μL system (7.3 μL ddH$_2$O; 1 μL 10× Taq buffer; 0.8 μL dNTPs; 0.2 μL primers; 0.1 μL Taq enzyme; 0.4 μL DNA template) using the following PCR program: 5 min of full denaturation at 95 °C; 29 cycles (95 °C, 30 s; 56 °C, 30 s; 72°C, 30 s); 72 °C extension for 7 min. After the standard PCR, 3 μL DNA loading buffer was added to the PCR product. Then, the mixture was separated in 3.5% agarose gel.

3. Results

3.1. Genome Resequencing and Sequence Polymorphism Identification

The successful mapping of QTLs relies on the genetic maps with high-density molecular markers between the accessions. Previously, the genome sequence data of *G. paniculata*, a wild-type accession with pink flowers (WT-P), was assembled onto 17 chromosomes [25]. To develop sufficient molecular markers for *G. paniculata* genetic research, we detected sequence polymorphisms between WT-P and another wild-type accession with white flowers (WT-W) through genome resequencing by the high-throughput sequencing platform MGISEQ-2000 PE150. After filtering, a total of 8.05 Gb of clean reads was generated, 82.23% of which were mapped to the reference genome, displaying an average sequencing depth of 5.70 (Figure 2B). Different kinds of natural genetic variations were detected between the reference and resequencing genome, including 2,377,499 SNPs, 1,366,056 InDels, 1403 SVs, and 28 CNVs, whose densities were shown on the circus map (Figure 2A). Interestingly, the InDels preferred to distribute at the end of the chromosomes rather than the centromeric region, as shown by the circus map. Meanwhile, the length of most InDels (>80%) was less than 5 bp, and the InDels between 5 and 10 bp accounted for 10% of this variation. There were about ~5% (68,302/1,366,056) InDels over 10 bp which are suitable for genome-wide marker construction (Figure 2C).

3.2. Construction of InDel Markers for Polymorphism Analysis

To develop InDel markers that can discriminate alleles between WT-P and WT-W, insertions or deletions over 10 bp were chosen as candidates with the interval of the neighbouring markers set as ~2 Mb. Sequence fragments about 400 bp long that contained either the insertions or deletions were used as templates to design primers. In total, 407 pairs of primers were designed for 17 chromosomes (Figure 3). To validate the newly designed markers, PCR analysis was conducted and the products were analysed by gel electrophoresis. Of the 407 markers, 289 markers distinguished the alleles of WT-P and WT-W clearly. Another 34 markers produced close bands on the 3.5% gel, but could still discriminate the alleles of WT-P and WT-W. These markers can be used when the chromosome region has limited markers, probably separated by gel with higher concentration. The success rates of the designed primers varied across the chromosomes from 40% to 92.9%, and the average success rate was as high as 71.0% (Table 1). Our data provided the successful establishment of genome-wide InDel markers based on a genetic map for *G. paniculata*. Nevertheless, it has to be acknowledged that for some chromosomes, such as Chr.4, Chr.7, Chr.12 and Chr.14, there were obvious gaps between two available markers, which was probably due to the low density of InDels on the centromeric region of these chromosomes. Thus, it might be essential to develop other molecular markers such as SNPs to compensate for these empties in the future.

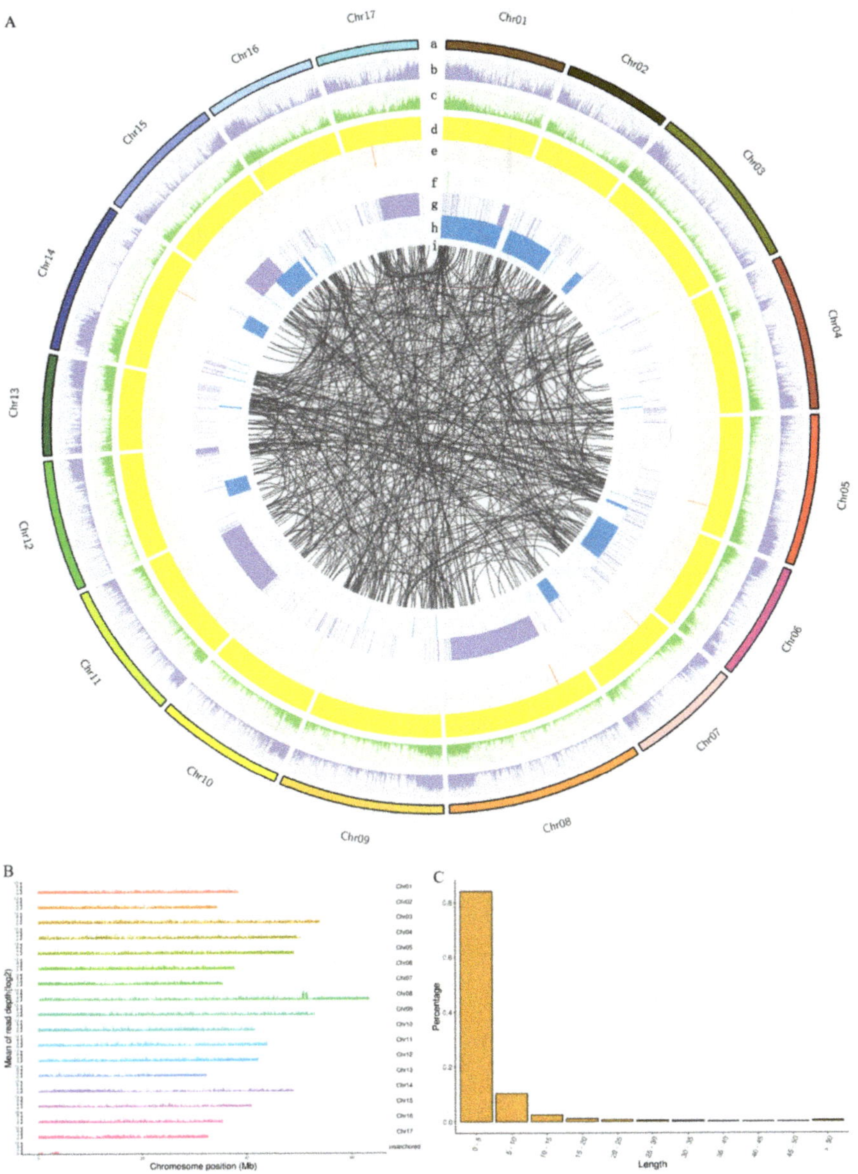

Figure 2. Resequencing of WT-W based on the WT-P genome sequence. (**A**) Genomic structure variation distribution between the two *G. paniculata* wild-type accessions. a: reference sequence. b: SNP density distribution. c: InDel distribution density. d: CNV duplication. e: CNV deletion. f: SV insertion. g: SV deletion. h: SV inversion. i: SV translocation. Abbreviations include SNP: Single Nucleotide Polymorphism; InDel: Insertion/Deletion; CNV: Copy Number Variations; SV: Structure Variation. (**B**) The sequencing coverage depth distribution map of each chromosome of *G. paniculata*. The mean of read depth was calculated using the coverage depth (10,000 bp as the statistical window) by logarithm (log2). (**C**) The distribution of the InDel length between WT-P and WT-W.

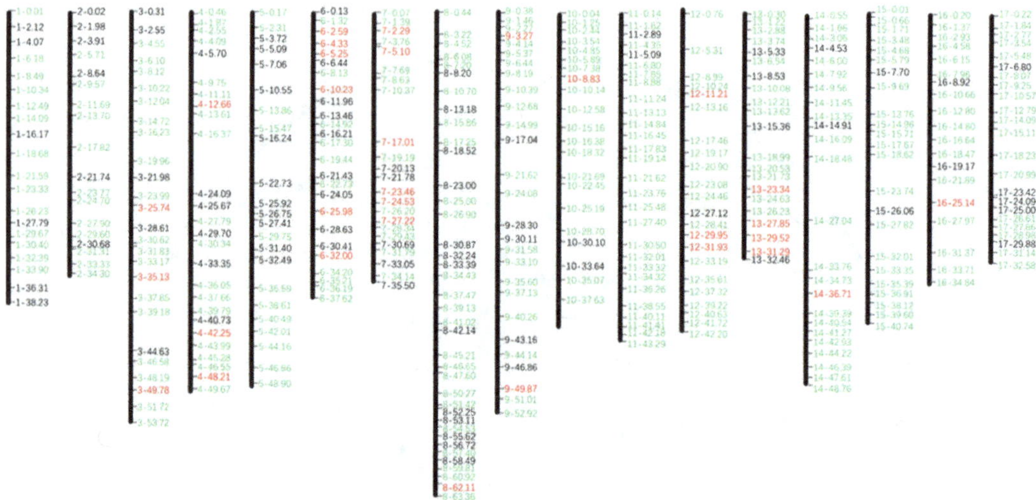

Figure 3. The physical map of 407 InDel markers distributed across all 17 chromosomes of *G. paniculata* genome. The name code of the InDel marker was presented as a chromosome number with the physical distance. Green markers discriminate alleles between WT-P and WT-W. Red markers amplified close bands on gel, and black markers were unavailable.

Table 1. The successful rates of InDel markers for all 17 chromosomes.

Chromosome	Number of Markers	Number of Green Markers	Number of Red Markers	Successful Rate (%)
Chr.1	20	14		70.0
Chr.2	18	12		66.7
Chr.3	26	18	3	69.2
Chr.4	26	17	3	65.4
Chr.5	23	12		52.2
Chr.6	25	10	6	40.0
Chr.7	23	12	6	52.2
Chr.8	38	24	1	63.2
Chr.9	27	20	2	74.1
Chr.10	21	18	1	85.7
Chr.11	28	26		92.9
Chr.12	22	18	3	81.8
Chr.13	21	13	4	61.9
Chr.14	24	21	1	87.5
Chr.15	23	21		91.3
Chr.16	19	16	1	84.2
Chr.17	23	17		73.9
Total	407	289	31	71.0

Note: Green markers discriminate alleles between WT-P and WT-W. Red markers amplified close bands on gel.

3.3. InDel Marker Polymorphisms among Commercial Cultivars

The wild and commercial cultivars possess excellent agronomic traits, for example, wild types are generally more resistant, while the commercial varieties display larger flowers and more petals. However, limited research focuses on the genetic regulators underlying these traits, causing the relative mechanisms to remain unknown. To explore the applicability of the InDel markers designed in distinguishing the alleles between wild-type and commercial varieties, PCR amplification was conducted using the genomic DNA of WT-P and four commercial varieties (YX1-4) as templates. Out of the 407 pairs of primers, 191 were able to discriminate alleles between WT-P and commercial cultivars. The polymorphism of the InDel markers between WT-P and commercial cultivars was then analysed by pairwise comparisons (Table 2). In total, the number of available markers for each pair of accessions ranged from 31 (YX1 vs. YX4) to 173 (YX1 vs. WT-P), with an average of 92. The InDel markers were suitable to discriminate alleles between WT-P and commercial cultivars (an average of 171 markers available) since a high degree of polymorphism was observed (Figure 4), whereas the markers available between the commercial cultivars were no more than 50. This implies that the commercial cultivars are closely related, which is consistent with the observation that all four commercial cultivars bloom white flowers but differ only in flower size.

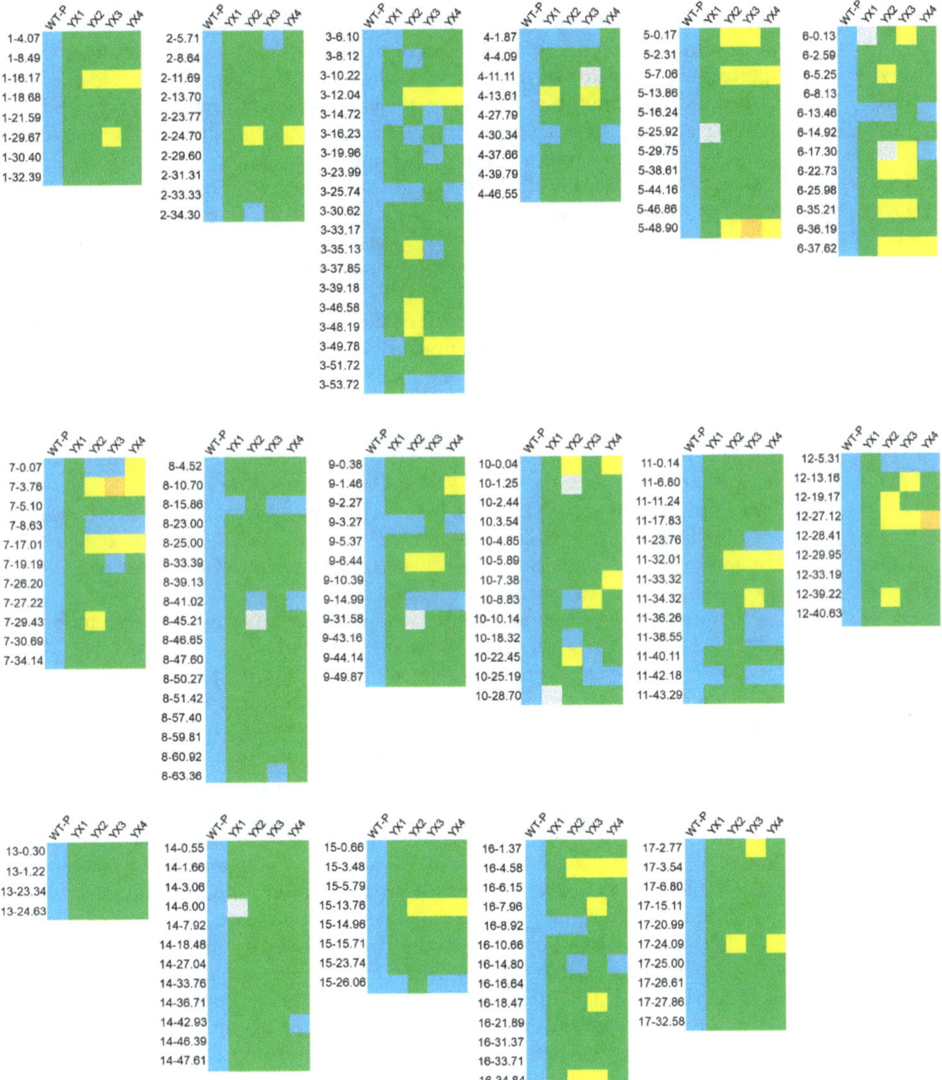

Figure 4. Matrix of the polymorphisms using the InDel markers among the five accessions of *G. paniculata*. Blue squares are WT-P bands, green, yellow and orange squares represent bands different from WT-P, and grey squares mean no bands detected.

Table 2. Number of InDel markers that were polymorphic in pairwise comparison of five *G. paniculata* accessions.

Cultivars	WT-P	YX1	YX2	YX3
YX1	173			
YX2	172	49		
YX3	170	43	45	
YX4	171	31	33	38

4. Discussion

The elaboration of the key regulatory mechanism underlying one or several traits as well as the fast selection of elite progenies is crucial for plant breeding. When obtaining a certain mutant, the identification of the allele(s) related to the phenotype is usually performed by the forward genetics, in which F2 rough mapping provides an approximate location of the mutation causative allele(s) on chromosomes without the requirement of a large amount of samples and high-throughput sequencing, narrowing down the targets for further fine mapping. In the last decades, the development of PCR-based markers such as RAPD, SSR and amplified fragment length polymorphisms (AFLPs) have fulfilled the shortage of map-based cloning [34]. In the field of ornamental breeding, these sequence-related amplified polymorphisms (SRAP) markers have been applied to the vase life-associated or disease-resistant QTL mapping and analyse chrysanthemum, carnation and lily, to cite a few [35].

Nevertheless, the development of such markers is labour intensive, and their application is limited in certain situations, since they are usually not genome-wide. Earlier, benefiting from the availability of an annotated reference genome and sequenced accessions, genetic markers based on InDels have been developed in *Arabidopsis*, accelerating the identification of the mutated allele(s) [36]. With the booming of sequencing technology and the following drop in sequencing expense, plentiful plant genomes were released for crops and horticultural plants [37–40]. The resequencing-based InDel makers were then developed in cotton [41], rice [42], Brassica [43], buckwheat [44], jute [45], melon [46], chickpea [47], cucumber [48], etc., used for research such as disease-related gene identification or accession discrimination. However, the systematic development of such markers has not been reported in ornamental species.

In this study, we constructed a genome-wide InDel marker system for *G. paniculata* through genome resequencing. Similar to the early report in jute [45], InDels detected in the *G. paniculata* genome are quite abundant, but most of them are shorter than 5 bp, which makes them hard to use as markers. Regardless, the number of the ~5% InDels that are longer than 10 bp is as large as 68302, equivalent to 91 InDels per Mb, which is more than needed. To meet the demand for mapping (1 maker/2 Mb), 409 InDels distributed on 17 chromosomes were selected, and the relative primers were then designed. Of these, 289 can discriminate alleles from 2 wild types donating the genome sequence data, coming to a success rate of 70.6%. Although we expected to obtain an available marker every 2 Mb, the outcome was barely satisfactory. There were usually missing available InDel markers in the middle (calculated by physical distance) of the chromosome, such as Chr. 4, 7 and 14 (Figure 3). The same situation happened during the development of InDel markers in rice [42] and *Capsicum* spp. [49]. It might be dissolved by adding other molecular markers when mapping a certain QTL, or the InDels shorter than 10 bp can also be developed as markers based on a high-resolution melting curve, as reported [50]. Since the discrimination of genetic resources and extension of the application of markers are crucial in the breeding process [44], we then detected the polymorphisms of the designed InDel markers in four best-selling cultivars. Over 170 makers were available to differentiate each commercial cultivar and WT-P, whereas only less than 50 markers worked for discrimination between the 4 commercial cultivars. It makes sense, since all the commercial cultivars bloom with white flowers and may share more common genetic information rather than WT-P.

Used not only as the filler flower but also as a preserved flower which decorates the environment after colourful staining, the status of cut flower *G. paniculata* is rising, leading to a massive demand for the innovation of this species. Molecular genetics play a more and more important role in floricultural breeding, in which a molecular marker system covering the whole genome is the basis for genetic molecular research in the era of genomics. However, it is still a gap for *G. paniculata*. Here, we provide the first genetic map of *G. paniculata* in this study, consisting of a comprehensive set of InDel markers for the molecular research of *G. paniculata*. The success in our case also implies that the development of InDel markers covering the whole genome is cost- and labour-effective with a high success rate, deserving to be applied in other ornamental species for which cross-breeding is the main method for cultivar innovation.

Supplementary Materials: The following supporting information can be downloaded at: https://www.mdpi.com/article/10.3390/horticulturae8100921/s1, Table S1: Primers used in this study.

Author Contributions: C.J. and F.L. conceived and designed the research; B.L. performed the experiments and analysed the data; C.J. and F.L. wrote the manuscript and revised the manuscript; J.R. and C.Y. provided the wildtype plants and commercial cultivars. F.L. supervised the project. All authors have read and agreed to the published version of the manuscript.

Funding: This research was funded by National Natural Science Foundation of China (31960608), Yunnan Fundamental Research Projects (202101AT070147) and High-level Talent Introduction Program of Yunnan Province—Industrial Talent Special Project (YNQR-CYRC-2020-004).

Institutional Review Board Statement: Not applicable.

Informed Consent Statement: Not applicable.

Data Availability Statement: Not applicable.

Conflicts of Interest: The authors declare no conflict of interest.

References

1. Lu, D.; Nicholas, J.T. *Gypsophila Linnaeus*; Science Press: St. Louis, MA, USA, 2001; Volume 6.
2. Li, F.; Wang, G.; Yu, R.; Wu, M.; Shan, Q.; Wu, L.; Ruan, J.; Yang, C. Effects of Seasonal Variation and Gibberellic Acid Treatment on the Growth and Development of Gypsophila paniculata. *HortScience* **2019**, *54*, 1370–1374. [CrossRef]
3. Kuligowska, K.; Lütken, H.; Müller, R. Towards development of new ornamental plants: Status and progress in wide hybridization. *Planta* **2016**, *244*, 1–17. [CrossRef] [PubMed]
4. Zvi, M.M.B.; Zuker, A.; Ovadis, M.; Shklarman, E.; Ben-Meir, H.; Zenvirt, S.; Vainstein, A. Agrobacterium-mediated transformation of gypsophila (*Gypsophila paniculata* L.). *Mol. Breed.* **2008**, *22*, 543–553.
5. Li, F.; Mo, X.; Wu, L.; Yang, C. A Novel Double-flowered Cultivar of Gypsophila paniculata Mutagenized by 60Co γ-Ray. *HortScience* **2020**, *55*, 1531–1532. [CrossRef]
6. Xue, Y.; Liu, R.; Xue, J.; Wang, S.; Zhang, X. Genetic Diversity and Relatedness Analysis of Nine Wild Species of Tree Peony Based on Simple Sequence Repeats Markers. *Hortic. Plant J.* **2021**, *7*, 579–588. [CrossRef]
7. Sousa, A.; Souza, M.M.; Melo, C.; Sodré, G. ISSR markers in wild species of Passiflora L. (*Passifloraceae*) as a tool for taxon selection in ornamental breeding. *Genet. Mol. Res. Gmr* **2015**, *14*, 18534. [CrossRef]
8. Conceição, L.; Belo, G.; Souza, M.; Santos, S.; Cerqueira-Silva, C.; Corrêa, R. Confirmation of cross-fertilization using molecular markers in ornamental passion flower hybrids. *Genet. Mol. Res.* **2011**, *10*, 47–52. [CrossRef] [PubMed]
9. Zheng, S.-g.; Hu, Y.-d.; Zhao, R.-x.; Yan, S.; Zhang, X.-q.; Zhao, T.-m.; Chun, Z. Genome-wide researches and applications on Dendrobium. *Planta* **2018**, *248*, 769–784. [CrossRef]
10. Yang, C.; Ma, Y.; Cheng, B.; Zhou, L.; Yu, C.; Luo, L.; Pan, H.; Zhang, Q. Molecular Evidence for Hybrid Origin and Phenotypic Variation of Rosa Section Chinenses. *Genes* **2020**, *11*, 996. [CrossRef]
11. Fan, Y.; Wang, Q.; Dong, Z.; Yin, Y.; Teixeira da Silva, J.A.; Yu, X. Advances in molecular biology of *Paeonia* L. *Planta* **2020**, *251*, 23. [CrossRef]
12. Semagn, K.; Babu, R.; Hearne, S.; Olsen, M. Single nucleotide polymorphism genotyping using Kompetitive Allele Specific PCR (KASP): Overview of the technology and its application in crop improvement. *Mol. Breed.* **2014**, *33*, 1–14. [CrossRef]
13. Cao, T.X.; Piao, X.C.; Wu, S.Q.; Wang, S.M.; Lian, M.L. Analysis of RAPD Fingerprint of Shoots and Its Vitrification Shoots in vitro of Gypsophila paniculata L. *Plant Physiol. Commun.* **2007**, *43*, 288–290.
14. Calistri, E.; Buiatti, M.; Bogani, P. Characterization of Gypsophila species and commercial hybrids with nuclear whole-genome and cytoplasmic molecular markers. *Plant Biosyst. Int. J. Deal. All Asp. Plant Biol.* **2016**, *150*, 11–21.

15. Li, M.; Wen, Z.; Meng, J.; Cheng, T.; Zhang, Q.; Sun, L. The genomics of ornamental plants: Current status and opportunities. *Ornam. Plant Res.* **2022**, *2*, 6. [CrossRef]

16. Rouet, C.; O'Neill, J.; Banks, T.; Tanino, K.; Derivry, E.; Somers, D.; Lee, E.A. Mapping Winterhardiness in Garden Roses. *J. Am. Soc. Hortic. Sci.* **2022**, *147*, 216–238. [CrossRef]

17. Mieulet, D.; Aubert, G.; Bres, C.; Klein, A.; Droc, G.; Vieille, E.; Rond-Coissieux, C.; Sanchez, M.; Dalmais, M.; Mauxion, J.-P.; et al. Unleashing meiotic crossovers in crops. *Nat. Plants* **2018**, *4*, 1010–1016. [CrossRef]

18. Li, F. Meiotic Recombination Suppressors of Arabidopsis Thaliana. Ph.D. Thesis, Ghent University, Ghent, Belgium, 2018.

19. De Maagd, R.A.; Loonen, A.E.H.M.; Chouaref, J.; Pele, A.; Meijerdekens, F.; Fransz, P.; Bai, Y. CRISPR/Cas inactivation of RECQ4 increases homeologous crossovers in an interspecific tomato hybrid. *Plant Biotechnol. J.* **2020**, *18*, 805–813. [CrossRef]

20. Song, X.; Xu, Y.; Gao, K.; Fan, G.; Zhang, F.; Deng, C.; Dai, S.; Huang, H.; Xin, H.; Li, Y. High-density genetic map construction and identification of loci controlling flower-type traits in Chrysanthemum (Chrysanthemum × morifolium Ramat.). *Hortic. Res.* **2020**, *7*, 108. [CrossRef] [PubMed]

21. Rouet, C.; Lee, E.A.; Banks, T.; O'Neill, J.; LeBlanc, R.; Somers, D.J. Identification of a polymorphism within the Rosa multiflora muRdr1A gene linked to resistance to multiple races of Diplocarpon rosae W. in tetraploid garden roses (Rosa × hybrida). *Theor. Appl. Genet.* **2020**, *133*, 103–117. [CrossRef] [PubMed]

22. Saint-Oyant, L.H.; Ruttink, T.; Hamama, L.; Kirov, I.; Lakhwani, D.; Zhou, N.-N.; Bourke, P.; Daccord, N.; Leus, L.; Schulz, D. A high-quality genome sequence of Rosa chinensis to elucidate ornamental traits. *Nat. Plants* **2018**, *4*, 473–484. [CrossRef]

23. Wang, Q.; Zhang, X.; Lin, S.; Yang, S.; Yan, X.; Bendahmane, M.; Bao, M.; Fu, X. Mapping a double flower phenotype-associated gene DcAP2L in Dianthus chinensis. *J. Exp. Bot.* **2020**, *71*, 1915–1927. [CrossRef] [PubMed]

24. Gull, S.; Haider, Z.; Gu, H.; Raza Khan, R.A.; Miao, J.; Wenchen, T.; Uddin, S.; Ahmad, I.; Liang, G. InDel marker based estimation of multi-gene allele contribution and genetic variations for grain size and weight in rice (*Oryza sativa* L.). *Int. J. Mol. Sci.* **2019**, *20*, 4824. [CrossRef]

25. Li, F.; Gao, Y.; Jin, C.; Wen, X.; Geng, H.; Cheng, Y.; Qu, H.; Liu, X.; Feng, S.; Zhang, F.; et al. The chromosome-level genome of Gypsophila paniculata reveals the molecular mechanism of floral development and ethylene insensitivity. *Hortic. Res.* **2022**, *9*, uhac176. [CrossRef]

26. Li, H.; Durbin, R. Fast and accurate short read alignment with Burrows–Wheeler transform. *Bioinformatics* **2009**, *25*, 1754–1760. [CrossRef]

27. Li, H.; Handsaker, B.; Wysoker, A.; Fennell, T.; Ruan, J.; Homer, N.; Marth, G.; Abecasis, G.; Durbin, R. The sequence alignment/map format and SAMtools. *Bioinformatics* **2009**, *25*, 2078–2079. [CrossRef] [PubMed]

28. McKenna, A.; Hanna, M.; Banks, E.; Sivachenko, A.; Cibulskis, K.; Kernytsky, A.; Garimella, K.; Altshuler, D.; Gabriel, S.; Daly, M. The Genome Analysis Toolkit: A MapReduce framework for analyzing next-generation DNA sequencing data. *Genome Res.* **2010**, *20*, 1297–1303. [CrossRef]

29. Abyzov, A.; Urban, A.E.; Snyder, M.; Gerstein, M. CNVnator: An approach to discover, genotype, and characterize typical and atypical CNVs from family and population genome sequencing. *Genome Res.* **2011**, *21*, 974–984. [CrossRef] [PubMed]

30. Chen, X.; Schulz-Trieglaff, O.; Shaw, R.; Barnes, B.; Schlesinger, F.; Källberg, M.; Cox, A.J.; Kruglyak, S.; Saunders, C.T. Manta: Rapid detection of structural variants and indels for germline and cancer sequencing applications. *Bioinformatics* **2016**, *32*, 1220–1222. [CrossRef] [PubMed]

31. Wang, K.; Li, M.; Hakonarson, H. ANNOVAR: Functional annotation of genetic variants from high-throughput sequencing data. *Nucleic Acids Res.* **2010**, *38*, e164. [CrossRef] [PubMed]

32. Krzywinski, M.; Schein, J.; Birol, I.; Connors, J.; Gascoyne, R.; Horsman, D.; Jones, S.J.; Marra, M.A. Circos: An information aesthetic for comparative genomics. *Genome Res.* **2009**, *19*, 1639–1645. [CrossRef] [PubMed]

33. Li, F.; Cheng, Y.; Zhao, X.; Yu, R.; Li, H.; Wang, L.; Li, S.; Shan, Q. Haploid induction via unpollinated ovule culture in Gerbera hybrida. *Sci. Rep.* **2020**, *10*, 1702. [CrossRef] [PubMed]

34. Peters, J.L.; Cnudde, F.; Gerats, T. Forward genetics and map-based cloning approaches. *Trends Plant Sci.* **2003**, *8*, 484–491. [CrossRef] [PubMed]

35. Van Huylenbroeck, J. *Ornamental Crops*; Springer: Amsterdam, The Netherlands, 2018; Volume 11.

36. Hou, X.; Li, L.; Peng, Z.; Wei, B.; Tang, S.; Ding, M.; Liu, J.; Zhang, F.; Zhao, Y.; Gu, H. A platform of high-density INDEL/CAPS markers for map-based cloning in Arabidopsis. *Plant J.* **2010**, *63*, 880–888. [CrossRef] [PubMed]

37. Chen, F.; Su, L.; Hu, S.; Xue, J.-Y.; Liu, H.; Liu, G.; Jiang, Y.; Du, J.; Qiao, Y.; Fan, Y.; et al. A chromosome-level genome assembly of rugged rose (*Rosa rugosa*) provides insights into its evolution, ecology, and floral characteristics. *Hortic. Res.* **2021**, *8*, 141. [CrossRef]

38. Liang, Y.; Li, F.; Gao, Q.; Jin, C.; Dong, L.; Wang, Q.; Xu, M.; Sun, F.; Bi, B.; Zhao, P.; et al. The genome of Eustoma grandiflorum reveals the whole-genome triplication event contributing to ornamental traits in cultivated lisianthus. *Plant Biotechnol. J.* **2022**, *20*, 1856–1858. [CrossRef] [PubMed]

39. Zheng, T.; Li, P.; Li, L.; Zhang, Q. Research advances in and prospects of ornamental plant genomics. *Hortic. Res.* **2021**, *8*, 65. [CrossRef] [PubMed]

40. Zhang, L. Advance of Horticultural Plant Genomes. *Hortic. Plant J.* **2019**, *5*, 229–230. [CrossRef]

41. Feng, J.; Zhu, H.; Zhang, M.; Zhang, X.; Guo, L.; Qi, T.; Tang, H.; Wang, H.; Qiao, X.; Zhang, B. Development and utilization of an InDel marker linked to the fertility restorer genes of CMS-D8 and CMS-D2 in cotton. *Mol. Biol. Rep.* **2020**, *47*, 1275–1282. [CrossRef]

42. Hechanova, S.L.; Bhattarai, K.; Simon, E.V.; Clave, G.; Karunarathne, P.; Ahn, E.-K.; Li, C.-P.; Lee, J.-S.; Kohli, A.; Hamilton, N. Development of a genome-wide InDel marker set for allele discrimination between rice (*Oryza sativa*) and the other seven AA-genome Oryza species. *Sci. Rep.* **2021**, *11*, 8962. [CrossRef] [PubMed]

43. Liu, B.; Wang, Y.; Zhai, W.; Deng, J.; Wang, H.; Cui, Y.; Cheng, F.; Wang, X.; Wu, J. Development of InDel markers for Brassica rapa based on whole-genome re-sequencing. *Theor. Appl. Genet.* **2013**, *126*, 231–239. [CrossRef] [PubMed]

44. Sohn, H.-B.; Kim, S.-J.; Hong, S.-Y.; Park, S.-G.; Oh, D.-H.; Lee, S.; Nam, H.Y.; Nam, J.H.; Kim, Y.-H. Development of 50 InDel-based barcode system for genetic identification of tartary buckwheat resources. *PLoS ONE* **2021**, *16*, e0250786. [CrossRef]

45. Yang, Z.; Dai, Z.; Xie, D.; Chen, J.; Tang, Q.; Cheng, C.; Xu, Y.; Wang, T.; Su, J. Development of an InDel polymorphism database for jute via comparative transcriptome analysis. *Genome* **2018**, *61*, 323–327. [CrossRef]

46. Islam, M.R.; Hossain, M.R.; Jesse, D.M.I.; Jung, H.-J.; Kim, H.-T.; Park, J.-I.; Nou, I.-S. Development of molecular marker linked with bacterial fruit blotch resistance in melon (*Cucumis melo* L.). *Genes* **2020**, *11*, 220. [CrossRef] [PubMed]

47. Jain, A.; Roorkiwal, M.; Kale, S.; Garg, V.; Yadala, R.; Varshney, R.K. InDel markers: An extended marker resource for molecular breeding in chickpea. *PLoS ONE* **2019**, *14*, e0213999. [CrossRef] [PubMed]

48. Adedze, Y.M.N.; Lu, X.; Xia, Y.; Sun, Q.; Nchongboh, C.G.; Alam, M.; Liu, M.; Yang, X.; Zhang, W.; Deng, Z. Agarose-resolvable InDel markers based on whole genome re-sequencing in cucumber. *Sci. Rep.* **2021**, *11*, 3872. [CrossRef] [PubMed]

49. Guo, G.; Zhang, G.L.; Pan, B.G.; Diao, W.P.; Liu, J.B.; Ge, W.; Gao, C.Z.; Zhang, Y.; Jiang, C.; Wang, S.B. Development and application of InDel markers for Capsicum spp. based on whole-genome re-sequencing. Scientific reports. *Sci. Rep.* **2019**, *9*, 3691.

50. Chen, R.; Chang, L.C.; Cai, X.; Wu, J.; Liang, J.L.; Lin, R.M.; Song, Y.; Wang, X.W. Development of InDel markers for Brassica rapa based on a high-resolution melting curve. *Hortic. Plant J.* **2021**, *7*, 31–37.

Article

Genetic Diversity and Genome-Wide Association Study of Architectural Traits of Spray Cut Chrysanthemum Varieties

Daojin Sun, Luyao Zhang, Jiangshuo Su, Qi Yu, Jiali Zhang, Weimin Fang, Haibin Wang, Zhiyong Guan, Fadi Chen and Aiping Song *

State Key Laboratory of Crop Genetics and Germplasm Enhancement, Key Laboratory of Landscaping, Ministry of Agriculture and Rural Affairs, Key Laboratory of Biology of Ornamental Plants in East China, National Forestry and Grassland Administration, College of Horticulture, Nanjing Agricultural University, Nanjing 210095, China; 2020104106@njau.edu.cn (D.S.); 2019804185@njau.edu.cn (L.Z.); sujiangshuo@njau.edu.cn (J.S.); 2019104103@njau.edu.cn (Q.Y.); 2020804199@stu.njau.edu.cn (J.Z.); fangwm@njau.edu.cn (W.F.); hb@njau.edu.cn (H.W.); guanzhy@njau.edu.cn (Z.G.); chenfd@njau.edu.cn (F.C.)
* Correspondence: aiping_song@njau.edu.cn

Abstract: The architecture of spray cut chrysanthemum is crucial for the quality and quantity of cut flower production. However, the mechanism underlying plant architecture still needs to be clarified. In this study, we measured nine architecture-related traits of 195 spray cut chrysanthemum varieties during a two-year period. The results showed that the number of upper primary branches, number of lateral flower buds and primary branch length widely varied. Additionally, plant height had a significant positive correlation with number of leaf nodes and total number of lateral buds. Number of upper primary branches had a significant negative correlation with primary branch diameter, primary branch angle and primary branch length. Plant height, total number of lateral buds, number of upper primary branches, stem diameter, primary branch diameter and primary branch length were vulnerable to environmental impacts. All varieties could be divided into five categories according to cluster analysis, and the typical plant architecture of the varieties was summarized. Finally, a genome-wide association study (GWAS) was performed to find potential functional genes.

Keywords: spray cut chrysanthemum; GWAS; plant architecture; statistical analysis

Citation: Sun, D.; Zhang, L.; Su, J.; Yu, Q.; Zhang, J.; Fang, W.; Wang, H.; Guan, Z.; Chen, F.; Song, A. Genetic Diversity and Genome-Wide Association Study of Architectural Traits of Spray Cut Chrysanthemum Varieties. *Horticulturae* **2022**, *8*, 458. https://doi.org/10.3390/horticulturae8050458

Academic Editor: Luigi De Bellis

Received: 26 April 2022
Accepted: 18 May 2022
Published: 19 May 2022

Publisher's Note: MDPI stays neutral with regard to jurisdictional claims in published maps and institutional affiliations.

1. Introduction

Chrysanthemum (*Chrysanthemum morifolium* Ramat.) is one of the four most popular cut flowers worldwide and is an important component in the floral industry [1]. Branching is one of the most important agricultural traits of chrysanthemum, playing an important role in morphological formation, and affecting ornamental quality and economic value. Operations involving decapitation and/or removal of lateral buds constitute nearly 1/3 of production costs [2]. The growth and development of chrysanthemum are largely affected by various environmental factors. As a quantitative trait controlled by multiple genes, branching is affected by both the environment and the genetic background [3,4], and the underlying molecular mechanism that governs branching still needs to be elucidated.

Shoot branching is controlled by various hormone signaling pathways, including auxin, strigolactones (SLs), cytokinins (CKs) and brassinosteroids (BRs) [5]. Apical dominance is a universal phenomenon in plants and is mainly maintained by auxin. According to the auxin canalization model, auxin can act as a second messenger to regulate downstream signals or functions [6]. Moreover, auxin acts upstream of SLs and CKs, which promote and inhibit shoot branching, respectively. The biosynthesis of CKs is repressed by auxin, as the key CK biosynthesis-related enzyme Isopentenyltransferase (IPT) is downregulated by auxin [7,8]. The biosynthesis of SLs is activated by auxin, as the key SL biosynthesis-related genes *carotenoid cleavage dioxygenase7* (*CCD7*) and *carotenoid cleavage dioxygenase8* (*CCD8*) are upregulated by auxin [9]. Through the BR signaling component brassinazole-resistant1 (BZR1), BRs can promote increased tillering in rice [10] and bud outgrowth of tomato [11]. Sugars are a major source of carbon and energy in plants. A recent study indicated that sugars can promote initial bud outgrowth and downregulate the expression levels of *BRANCHED1* (*BRC1*) [12]. However, the gene regulatory network governing plant architecture still needs to be elucidated, and new key genes and pathways need to be identified for continued research.

The environment also plays an important role in determining plant architecture [13]. Treatments involving drought, heat and drought plus heat were shown to reduce the shoot outgrowth of *Pinus edulis* [14]. Leaf distribution, branch distribution and canopy photosynthetic rate were also influenced by temperature in potato [15]. Light is a pivotal environmental factor that influences the growth of shoots, and increasing light intensity can promote the growth of branches in herbaceous and tree species [16–18]. Nitrogen is an important nutrient element in the soil and can alter the amino acid content and influence the branch growth of plants [19]. *TaNAC2-5A* is a nitrate-inducible gene and can increase tiller numbers and spikelet numbers of wheat [20]. Various environmental factors influence the architecture of plants, which reflects their adaptation and evolution.

Genome-wide association studies (GWASs) are efficient tools to exploit complex genetic mechanisms through associations of agronomic traits with single-nucleotide polymorphisms (SNPs) within a group of individuals or natural inbred lines [21,22]. For chrysanthemum, GWASs have been used to identify genes related to waterlogging resistance and flower color [23–25], providing a reference for transgenic breeding. GWASs have also been used to identify key regulatory genes controlling plant architecture. In *Brassica napus*, plant height, branch initiation height and branch number have been used to identify functional loci [26–29]. In rice, plant height, tillering, and panicle morphology were examined, and the gibberellic acid (GA) signaling-related gene *OsSPY* was found to be associated with semidwarfism and small panicles [30]. However, no architecture-related research has been conducted in spray cut chrysanthemum, and the molecular mechanism controlling architecture still needs to be elucidated.

In this study, we performed phenotypic measurements and a statistical analysis on nine architectural traits of 195 spray cut chrysanthemum varieties in two continuous years (2019 and 2020). Because the environments of these two years were different, we defined the environment in 2019 as EN2019 and the environment in 2020 as EN2020. The effect of different environmental factors, variation and relationships of different traits and cluster analysis of all spray cut chrysanthemum were analyzed. GWAS was also performed to find latent functional genes controlling architectural traits in chrysanthemums.

2. Materials and Methods

2.1. Plant Materials

A total of 195 spray cut chrysanthemum varieties were used in this study, including varieties developed by Nanjing Agricultural University and those collected from around the world. These varieties were maintained at the Chrysanthemum Germplasm Resource Preserving Centre of Nanjing Agricultural University, China (E118°85′, N31°95′). The 195 varieties evaluated during the two years are listed in Table S1.

2.2. Phenotypic Evaluation of Architectural Traits

Seedlings were planted in seedbeds in June 2019 and June 2020. Vigorously growing and similarly appearing rooted seedlings were selected and transplanted into a greenhouse in July of the same year. Fifty seedlings of each variety were planted in accordance with a row spacing of 10 cm × 10 cm, and conventional field management practices were performed. Flowering occurred from late October to late November. The monthly average temperature, monthly precipitation and monthly average relative humidity of EN2019 and EN2020 in planting location, Jiangning District, Nanjing, China (E118°85′, N31°95′) were shown in Table S2.

Nine phenotypic traits were measured. 1. For plant height, the height of the above-ground part of the plant was measured with a ruler, with a precision of 0.1 cm; 2. for number of leaf nodes, the number of leaf nodes on the trunk of the aboveground part of the plant was counted visually; 3. for total number of lateral buds, the number of nodes of all germinating buds or sprouted branches on the stem of the plant was counted visually; 4. for number of upper lateral branches, the number of all primary branches within 15 cm from the top of the plant was counted visually; 5. for number of lateral flower buds, the total number of flower buds on all primary branches was counted visually; 6. for stem diameter, the diameter at 40 cm below the top of the plant was measured with a digital Vernier caliper with precision of 0.01 mm; 7. for branch diameter, the diameter at 1/2 of the three nearest primary branches around the main bud was measured with a digital Vernier caliper with a precision of 0.01 mm; 8. for branch angle, the angle of the three nearest primary branches around the main bud was measured with a protractor; 9. for branch length, the length of all primary branches was measured with a ruler, with a precision of 0.1 cm. At the full-flowering stage, the measurement was performed on six plants for each variety, and the mean values were taken.

2.3. Phenotypic Data Analysis

Microsoft Excel 2019 was used for basic descriptive statistical analysis of the 9 architecture-related phenotypic traits of the 195 cut chrysanthemum varieties in EN2019 and EN2020 environments, and IBM SPSS 25.0 statistical software was used for correlation analysis of the EN2019 and EN2020 data. Significant differences (paired-sample t tests) were assessed and violin mapping and cluster analysis of two environmental data were conducted by R 4.0.4 (https://www.r-project.org/, accessed on 7 December 2021).

2.4. GWAS and Candidate Gene Annotation

In our previous study [31], 199 chrysanthemum accessions were sequenced, of which forty-four spray cut chrysanthemum varieties were also measured in our study and the list is shown in Table S3. In order to obtain more meaningful information, we performed GWAS using these raw sequencing data. SLAF-seq raw reads whose quality scores were <30 and separated by barcodes were discarded. The highest depth tag in each SLAF was chosen as a reference due to the lack of a reference genome sequence. The qualified sequencing data of the samples were aligned to the genome reference sequence of chrysanthemum using Burrow–Wheeler Aligner (BWA) V0.7 (http://bio-bwa.sourceforge.net/, accessed on 7 December 2021) [32], and then SNP sites were detected by SAMtools V1.4 (http://samtools.sourceforge.net/, accessed on 7 December 2021) [33]. After removing the SNPs with a sequencing depth less than 3, a data loss percentage greater than 20%, and a minor allele frequency (MAF) less than 5%, 191,417 high-quality SNPs were ultimately identified for further analysis.

GCTA software V1.93 (https://yanglab.westlake.edu.cn/software/gcta/#Overview, accessed on 7 December 2021) [34] was used for principal component analysis (PCA) and construction of a kinship matrix, yielding an eigenvector principal component (PC) matrix of all the individuals and a kinship matrix comprising data between every pair of individuals. Combining the data of the nine phenotypic traits and the SNP sequencing data, a GWAS was conducted via the compressed mixed linear model (cMLM) of GAPIT software V3 (https://www.zzlab.net/GAPIT/, accessed on 7 December 2021) [35] and via the cMLM and mixed linear model (MLM) of TASSEL software V5.0 (https://www.maizegenetics.net/tassel, accessed on 7 December 2021) [36]. The mean values were used for the GWAS, and the significance threshold was set at $p \leq 0.001$. As a result, the SNPs found to be significantly associated with the phenotypic data and the phenotypic variance explained (PVE) were identified for gene mining.

According to the significant SNP sites detected by cMLM model of TASSEL, candidate genes within 300 k of SNP sites were found. The function of genes was annotated via The Arabidopsis Information Resource (TAIR) website (https://www.arabidopsis.org/, accessed on 7 May 2022) by BLASTX [37]. Through the functional annotations of Arabidopsis and other function reported in other plants, the genes related to plant architecture, hormone signaling pathways or plant development regulation were further selected as final candidate functional genes.

3. Results

3.1. Analysis of Significantly Different Architectural Traits of Spray Cut Chrysanthemum between EN2019 and EN2020

To explore the effects of different years on architectural traits, nine architectural traits in EN2019 and EN2020 were compared, and their significance was assessed (paired-sample t tests). As shown in Figure 1, there were significant differences in plant height, total number of lateral buds, number of upper primary branches, stem diameter, primary branch diameter and primary branch length between EN2019 and EN2020, while there were no significant differences in number of leaf nodes, number of lateral flower buds or primary branch angle. Among these traits, the median plant height, stem diameter, primary branch diameter and primary branch length in EN2019 were larger than those in EN2020. The median total number of lateral buds in EN2019 was similar to that in EN2020, while the median number of upper primary branches in EN2019 was smaller than that in EN2020.

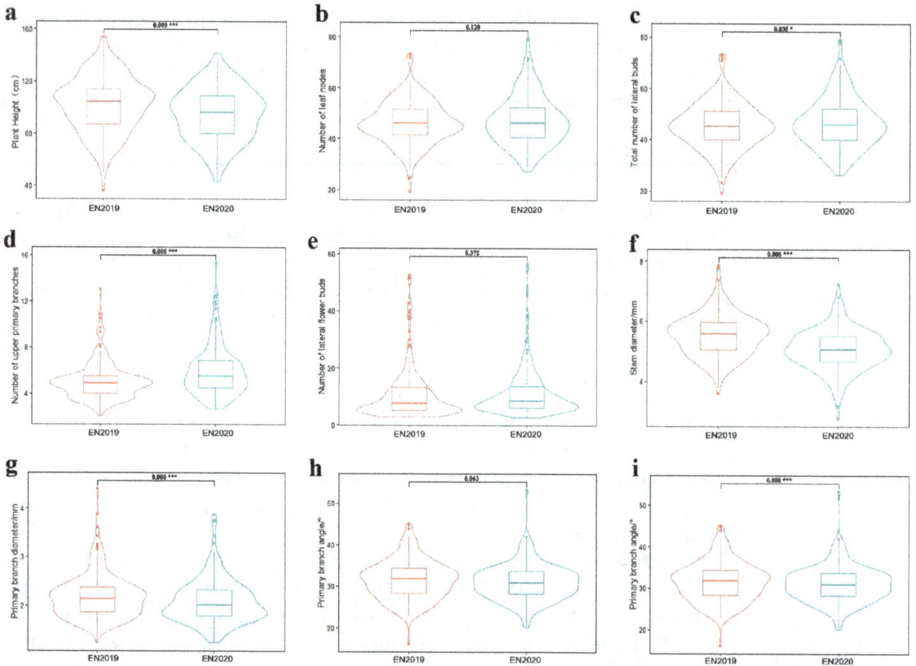

Figure 1. Violin plots indicating variation in architectural traits of 195 spray cut chrysanthemum varieties in EN2019 and EN2020. (**a**) Plant height; (**b**) Number of leaf nodes; (**c**) Total number of lateral buds; (**d**) Number of upper primary branches; (**e**) Number of lateral flower buds; (**f**) Stem diameter; (**g**) Primary branch diameter; (**h**) Primary branch angle; (**i**) Primary branch length. Note: ***, ** and * indicate significant differences at the 0.001, 0.01 and 0.05 probability levels, respectively.

3.2. Descriptive Statistics of the Architectural Characteristics of Spray Cut Chrysanthemum in EN2019 and EN2020

The basic statistical analysis results of the data of the nine phenotypic traits of the 195 spray cut chrysanthemum species in EN2019 and EN2020 are shown in Table 1. Although there were different coefficients of variation (CVs) for all the traits, the same traits in both years showed a significant positive correlation. In EN2019, the CV varied from 12.67% to 85.84%, among which the CV of stem diameter was the smallest and the CV of number of lateral flower buds was the largest. In EN2020, the CV varied from 13.30% to 73.81%, and traits with extreme values were the same as those in EN2019. These results showed that stem diameter trait was rather stable, while number of lateral flower buds varied among the different spray cut chrysanthemum varieties. In addition, compared with the other traits, primary branch length and number of upper primary branches also was associated with larger CVs, which were more than 30%.

Table 1. Phenotypic characteristics of the architectural traits of 195 spray cut chrysanthemum varieties.

Trait	Environment	Max	Min	Rang	Mean	SD	CV/%	Skewness	Kurtosis	r
Plant height/cm	EN2019	153.67	36.10	117.57	101.84	21.20	20.82	−0.22	−0.16	0.794 **
	EN2020	141.42	42.98	98.43	94.67	19.89	21.01	−0.15	−0.55	
Number of leaf nodes	EN2019	73.33	19.17	54.17	46.51	8.62	18.52	0.24	0.93	0.701 **
	EN2020	79.50	27.17	52.33	47.32	9.82	20.75	0.63	0.41	
Total number of lateral buds	EN2019	73.33	19.17	54.17	45.40	8.94	19.68	0.20	0.81	0.649 **
	EN2020	79.17	26.33	52.83	46.60	9.77	20.96	0.57	0.43	
Number of upper primary branches	EN2019	13.00	2.00	11.00	5.04	1.60	31.75	1.56	4.16	0.593 **
	EN2020	15.33	2.67	12.67	5.95	2.13	35.82	1.26	2.16	
Number of lateral flower buds	EN2019	52.67	2.83	49.83	10.83	9.30	85.84	2.45	6.69	0.738 **
	EN2020	56.50	2.67	53.83	11.25	8.30	73.81	2.71	9.85	
Stem diameter/mm	EN2019	7.87	3.62	4.24	5.54	0.70	12.67	0.25	0.58	0.618 **
	EN2020	7.21	2.77	4.44	5.07	0.67	13.30	−0.07	0.86	
Primary branch diameter/mm	EN2019	4.40	1.24	3.16	2.18	0.46	20.90	1.34	3.35	0.712 **
	EN2020	3.88	1.25	2.62	2.09	0.45	21.40	1.14	2.26	
Primary branch angle/°	EN2019	45.00	16.22	28.78	31.49	4.72	14.99	0.22	0.28	0.718 **
	EN2020	53.38	20.11	33.17	31.03	4.30	13.85	0.80	3.11	
Primary branch length/cm	EN2019	78.08	2.22	75.86	16.47	8.75	53.11	3.15	15.82	0.521 **
	EN2020	54.96	1.63	53.33	13.39	7.77	58.06	3.04	11.42	

Note: r refers to the Pearson correlation coefficient between the two environments (EN2019 and EN2020). ** indicates a significant difference at the 0.01 probability level.

3.3. Correlation Analysis of Architectural Traits in EN2019 and EN2020

Table 2 shows the Pearson correlations between the nine phenotypic traits in EN2019 and EN2020. In total, there were 72 pairs of architecture correlations in EN2019 and EN2020. Among all the architectural traits, the positive correlation between the number of leaf nodes and total number of lateral buds was the strongest, whose correlation coefficients in EN2019 and EN2020 were 0.913 and 0.986, respectively, followed by the positive correlation between primary branch length and primary branch diameter, whose correlation coefficients were 0.639 and 0.661, respectively. Similarly, the negative correlation between primary branch length and number of upper primary branches in EN2019 was the strongest, with a correlation coefficient of −0.317; in EN2020, the negative correlation between primary branch diameter and number of upper primary branches was the strongest, with a correlation coefficient of −0.466.

3.4. Cluster Analysis of the Architectural Traits of 195 Spray Cut Chrysanthemum Species

Cluster analysis was performed using both the K-means clustering algorithm and the Pedigree clustering algorithm based on the nine architectural traits during the two-year period (Figure 2). The best preset number of K calculated by K-means in EN2019 and EN2020 was 5 (Figure 2a,b), which was verified by pedigree clustering (Figure 2c,d). However, in EN2019, the 195 spray cut chrysanthemum varieties were divided into six categories according to the pedigree clustering diagram, which was inconsistent with the results of the K-means clustering; in EN2020, the distribution of varieties in the pedigree clustering diagram was nearly the same as that in the K-means clustering diagram.

Table 2. Pearson correlations between the architectural traits of 195 spray cut chrysanthemum varieties.

EN2020 / EN2019	Plant Height	Number of Leaf Nodes	Total Number of Lateral Buds	Number of Upper Primary Branches	Number of Lateral Flower Buds	Stem Diameter	Primary Branch Diameter	Primary Branch Angle	Primary Branch Length
Plant height	1	0.196 **	0.188 **	−0.298 **	−0.195 **	0.153 *	0.027	0.112	0.142 *
Number of leaf nodes	0.250 **	1	0.913 **	0.197 **	0.079	0.388 **	0.012	−0.154 *	−0.059
Total number of lateral buds	0.253 **	0.986 **	1	0.133	0.041	0.334 **	−0.016	−0.160 *	−0.138
Number of upper primary branches	−0.369 **	0.134	0.126	1	0.231 **	0.110	−0.200 **	−0.293 **	−0.317 **
Number of lateral flower buds	−0.232 **	0.351 **	0.341 **	0.240 **	1	0.109	0.579 **	−0.074	0.534 **
Stem diameter	0.322 **	0.250 **	0.257 **	−0.019	−0.191 **	1	0.394 **	−0.041	0.044
Primary branch diameter	0.096	0.028	0.026	−0.466 **	0.368 **	0.147 *	1	0.147 *	0.639 **
Primary branch angle	0.063	−0.215 **	−0.242 **	−0.253 **	−0.108	−0.031	0.264 **	1	0.148 *
Primary branch length	0.113	0.063	0.057	−0.402 **	0.513 **	−0.279 **	0.661 **	0.160 *	1

Note: ** and * indicate significant differences at the 0.01 and 0.05 probability levels, respectively.

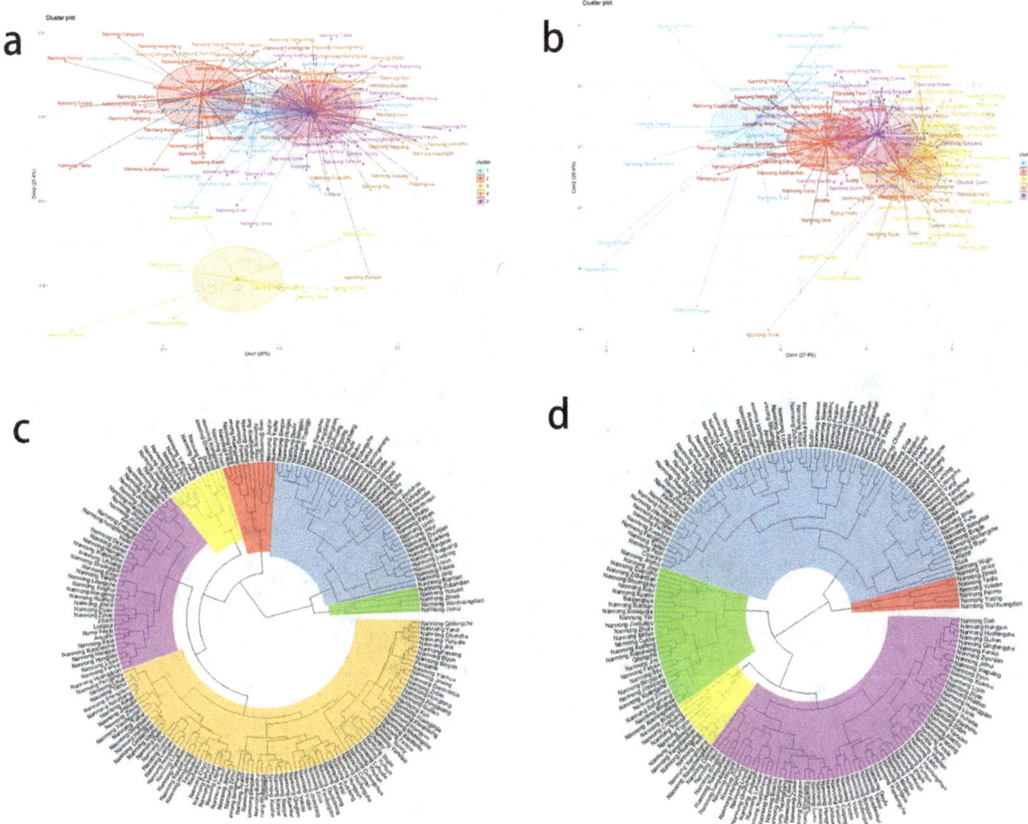

Figure 2. Cluster analysis map of 195 spray cut chrysanthemum varieties in EN2019 and EN2020. (**a**) K-means cluster of 195 spray cut chrysanthemum varieties in EN2019; (**b**) K-means cluster of 195 spray cut chrysanthemum varieties in EN2020; (**c**) Pedigree cluster of 195 spray cut chrysanthemum varieties in EN2019; (**d**) Pedigree cluster of 195 spray cut chrysanthemum varieties in EN2020. Different colors in every figure refer to different clusters.

Therefore, the categories in EN2020 were more suitable for summarizing the 195 spray cut chrysanthemum varieties. As shown in Figure 3, the 195 were divided into five categories, accounting for 3.08%, 12.31%, 29.74%, 13.85% and 41.45% of all the varieties, of which the typical architecture types were summarized. In the first category, there were only six varieties, represented by Nannong Taoliu (Figure 3a), which had a low starting-branch height, long primary branches, a semispreading plant growth habit and a large number of secondary branches. There were tertiary branches on the plants only in the first category. The second category, represented by Nannong Meifengche (Figure 3b), included 24 varieties, which had loosely distributed branches and longer primary branches. There were fewer leaf nodes and secondary branches on the plants in the second category. The third category, represented by Nannong Cuilongzhao (Figure 3c), included 58 varieties whose starting-branch height was lower than 1/3 of the total plant height. The plants in this category had a relatively large distribution of flowering branches and secondary branches. The fourth category, represented by Nannong Songmang (Figure 3d), included 27 varieties, which had a higher starting-branch height and shorter flowering branches. These compact flowering branches displayed a nearly spherical appearance. The fifth category, represented by Nannong Bingqing (Figure 3e), included 80 varieties, the plants of which had a relatively large amount of leaf nodes. The flowering branches of the plants in this category displayed a tower-like shape.

Figure 3. Five representative varieties with different architectural traits among 195 spray cut chrysanthemum. (**a**) Nannong Taoliu; (**b**) Nannong Meifengche; (**c**) Nannong Cuilongzhua; (**d**) Nannong Songmang; (**e**) Nannong Bingqing. Bars = 10 cm. In each sub figure, the left one refers to an intact plant, middle one refers to an intact plant without leaves and the right one refers to an intact plant without leaves and flowers.

3.5. GWAS and Mining of Genes Controlling Plant Architecture

Combining the data concerning 191,417 high-quality SNPs and the data of the nine phenotypic traits in EN2019 and EN2020, we performed a GWAS via the cMLM method of GAPIT software and the cMLM and MLM methods of TASSEL software, with the PC matrix and kinship matrix serving as covariates. When the significance threshold was 1×10^{-3}, 281 SNPs associated with each trait and corresponding PVE values were obtained. According to the data in Tables S4–S16, GAPIT software revealed 113 SNPs associated with plant architecture in EN2019 and 93 SNPs in EN2020, and the PVE values ranged from 24.06% to 44.46% in EN2019 and from 25.20% to 44.48% in EN2020. The cMLM model of TASSEL software revealed 35 SNPs associated with plant architecture in EN2019 and 40 SNPs in EN2020, and the PVE values ranged from 23.33% to 49.63% in EN2019 and from 22.37% to 59.73% in EN2020.

Among the identified SNPs, 18 were detected by GAPIT. 5__55325230, 5__55325289 and 10__268875261 were associated with plant height in EN2019 and EN2020; 27__13304666 was associated with stem diameter in EN2019 and EN2020; 9_193339518 was associated with primary branch diameter in EN2019 and EN2020. Eight SNPs (17__232431589, 27__194241646, 27__194241707, 17__46614191, 8__66464970, 27__55624912, 27__112009133 and 14__58564373) were associated with number of leaf nodes and total number of lateral buds in EN2019. Five SNPs (3__163391415, 25__13632558, 25__13632658, 22__108888654 and 7__185418131) were associated with number of leaf nodes and total number of lateral buds in EN2020. According to the cMLM model of TASSEL, the same 10 SNPs were detected: 23__171200599 was associated with plant height in EN2019 and EN2020, 8 SNPs (11__125802422, 17__280870788, 21__242369101, 24__15091361, 23__293787130, 23__293967285, 23__294075102 and 23__308947968) were associated with number of leaf nodes and total number of lateral buds in EN2019, and 9__215337463 was associated with number of leaf nodes and total number of lateral buds in EN2020. Additionally, 19__104723464 was associated with number of lateral flower buds in EN2020. According to the MLM method of TASSEL, 23__171200599 was also associated with plant height in EN2020; 11__125802422 was associated with total number of lateral buds in EN2019; 17__280870788 was associated with number of leaf nodes in EN2019; 21__242369101, 24__15091361, 23__293787130, 23__293967285, 23__294075102 and 23__308947968 were associated with the number of leaf nodes and total number of lateral buds in EN2019; 9__215337463 was associated with number of leaf nodes in EN2020.

After comparing genes related to SNP loci of three models, the cMLM model of TASSEL software was chosen finally. Combining the annotation of TAIR, candidate genes are shown in Table 3, and the Manhattan plots of cMLM model of TASSEL can be found in supplementary file S1. We identified four candidate genes: *phyB*, *BRH1*, *CPC* and *bZIP16*.

Table 3. List of SNP sites, candidate genes, and functional annotation for selected architectural traits in spray cut chrysanthemums identified with cMLM model of TASSEL.

SNP Site	Significantly Associated Traits	Candidate Genes	Homologs in Arabidopsis
19__104723464	Number of lateral flower buds/Number of upper primary branches	evm.model.scaffold_940.421	AT2G18790.1 (phytochrome B, phyB)
9__215337463	Number of leaf nodes/Total number of lateral buds	evm.model.scaffold_11169.22	AT3G61460.1 (brassinosteroid-responsive RING-H2, BRH1)
17__280870788	Total number of lateral buds	evm.model.scaffold_3682.68	AT2G46410.1 (Homeodomain-like superfamily protein, CPC)
21__242369101	Total number of lateral buds	evm.model.scaffold_881.80	AT2G35530.1 (basic region/leucine zipper transcription factor 16, bZIP16)

4. Discussion

In the two consecutive years of EN2019 and EN2020, the number of upper primary branches, number of lateral flower buds and primary branch length presented the highest CV values; these traits were also the key traits used to determine the output of spray cut chrysanthemum. These results indicated that the architectural traits of the 195 spray cut chrysanthemum varieties selected by artificial breeding were diverse and controlled by complex gene pathways. As such, when selecting the appropriate spray cut chrysanthemum, breeders should consider these three traits to have high priority.

During their growth, plants have limited resources. They can allocate resources reasonably through different gene pathways to complete their life cycle. In the present study, plant height had significant slightly large positive Pearson coefficients with number of leaf nodes and total number of lateral buds, but had significant large negative Pearson coefficients with number of upper flower branches and number of lateral flower buds. Plant height was also found to positively correlate with tiller number in sorghum [38], and negatively correlated with fruit branch length in Chinese upland cotton [39]. The correlation relationships in chrysanthemums reflect the mutual negative relationship between vegetative growth and reproductive growth. The number of upper primary branches had very significant negative correlations with primary branch diameter, primary branch angle and primary branch length, which are key traits determining the quality of spray cut chrysanthemum. Therefore, balancing the number of upper primary branches and their quality is highly important.

Plant architecture is species specific, and influenced by environmental conditions such as light, temperature, humidity and nutrient status [40]. Low temperature can lead to the dwarfed rosette and leaves with increased thickness in Arabidopsis [41]. Main differences between EN2019 and EN2020 were monthly precipitation and monthly average relative humidity according to Table S2. Monthly precipitation and monthly average relative humidity of EN2020 in nearly each month were larger than that of EN2019. The mean values of plant height, stem diameter, primary branch diameter, primary branch angle and primary branch length decreased in EN2020 compared with EN2019, indicating a weaker growth state in EN2020. These differences in growth might be attributed to three main factors. First, the growth of chrysanthemum is sensitive to continuously cropped soils, which is related to changes in physicochemical properties, soil microorganisms and allelopathy of plants [42]. Second, the plum rain season in Nanjing was longer in 2020 than in 2019, as mentioned above, which increased the air humidity during the rooting period of the seedlings and the initial root growth stages. Changes in vapor pressure deficit (VPD) and relative humidity (RH) affect the height and flowering time of chrysanthemum plants [43,44]. This explained why the growth of seedlings in EN2020 was rather poor to some extent. Lastly, the seedlings of EN2020 were collected from mother plants overwintering in EN2019, whose growth state might be worse than that in EN2019. Number of leaf nodes, number of lateral flower buds and primary branch angle showed no significant difference in two years, which means that they might not be significantly influenced by a changed environment.

In this experiment, due to the differences in each trait in the two years, two cluster methods were used to cluster the data from the two years. After the trials, the best K value was set as 5. According to the clustering results, we divided the spray cut chrysanthemum varieties into five categories; these categories could be used as typical architecture types for summarize the architectural traits of spray cut chrysanthemum.

A low red light:far-red light ratio (R:FR) can lead to shade avoidance syndrome of plants, resulting in enhanced shoot elongation and reduced branching, and phyB is a major sensor of R:FR signal [45]. phyB has been found to control shooting branching together with photosynthetic photon flux density (PPFD) and PIF4/PIF5 in Arabidopsis, regulating related hormone pathway and expression levels of downstream genes such as *BRC1* [46,47]. The *phyB* mutant in sorghum also showed enhanced apical dominance and shortened bud length and the expression levels of *TEOSINTE BRANCHED1* (*TB1*), *Dormancy-associated gene-1* (*DRM1*) and *MORE AXILLARY BRANCHES2* (*MAX2*) were found to increase in the axillary buds [48,49]. The number of upper primary branches in chrysanthemums was found to correlate with *phyB* in our study, indicating the functional role of controlling plant architecture. Three other genes were also found to participate in plant development regulation and hormone signal pathway. *BRH1* is a BR-responsive gene, and overexpression of *BRH1* results in the production of rounded leaves and may result in the growth and development of rosette leaves [50]. By promoting the conversion of nonhair cells to root hair cells, the R3-type MYB transcription factor protein CAPRICE (CPC) was shown to induce root hair formation in the root epidermis [51]. bZIP16 can promote seed germination and hypocotyl elongation in the initial stages of seedling development [52]. These three genes also play a role in plant development in other plants, which might control plant architecture in chrysanthemums.

5. Conclusions

In this study, we found that the number of leaf nodes, number of lateral flower buds and primary branch angle were less influenced by environmental factors, while plant height, stem diameter, total number of lateral buds, number of upper primary branches, primary branch diameter and primary branch length were significantly influenced by environmental factors. The number of upper primary branches, number of lateral flower buds and primary branch length presented larger variation degree in 195 species. The number of upper primary branches had very significant negative correlations with primary branch diameter, primary branch angle and primary branch length. We also summarized five clusters with typical architecture and predicted four candidate functional genes (*phyB*, *BRH1*, *CPC* and *bZIP16*) which might control plant architecture in chrysanthemums.

Supplementary Materials: The following supporting information can be downloaded at: https://www.mdpi.com/article/10.3390/horticulturae8050458/s1, Table S1: 195 spray cut chrysanthemum varieties tested; Table S2. The monthly average temperature, monthly precipitation and monthly average relative humidity of EN2019 and EN2020 in planting location, Jiangning District, Nanjing, China (E118°85′, N31°95′) Table S3: 44 spray cut chrysanthemum varieties sequencing completed; Table S4: SNPs (GAPIT-cMLM) for the architecture traits of plant height and number of leaf nodes; Table S5: SNPs (GAPIT-cMLM) for the architecture traits of number of upper primary branches, number of lateral flower buds and stem diameter; Table S6: SNPs (GAPIT-cMLM) for the architecture traits of stem diameter, primary branch diameter and primary branch angle; Table S7: SNPs (GAPIT-cMLM) for the architecture traits of primary branch length; Table S8: SNPs (GAPIT-cMLM) for the architecture traits of primary branch length; Table S9: SNPs (Tassel-cMLM) for the architecture traits of plant height, number of leaf nodes, total number of lateral buds, number of upper primary branches and number of lateral flower buds; Table S10: SNPs (Tassel-cMLM) for the architecture traits of number of lateral flower buds, stem diameter, primary branch diameter and primary branch length; Table S11: SNPs (Tassel-MLM) for the architecture traits of plant height, number of leaf nodes and total number of lateral buds; Table S12: SNPs (Tassel-MLM) for the architecture traits of number of upper primary branches number of lateral flower buds and stem diameter; Table S13: SNPs (Tassel-MLM) for the architecture traits of stem diameter, primary branch diameter, primary branch angle and primary branch length; Table S14: SNPs (Tassel-MLM) for the architecture traits of primary branch length; Table S15: SNPs (Tassel-MLM) for the architecture traits of primary branch length; Table S16: SNPs (Tassel-MLM) for the architecture traits of primary branch length. Supplementary Figures S1–S18: The Manhattan plots of SNPs detected by the cMLM model of TASSEL software.

Author Contributions: Conceptualization, A.S.; methodology, A.S. and J.S.; software, A.S., J.S. and D.S.; validation, D.S., L.Z. and Q.Y.; formal analysis, D.S. and L.Z.; writing—original draft preparation, D.S. and L.Z.; writing—review and editing, D.S., L.Z., J.S., Q.Y., J.Z., W.F., H.W., Z.G., F.C. and A.S.; funding acquisition, A.S., F.C., W.F. and H.W. All authors have read and agreed to the published version of the manuscript.

Funding: This research was funded by China Agriculture Research System (CARS-23-A18), National Natural Science Foundation of China (32172609, 31870694, 31872149), the earmarked fund for Jiangsu Agricultural Industry Technology System (JATS [2021]454), and a project Funded by the Priority Academic Program Development of Jiangsu Higher Education Institution.

Institutional Review Board Statement: Not applicable.

Informed Consent Statement: Not applicable.

Data Availability Statement: Not applicable.

Acknowledgments: Data analysis was supported by the high-performance computing platform of Bioinformatics Center, Nanjing.

Conflicts of Interest: The authors declare no conflict of interest.

References

1. Su, J.; Jiang, J.; Zhang, F.; Liu, Y.; Ding, L.; Chen, S.; Chen, F. Current achievements and future prospects in the genetic breeding of chrysanthemum: A review. *Hortic. Res.* **2019**, *6*, 109. [CrossRef] [PubMed]
2. Chen, X.; Zhou, X.; Xi, L.; Li, J.; Zhao, R.; Ma, N.; Zhao, L. Roles of *DgBRC1* in regulation of lateral branching in chrysanthemum (*Dendranthema × grandiflora* cv. Jinba). *PLoS ONE* **2013**, *8*, e61717. [CrossRef] [PubMed]
3. Yang, J.; Jeong, B.R. Side lighting enhances morphophysiology by inducing more branching and flowering in chrysanthemum grown in controlled environment. *Int. J. Mol. Sci.* **2021**, *22*, 12019. [CrossRef] [PubMed]
4. Sun, D.; Zhang, L.; Yu, Q.; Zhang, J.; Li, P.; Zhang, Y.; Xing, X.; Ding, L.; Fang, W.; Chen, F.; et al. Integrated Signals of Jasmonates, Sugars, Cytokinins and Auxin Influence the Initial Growth of the Second Buds of Chrysanthemum after Decapitation. *Biology* **2021**, *10*, 440. [CrossRef]
5. Barbier, F.F.; Dun, E.A.; Kerr, S.C.; Chabikwa, T.G.; Beveridge, C.A. An update on the signals controlling shoot branching. *Trends Plant Sci.* **2019**, *24*, 220–236. [CrossRef]

6. Rameau, C.; Bertheloot, J.; Leduc, N.; Andrieu, B.; Foucher, F.; Sakr, S. Multiple pathways regulate shoot branching. *Front. Plant Sci.* **2015**, *5*, 741. [CrossRef]

7. Nordström, A.; Tarkowski, P.; Tarkowska, D.; Norbaek, R.; Åstot, C.; Dolezal, K.; Sandberg, G. Auxin regulation of cytokinin biosynthesis in *Arabidopsis thaliana*: A factor of potential importance for auxin–cytokinin-regulated development. *Proc. Natl. Acad. Sci. USA* **2004**, *101*, 8039–8044. [CrossRef]

8. Tanaka, M.; Takei, K.; Kojima, M.; Sakakibara, H.; Mori, H. Auxin controls local cytokinin biosynthesis in the nodal stem in apical dominance. *Plant J.* **2006**, *45*, 1028–1036. [CrossRef]

9. Alder, A.; Jamil, M.; Marzorati, M.; Bruno, M.; Vermathen, M.; Bigler, P.; Ghisla, S.; Bouwmeester, H.; Beyer, P.; Al-Babili, S. The path from β-carotene to carlactone, a strigolactone-like plant hormone. *Science* **2012**, *335*, 1348–1351. [CrossRef]

10. Wu, C.; Trieu, A.; Radhakrishnan, P.; Kwok, S.F.; Harris, S.; Zhang, K.; Wang, J.; Wan, J.; Zhai, H.; Takatsuto, S.; et al. Brassinosteroids regulate grain filling in rice. *Plant Cell* **2008**, *20*, 2130–2145. [CrossRef]

11. Xia, X.; Dong, H.; Yin, Y.; Song, X.; Gu, X.; Sang, K.; Zhou, J.; Shi, K.; Zhou, Y.; Foyer, C.H.; et al. Brassinosteroid signaling integrates multiple pathways to release apical dominance in tomato. *Proc. Natl. Acad. Sci. USA* **2021**, *118*, e2004384118. [CrossRef]

12. Mason, M.G.; Ross, J.J.; Babst, B.A.; Wienclaw, B.N.; Beveridge, C.A. Sugar demand, not auxin, is the initial regulator of apical dominance. *Proc. Natl. Acad. Sci. USA* **2014**, *111*, 6092–6097. [CrossRef]

13. Wang, B.; Smith, S.M.; Li, J. Genetic regulation of shoot architecture. *Annu. Rev. Plant Biol.* **2018**, *69*, 437–468. [CrossRef]

14. Adams, H.D.; Collins, A.D.; Briggs, S.P.; Vennetier, M.; Dickman, L.T.; Sevanto, S.A.; Garcia-Forner, N.; Powers, H.H.; McDowell, N.G. Experimental drought and heat can delay phenological development and reduce foliar and shoot growth in semiarid trees. *Glob. Chang. Biol.* **2015**, *21*, 4210–4220. [CrossRef]

15. Fleisher, D.H.; Timlin, D.J.; Reddy, V.R. Temperature influence on potato leaf and branch distribution and on canopy photosynthetic rate. *Agron. J.* **2006**, *98*, 1442–1452. [CrossRef]

16. Leduc, N.; Roman, H.; Barbier, F.; Péron, T.; Huché-Thélier, L.; Lothier, J.; Demotes-Mainard, S.; Sakr, S. Light signaling in bud outgrowth and branching in plants. *Plants* **2014**, *3*, 223–250. [CrossRef]

17. Bahmani, I.; Hazard, L.; Varlet-Grancher, C.; Betin, M.; Lemaire, G.; Matthew, C.; Thom, E. Differences in tillering of long- and short-leaved perennial ryegrass genetic lines under full light and shade treatments. *Crop Sci.* **2000**, *40*, 1095–1102. [CrossRef]

18. Evers, J.B.; Vos, J.; Andrieu, B.; Struik, P.C. Cessation of tillering in spring wheat in relation to light interception and red: Far-red ratio. *Ann. Bot.* **2006**, *97*, 649–658. [CrossRef]

19. Luo, L.; Zhang, Y.; Xu, G. How does nitrogen shape plant architecture? *J. Exp. Bot.* **2020**, *71*, 4415–4427. [CrossRef]

20. He, X.; Qu, B.; Li, W.; Zhao, X.; Teng, W.; Ma, W.; Ren, Y.; Li, B.; Li, Z.; Tong, Y. The nitrate-inducible NAC transcription factor TaNAC2-5A controls nitrate response and increases wheat yield. *Plant Physiol.* **2015**, *169*, 1991–2005. [CrossRef]

21. Tian, D.; Wang, P.; Tang, B.; Teng, X.; Li, C.; Liu, X.; Zou, D.; Song, S.; Zhang, Z. GWAS Atlas: A curated resource of genome-wide variant-trait associations in plants and animals. *Nucleic Acids Res.* **2020**, *48*, D927–D932. [CrossRef] [PubMed]

22. Cano-Gamez, E.; Trynka, G. From GWAS to function: Using functional genomics to identify the mechanisms underlying complex diseases. *Front. Genet.* **2020**, *11*, 424. [CrossRef]

23. Su, J.; Zhang, F.; Li, P.; Guan, Z.; Fang, W.; Chen, F. Genetic variation and association mapping of waterlogging tolerance in chrysanthemum. *Planta* **2016**, *244*, 1241–1252. [CrossRef]

24. Su, J.; Zhang, F.; Chong, X.; Song, A.; Guan, Z.; Fang, W.; Chen, F. Genome-wide association study identifies favorable SNP alleles and candidate genes for waterlogging tolerance in chrysanthemums. *Hortic. Res.* **2019**, *6*, 21. [CrossRef]

25. Sumitomo, K.; Shirasawa, K.; Isobe, S.; Hirakawa, H.; Hisamatsu, T.; Nakano, Y.; Yagi, M.; Ohmiya, A. Genome-wide association study overcomes the genome complexity in autohexaploid chrysanthemum and tags SNP markers onto the flower color genes. *Sci. Rep.* **2019**, *9*, 13947. [CrossRef]

26. He, Y.; Wu, D.; Wei, D.; Fu, Y.; Cui, Y.; Dong, H.; Tan, C.; Qian, W. GWAS, QTL mapping and gene expression analyses in *Brassica napus* reveal genetic control of branching morphogenesis. *Sci. Rep.* **2017**, *7*, 15971. [CrossRef]

27. Li, F.; Chen, B.; Xu, K.; Gao, G.; Yan, G.; Qiao, J.; Li, J.; Li, H.; Li, L.; Xiao, X.; et al. A genome-wide association study of plant height and primary branch number in rapeseed (*Brassica napus*). *Plant Sci.* **2016**, *242*, 169–177. [CrossRef]

28. Liu, H.; Wang, J.; Zhang, B.; Yang, X.; Hammond, J.P.; Ding, G.; Wang, S.; Cai, H.; Wang, C.; Xu, F. Genome-wide association study dissects the genetic control of plant height and branch number in response to low-phosphorus stress in *Brassica napus*. *Ann. Bot.* **2021**, *128*, 919–930. [CrossRef]

29. Zheng, M.; Peng, C.; Liu, H.; Tang, M.; Yang, H.; Li, X.; Liu, J.; Sun, X.; Wang, X.; Xu, J.; et al. Genome-wide association study reveals candidate genes for control of plant height, branch initiation height and branch number in rapeseed (*Brassica napus* L.). *Front. Plant Sci.* **2017**, *8*, 1246. [CrossRef] [PubMed]

30. Yano, K.; Morinaka, Y.; Wang, F.; Huang, P.; Takehara, S.; Hirai, T.; Ito, A.; Koketsu, E.; Kawamura, M.; Kotake, K.; et al. GWAS with principal component analysis identifies a gene comprehensively controlling rice architecture. *Proc. Natl. Acad. Sci. USA* **2019**, *116*, 21262–21267. [CrossRef] [PubMed]

31. Chong, X.; Zhang, F.; Wu, Y.; Yang, X.; Zhao, N.; Wang, H.; Guan, Z.; Fang, W.; Chen, F. A SNP-enabled assessment of genetic diversity, evolutionary relationships and the identification of candidate genes in chrysanthemum. *Genome Biol. Evol.* **2016**, *8*, 3661–3671. [CrossRef]

32. Li, H.; Durbin, R. Fast and accurate short read alignment with Burrows–Wheeler transform. *Bioinformatics* **2009**, *25*, 1754–1760. [CrossRef]

33. Li, H.; Handsaker, B.; Wysoker, A.; Fennell, T.; Ruan, J.; Homer, N.; Marth, G.; Abecasis, G.; Durbin, R. The sequence alignment/map format and SAMtools. *Bioinformatics* **2009**, *25*, 2078–2079. [CrossRef]
34. Yang, J.; Lee, S.H.; Goddard, M.E.; Visscher, P.M. GCTA: A tool for genome-wide complex trait analysis. *Am. J. Hum. Genet.* **2011**, *88*, 76–82. [CrossRef]
35. Lipka, A.E.; Tian, F.; Wang, Q.; Peiffer, J.; Li, M.; Bradbury, P.J.; Gore, M.A.; Buckler, E.S.; Zhang, Z. GAPIT: Genome association and prediction integrated tool. *Bioinformatics* **2012**, *28*, 2397–2399. [CrossRef]
36. Bradbury, P.J.; Zhang, Z.; Kroon, D.E.; Casstevens, T.M.; Ramdoss, Y.; Buckler, E.S. TASSEL: Software for association mapping of complex traits in diverse samples. *Bioinformatics* **2007**, *23*, 2633–2635. [CrossRef]
37. Altschul, S.F.; Gish, W.; Miller, W.; Myers, E.W.; Lipman, D.J. Basic local alignment search tool. *J. Mol. Biol.* **1990**, *215*, 403–410. [CrossRef]
38. Luo, F.; Pei, Z.; Zhao, X.; Liu, H.; Jiang, Y.; Sun, S. Genome-wide association study for plant architecture and bioenergy traits in diverse sorghum and sudangrass germplasm. *Agronomy* **2020**, *10*, 1602. [CrossRef]
39. Su, J.; Li, L.; Zhang, C.; Wang, C.; Gu, L.; Wang, H.; Wei, H.; Liu, Q.; Huang, L.; Yu, S. Genome-wide association study identified genetic variations and candidate genes for plant architecture component traits in Chinese upland cotton. *Theor. Appl. Genet.* **2018**, *131*, 1299–1314. [CrossRef]
40. Reinhardt, D.; Kuhlemeier, C. Plant architecture. *EMBO Rep.* **2002**, *3*, 846–851. [CrossRef]
41. Atkin, O.K.; Loveys, B.; Atkinson, L.J.; Pons, T.J. Phenotypic plasticity and growth temperature: Understanding interspecific variability. *J. Exp. Bot.* **2006**, *57*, 267–281. [CrossRef]
42. Li, P.; Chen, J.; Li, Y.; Zhang, K.; Wang, H. Possible mechanisms of control of Fusarium wilt of cut chrysanthemum by *Phanerochaete chrysosporium* in continuous cropping fields: A case study. *Sci. Rep.* **2017**, *7*, 15994. [CrossRef]
43. Mortensen, L.M. Effects of air humidity on growth, flowering, keeping quality and water relations of four short-day greenhouse species. *Sci. Hortic.* **2000**, *86*, 299–310. [CrossRef]
44. Körner, O.; Challa, H. Temperature integration and process-based humidity control in chrysanthemum. *Comput. Electron. Agric.* **2004**, *43*, 1–21. [CrossRef]
45. Casal, J.J. Shade avoidance. *Arab. Book/Am. Soc. Plant Biol.* **2012**, *10*, e0157. [CrossRef] [PubMed]
46. Su, H.; Abernathy, S.D.; White, R.H.; Finlayson, S.A. Photosynthetic photon flux density and phytochrome B interact to regulate branching in *Arabidopsis*. *Plant Cell Environ.* **2011**, *34*, 1986–1998. [CrossRef]
47. Holalu, S.V.; Reddy, S.K.; Blackman, B.K.; Finlayson, S.A. Phytochrome interacting factors 4 and 5 regulate axillary branching via bud abscisic acid and stem auxin signalling. *Plant Cell Environ.* **2020**, *43*, 2224–2238. [CrossRef]
48. Kebrom, T.H.; Burson, B.L.; Finlayson, S.A. Phytochrome B represses *Teosinte Branched1* expression and induces sorghum axillary bud outgrowth in response to light signals. *Plant Physiol.* **2006**, *140*, 1109–1117. [CrossRef]
49. Kebrom, T.H.; Mullet, J.E. Transcriptome profiling of tiller buds provides new insights into PhyB regulation of tillering and indeterminate growth in sorghum. *Plant Physiol.* **2016**, *170*, 2232–2250. [CrossRef]
50. Wang, X.; Chen, E.; Ge, X.; Gong, Q.; Butt, H.; Zhang, C.; Yang, Z.; Li, F.; Zhang, X. Overexpressed *BRH1*, a RING finger gene, alters rosette leaf shape in *Arabidopsis thaliana*. *Sci. China Life Sci.* **2018**, *61*, 79–87. [CrossRef]
51. Tominaga-Wada, R.; Wada, T. CPC-ETC1 chimeric protein localization data in *Arabidopsis* root epidermis. *Data Brief* **2018**, *18*, 1773–1776. [CrossRef]
52. Hsieh, W.-P.; Hsieh, H.-L.; Wu, S.-H. *Arabidopsis* bZIP16 transcription factor integrates light and hormone signaling pathways to regulate early seedling development. *Plant Cell* **2012**, *24*, 3997–4011. [CrossRef]

Article

C-CorA: A Cluster-Based Method for Correlation Analysis of RNA-Seq Data

Jianpu Qian [1,†], Wenli Liu [1,2,†], Yanna Shi [1], Mengxue Zhang [1], Qingbiao Wu [2], Kunsong Chen [1] and Wenbo Chen [1,*]

1   Zhejiang Provincial Key Laboratory of Horticultural Plant Integrative Biology, Zhejiang University, Hangzhou 310058, China; qianagnes@zju.edu.cn (J.Q.); ili@zju.edu.cn (W.L.); shiyanna@zju.edu.cn (Y.S.); zhangmengxue@zju.edu.cn (M.Z.); akun@zju.edu.cn (K.C.)
2   School of Mathematical Science, Zhejiang University, Hangzhou 310027, China; qbwu@zju.edu.cn
*   Correspondence: chenwenbo@zju.edu.cn; Tel.: +86-0571-88982224
†   These authors contributed equally to this work.

Abstract: Correlation analysis is a routine method of biological data analysis. In the process of RNA-Seq analysis, differentially expressed genes could be identified by calculating the correlation coefficients in the comparison of gene expression vs. phenotype or gene expression vs. gene expression. However, due to the complicated genetic backgrounds of perennial fruit, the correlation coefficients between phenotypes and genes are usually not high in fruit quality studies. In this study, a cluster-based correlation analysis method (C-CorA) is presented for fruit RNA-Seq analysis. C-CorA is composed of two main parts: the clustering analysis and the correlation analysis. The algorithm is described and then integrated into the MATLAB code and the C# WPF project. The C-CorA method was applied to RNA-Seq datasets of loquat (*Eriobotrya japonica*) fruit stored or ripened under different conditions. Low temperature conditioning or heat treatment of loquat fruit can alleviate the extent of lignification that occurs because of postharvest storage under low temperatures (0 °C). The C-CorA method generated correlation coefficients and identified many candidate genes correlated with lignification, including *EjCAD3* and *EjCAD4* and transcription factors such as *MYB* (*UN00328*). C-CorA is an effective new method for the correlation analysis of various types of data with different dimensions and can be applied to RNA-Seq data for candidate gene detection in fruit quality studies.

Keywords: correlation calculation; RNA-Seq; fruit quality; lignification

Citation: Qian, J.; Liu, W.; Shi, Y.; Zhang, M.; Wu, Q.; Chen, K.; Chen, W. C-CorA: A Cluster-Based Method for Correlation Analysis of RNA-Seq Data. *Horticulturae* **2022**, *8*, 124. https://doi.org/10.3390/horticulturae8020124

Academic Editors: Dilip R. Panthee and Diego Rivera-Nuñez

Received: 19 November 2021
Accepted: 26 January 2022
Published: 29 January 2022

Publisher's Note: MDPI stays neutral with regard to jurisdictional claims in published maps and institutional affiliations.

1. Introduction

High-throughput transcriptome sequencing (RNA-Seq) has become the main choice to measure gene expression levels. The correct identification of differentially expressed genes between specific conditions is a key to understanding phenotypic variation. The fold change of gene expression between samples and the absolute gene expression value are the main criteria for the identification of differentially expressed genes [1,2]. To mine more useful information, deep data analysis is necessary, such as the correlation analysis between gene expression value and phenotypic variation. Many different correlation-based methods, such as the WGCNA (weighted gene co-expression network analysis), have been used to find the correlation between gene expressions and phenotypic data, or between genes [3–5]. There are usually three different ways of ranking statistical correlation according to Spearman, Kendall and Pearson. Each coefficient will represent the result as 'r'. However, these statistical methods cannot deal directly with multi-dimensional data, such as the phenotypic data which is controlled by multiple genes.

Unlike model plants, perennial woody plants have a long-life cycle and a relatively large and complex genome (e.g., polyploidy and high heterogeneity). Furthermore, bud mutation makes the genetic background of the latter even more complicated [6]. Some traits of woody plants are controlled by multiple genes, and some fruit traits are quantitative, which could be easily affected by the environment. The fruit traits could even be different in the same tree due to different amounts of sunshine along the tree. These environmental factors introduce an amount of noise in the detection of correlation between these traits and the expression values of the genes if common approaches were used, resulting in decreased sensitivity for identifying trait-related genes. Therefore, a more inclusive correlation analysis approach is required.

In this study, we describe a new cluster-based correlation analysis (C-CorA) method which is applied to perennial plants with complicated genetic backgrounds. It could reduce the environmental effect on the traits when calculating the correlation between gene expression and phenotypic values, which is greatly helpful in fruit quality research. We describe the methodology and apply it to a set of RNA-seq data obtained from loquat, kiwifruit and persimmon.

Loquat fruits are sensitive to low temperatures and display chilling injuries, including lignin accumulation during low-temperature storage. Lignification causes an undesirable increase in fruit firmness, leading to a leathery texture [7,8]. Some transcription factors have been reported to be involved in the loquat fruit lignification process, including *EjMYB1/2* [9], *EjMYB8* [10] and *EjAP2-1* [11]. Chilling-induced lignification can be alleviated by an initial low-temperature conditioning (LTC; 5 °C for six days followed by transfer to 0 °C) or heat treatment (HT; 40 °C for four hours followed by transfer to 0 °C). In our previous study [12], we compared the transcriptome profiles of loquat fruit samples under LTC or HT with those stored at 0 °C at five points from day one to day eight after treatment. A total of 48 RNA-Seq samples, including controls and treatments, were analyzed. We identified 5824 differently expressed genes between the LTC and 0 °C samples and 3981 between the HT and 0 °C samples [12]. Correlation analysis was limited, however, due to the low correlation coefficients obtained from the Pearson calculations. Here, we carried out a more detailed analysis using the C-CorA method and identified additional genes related to the loquat lignification process during postharvest storage which were not detected by the Pearson calculation method in the previous study. Using the C-CorA method, we also detected additional genes related to the cell wall metabolism in kiwifruit and additional genes related to acetaldehyde production in persimmon.

2. Materials and Methods

2.1. Plant Materials and Treatments

The loquat fruits (*Eriobotrya japonica* Lindl. cv. Luoyangqing) were harvested at commercial maturity from the orchard of a lvyuanguopin cooperative in Luqiao, Zhejiang, China. For the identification and details of these samples, please refer to the following references [11,12]. Fruits were transported to the laboratory on the harvest day and screened for uniform size and maturity with no disease or mechanical damage. The fruit samples were divided into three pools with three biological replicates each. The first pool of fruit was stored at 0 °C. The second pool was subjected to HT (40 °C for 4 h and then transferred to 0 °C). The third pool was subjected to LTC (5 °C for 6 d then transferred to 0 °C for 2 d) [9]. Fruit flesh tissues were collected at days 0, 1, 2, 4, 6 and 8 during each treatment.

The lignin content of loquat fruits was determined using the method as described by Shan et al [13]. The specific parameter settings during each treatment are shown in the work of Xu et al. [9] and Liu et al [12].

2.2. RNA-Seq Analysis

Total RNA extraction from the flesh tissues was carried out with the QIAGEN RNeasy Plant Mini Kit following the manufacturer's instructions (QIAGEN, Chatsworth, CA, USA). The RNA quality was evaluated by electrophoresis on 1% agarose gels and quantity was determined by a NanoDrop 1000 spectrophotometer (Thermo Scientific, Wilmington, DE, USA). The construction of strand-specific RNA-Seq libraries was carried out using the protocol from Zhong et al. [14] and sequencing was completed on the Illumina HiSeq 2500 platform with single-end mode.

Raw RNA-Seq reads were first trimmed with Trimmomatic [15], and reads shorter than 40 bp were discarded. Reads were then aligned to the ribosomal RNA database [16] using bowtie [1] and aligned sequences were removed. The cleaned reads were assembled using Trinity [17] with minimum *k*mer coverage 10. The iAssembler was used to remove the redundant contigs [18]. The raw reads count of each contig was normalized to RPKM (reads per kilobase exon model per million mapped reads). The assembled contigs were blasted against three databases for gene annotation: TrEMBL, Swiss-Prot and Arabidopsis protein (TAIR), with an E-value cutoff of 1×10^{-5}.

2.3. Cluster-Based Correlation Method (C-CorA)

The cluster-based correlation method, named C-CorA, calculates the correlation coefficient using gene clustering results, based on gene expression and phenotype. In this study, we combined the basic *k*-means clustering method and the Pearson correlation coefficient to analyze the RNA-Seq of loquat fruit.

The variable *k* was set to 4 in the *k*-means clustering method in this study. The four clusters were then assembled into several 2-group combinations for the correlation coefficient calculations. There are C (4, 1) + C (4, 2)/2 = 7 different combinations in total for each of the two inputs which gave $7 \times 7 = 49$ correlation coefficients. Then the threshold was set from 0.7 to 0.9 to obtain the correlated and highly correlated candidate gene sets. The algorithm is written as follows, using MATLAB (R2018b, version 9.5, an environment developed by MathWorks) in Algorithm 1:

Algorithm 1: Cluster-Based Correlation Coefficient Calculation.

Input: S_{pheno}, S_{exp}, k, p
Output: Coe
1: $\vec{v} = \text{kmeans}\left(S_{pheno},\ \text{k}\right)$
2: **for** $i = 1, 2, \ldots, 7$ **do**
3:    $\vec{v_i} = f^i_{degeneration}\left(\vec{v}\right)$
4: **end for**
5: **while** readline S_{exp} **do**
6:    $\vec{w} = \text{kmeans}\left(S_{exp},\ \text{k}\right)$
7:    **for** $i = 1, 2, \ldots, 7$ **do**
8:      $\vec{w_i} = f^i_{degeneration}\left(\vec{w}\right)$
9:        **for** $j = 1, 2, \ldots, 7$ **do**
10:          $c = abs(corr(\vec{w_i}, \vec{v_j}), pearson)$
11:          **if** $c \geq p$ **then**
12:            store **c** in **Coe**
13:          **end if**
14:        **end for**
15:    **end for**
16: **end while**

---

3. Results

3.1. Method Implementation

The C-CorA method contains two modules: the cluster module and the correlation analysis module (Figure 1). This approach is highly flexible, as it can process various types of data and each of the two modules can be replaced with other suitable algorithms. The output is the correlation coefficients, which can be used in downstream analyses.

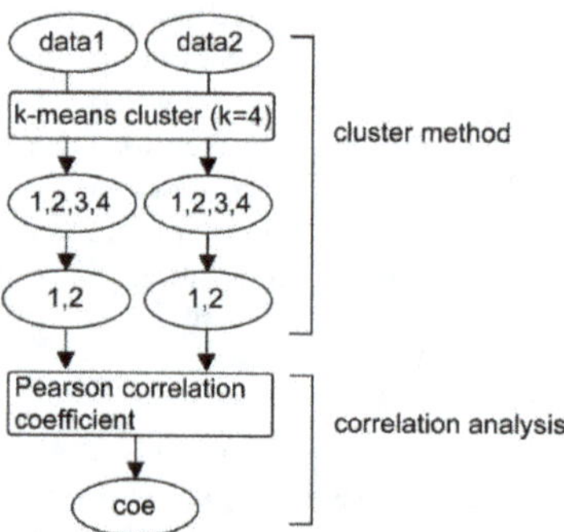

Figure 1. The C-CorA workflow showing the clustering module and a correlation analysis module. The numbers 1, 2, 3 and 4 represent the different clusters. When the k-mean is set as 4 for the cluster module, input data1 and input data 2 are first clustered into four groups. Then these groups are randomly combined into two groups for correlation analysis.

The algorithm used in the C-CorA method is written in two ways: MATLAB code and C# WPF (Windows Presentation Foundation). The algorithm written using MATLAB code is suitable for large-scale data sets, while the C# WPF is suitable for small-scale data sets and researchers with less experience in bioinformatics. The code and implementation results can be accessed on the website https://github.com/ili-4/C-corA (accessed: 20 January 2021). The MATLAB code can be run with MATLAB R2018b under the Windows or Linux systems. The cluster method used in the algorithm is the *k*-means function of MATLAB and the correlation calculation method is the corr function of MATLAB. The C# WPF is a visual program in windows that provides a convenient way to use the C-CorA method (Figure 2A). The program requires two paths for the input of data files, and the user can adjust the value of *k* used in the *k*-means clustering module of C-CorA. The output includes results of clustering and correlation coefficient calculations. An example of the output file is shown in Figure 2B, which was generated by the C# WPF application of C-CorA using the RNA-seq data of loquat fruit as mentioned above.

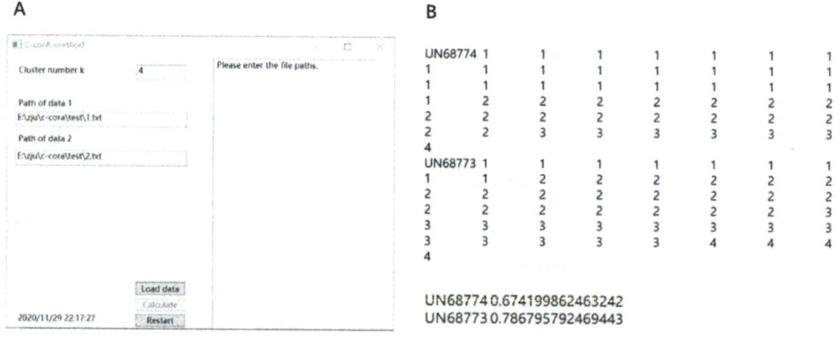

Figure 2. (**A**) The appearance of the C-CoA application written using C # WPF. (**B**) An example of an output file.

3.2. Transcriptome Analysis Using C-CorA Method

3.2.1. Clustering of Loquat Samples Based on Lignin Content

To get a more convincing 2-cluster pattern, the *k*-means method was used for the initial cluster. Setting the *k* parameter as 4, 48 loquat samples were randomly combined and clustered into four groups based on the lignin content (Figure 3). Then these clustered groups were randomly combined into seven 2-classes: ((1), (2, 3, 4)), ((2), (1, 3, 4)), ((3),

(1, 2, 4)), ((4), (1, 2, 3)), ((1, 2), (3, 4)), ((1, 3), (2, 4)) and ((1, 4), (2, 3)). For further correlation coefficient calculations, the lignin content value was replaced by the discrete variable 1 or 2 in each 2-class. For example, in ((1), (2, 3, 4)), the lignin content value was substituted as 1 in (1) and 2 in (2, 3, 4).

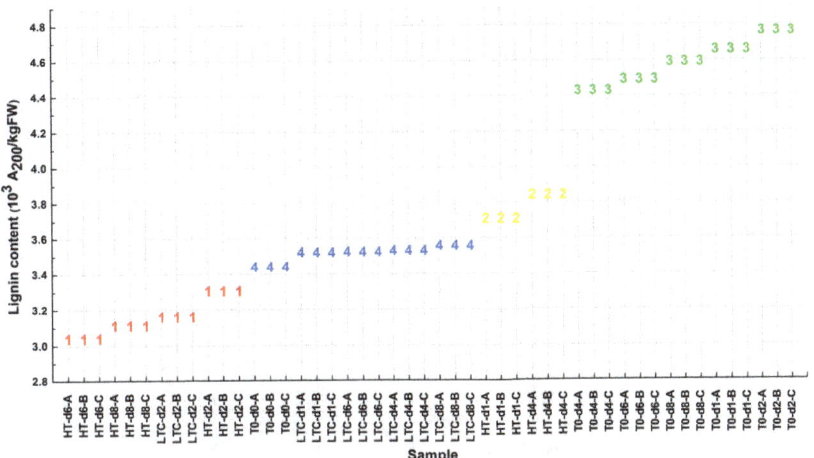

Figure 3. The *k*-means cluster result of the 48 loquat fruit samples from MATLAB. The number of groups was set to 4 in the *k*-means function. The data points are clustered based on the value of lignin content. (HT: heat treatment, LTC: Low temperature conditioning and T0: 0 °C).

3.2.2. Clustering of Gene Expression Data

The gene expression value was processed using the same cluster method as above. The *k* parameter was set to 4 as well. We divided the gene expression value into four groups and then randomly combined them into seven 2-classes for subsequent calculations. In each 2-class, the discrete variable 1 or 2 was substituted for the gene expression value, i.e., gene expression values were replaced as 1 in one class, and 2 in the other class.

3.2.3. Correlation Analysis of Gene Expression vs. Lignin Content

In correlation analysis, the strategy was to calculate all possible phenotype vs. gene expression combinations and then identify the most reliable pair. In this case, seven 2-class lignin content groups and seven 2-class gene expression groups were used for a Pearson correlation coefficient calculation. For each sample, there were seven patterns of phenotype and seven patterns of gene expression value. Be aware that these values only contain the discrete variables 1 and 2. Throughout the Pearson calculations for the seven groups of lignin content vs. the seven groups of gene expression, once the coefficient value appeared higher than the threshold given by the user, this gene was marked. The threshold was set as 0.7, 0.8 and 0.9. For each pattern of phenotype, the count of marked genes was summarized in Table 1. The 2-group pattern ((3), (1, 2, 4)) of phenotype had the most genes (Table 1). Moreover, the separation of group (3) and group (1, 2, 4) was consistent with the lignin content distribution for the 48 loquat fruit samples (Figure 3), thus, we used ((3), (1, 2, 4)) for further analysis.

For the 2-group pattern of phenotype ((3), (1, 2, 4)), 53 of the 71 highly correlated genes (correlation coefficient greater than 0.8) were well annotated (Table S1). The count of annotated genes with a correlation coefficient between 0.7 and 0.8 was 307. The correlated genes were more numerous than reported in our previous study [12]. The expression patterns of the 53 highly correlated genes are shown in Figure 4.

Table 1. The count of genes which highly correlated to lignin content in each of the seven 2-group patterns of phenotype data. The threshold of correlation coefficient was set as 0.7, 0.8 and 0.9.

2-Group Pattern	Correlation Coefficient		
	0.7	0.8	0.9
[{1}, {2, 3, 4}]	20	1	0
[{4}, {1, 2, 3}]	445	76	3
[{2}, {1, 3, 4}]	51	9	1
[{3}, {1, 2, 4}]	472	71	10
[{1, 4}, {2, 3}]	230	9	0
[{1, 2}, {3, 4}]	93	28	0
[{1, 3}, {2, 4}]	40	2	0

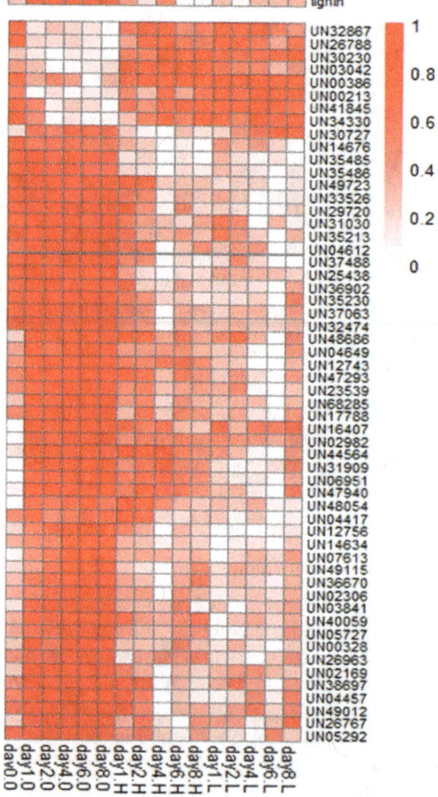

Figure 4. A heatmap of normalized gene expression values of the 53 highly correlated genes in gene expression vs. lignin content comparison, as calculated by the C-CorA method.

To assess the validity and accuracy of the C-CorA method, we compared the correlation coefficients calculated by C-CorA to those of the Pearson method and the Spearman method; the results are summarized in Figure 5. Genes highly related to the lignin content which were detected by the Pearson method and the Spearman method are all included in the result from C-CorA (Figure 5A). The C-CorA method identified a larger number of candidate genes, and the correlation coefficient values showed better discrimination. Using the Pearson or Spearman methods, 22 genes with correlation coefficients greater than 0.7 were screened (Figure 5B). For these 22 genes, the correlation coefficients calculated using C-CorA were consistent with at least one of the traditional methods, except for two genes

(UN68285 and UN35236). The correlation coefficients for these two genes calculated by the three methods are different. Comparison of the RPKM values for these two genes showed that the expression of UN68285 and UN35236 in all 48 samples (Figure 5C) was similar to trends in the lignin content, suggesting that these two genes are related to lignin content.

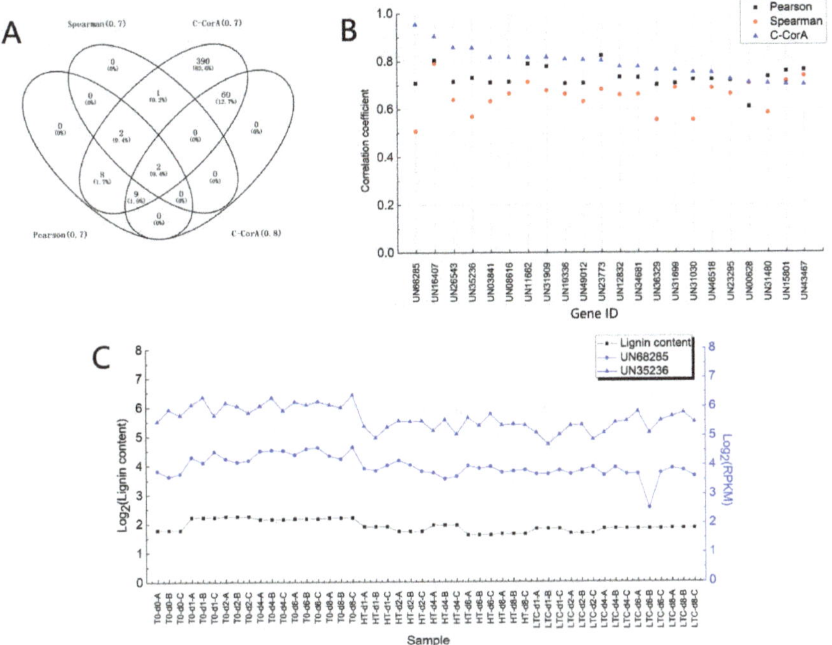

Figure 5. (**A**) A Venn diagram of the count of genes with high correlation coefficients in gene expression vs. lignin content comparison as calculated by the Pearson method, the Spearman method and the C-CorA method. (**B**) The correlation coefficients for 22 genes screened by the Pearson method or the Spearman method (>0.7) and the coefficients from the C-CorA method. (**C**) The base-2 logarithmic values of the RPKM values of genes UN68285 and UN35236.

In our previous study, we constructed co-expression networks between genes based on the Pearson product-moment correlation coefficient to determine lignin-related genes. Then *EjCAD3* (*UN68301*) and other genes were identified as candidate genes [12]. Using the C-CorA analysis, *EjCAD3* (*UN68301*) also showed a high correlation signal (Table S1). Other genes identified by C-CoA included a CBL-interacting protein kinase 08 (*UN00386*), an RNA binding protein (*UN30230*), a gibberellin 3-beta hydroxylase (*UN68191*) and so on, which was also consistent with the former work [12].

Among the newly identified genes were two lignin-related genes, encoding a CAD (cinnamyl alcohol dehydrogenase) (*UN26767*, *EjCAD4*) and a MYB transcription factor (*UN00328*). These two genes were not identified in the previous work [12]. The expression levels of *EjCAD4* and MYB (UN00328) decreased in LTC and HT treated samples compared with control samples (Figure 6), and they were positively correlated to the lignin content (the correlation coefficients with lignin content were 0.86 and 0.87, respectively). CAD is an enzyme that participates in the last step of monolignol biosynthesis. *AtCAD4* (AT3G19450), the homologous gene *EjCAD4* in Arabidopsis is involved in lignin biosynthesis and reported to act as an essential component in pathogen defense [19,20]. Moreover, the homologous gene *EjCAD4* in Chinese White Poplar (*Populus tomentosa* Carr.), JX986606.1, has also been identified as a candidate gene related to lignification by SSR markers [21].

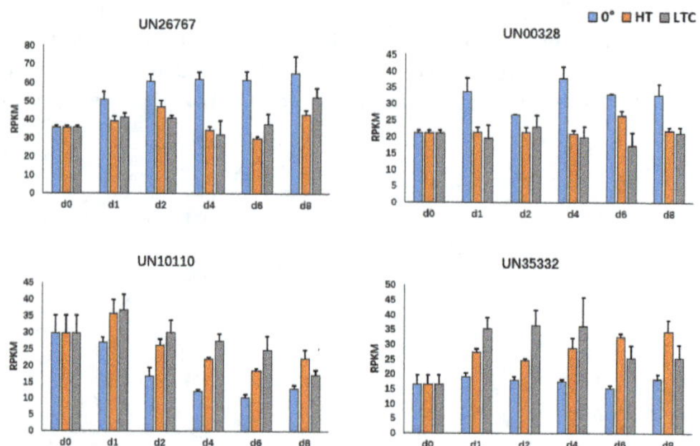

Figure 6. The RPKM of UN26767, UN00328, WRKY (UN10110) and BHLH (UN35332) at each time point under LTC, HT and 0 °C treatments from the RNA-Seq analysis of loquat fruit.

The expression of other transcription factors was also identified as being correlated with lignin content, including ERF (*UN41017*), BHLH (*UN35332*), MYB (*UN31768*, *UN02037*), WRKY (*UN10110* and *UN23662*), NAC (*UN48890*) and so on (Table S1).

3.2.4. Correlation Analysis for Newly Identified Candidate Genes

To better understand the regulatory network of genes newly identified by the C-CorA method above, a correlation analysis of gene vs. gene based on their expression values was performed. The C-CorA method was used to detect the genes correlated with *EjCAD4* (*UN26767*), and a total of 65 correlated genes were identified (Table S2). Interestingly, two transcription factors, WRKY (*UN10110*) and BHLH (*UN35332*), showed the opposite expression patterns when compared to *EjCAD4* (Figure 6), which suggested that they might be negative regulators of *EjCAD4*.

Multiple genes are involved in lignin accumulation. We used the *EjCAD4* as an example to perform a correlation analysis of multi-gene vs. lignin content. When searching for genes that act synergistically with *EjCAD4* to influence the fruit lignin content, we applied the *k*-means clustering method to a 2×48 matrix formed by the expression values of two genes, one of which was *EjCAD4*. The clustering results were then used as the input vector for the correlation analysis together with the lignin content. Based on the results produced by C-CorA, we obtained 38 annotated genes with a correlation coefficient cutoff value of 0.8. Six calcium-correlated genes were identified, including *EjCAD3* (Table S3). Cold shock elicits an immediate rise in cytosolic free calcium concentration [22] and can activate the expression of lignin-associated genes [23,24]. As we know, Ca^{2+} treatment could help to maintain fruit firmness via forming an "egg-box" [25]. Whether calcium-correlated genes interplay with loquat fruit lignification during postharvest and how they synergistically work with lignin-related gene *EjCAD4* requires further research.

3.3. Other Cases of Correlation Analysis Using C-CorA

The C-CorA method could be applied to any multi-dimensional data. It could detect more candidate genes in a correlation analysis of RNA-Seq. Two more cases are presented here.

3.3.1. Transcriptome Analysis in Kiwifruit Using C-CorA

In a recent study of fruit softening [26], the mechanism of postharvest cell wall metabolism was explored in kiwifruit. Six cell wall metabolism-related genes (*AdGAL1*, *AdMAN1*, *AdPL1*, *AdPL5*, *Adβ-Gal5* and *AdPME1*) were identified as candidate genes for pectin degradation by a correlation-based analysis as reported in Zhang et al. [26]. The correlation coefficients in this study were generated by the Pearson method. We applied the C-CorA method on the data provided by the author, and ten pectin degradation-related structural genes were highly related to the physiological traits. These ten genes included the six candidate genes mentioned above and four more genes (*Achn351951*, *Achn106231*, *Achn381701* and *Achn064441*). Among these four genes, two genes (*Achn351951* and *Achn106231*) were annotated as pectate lyase while the other two (*Achn381701* and *Achn064441*) were annotated as pectinesterase. Both pectate lyase and pectinesterase play an important role in pectin degradation [27]. Figure 7 illustrates the gene expression trends of these four genes in the firmness and the cell wall material pattern of kiwifruits. The pattern of physiological traits and the FPKM of these four genes showed the same or opposite trends during the treatment of kiwi fruit, which indicated that specific genes detected by C-CorA should also be considered as candidate genes in cell wall metabolism.

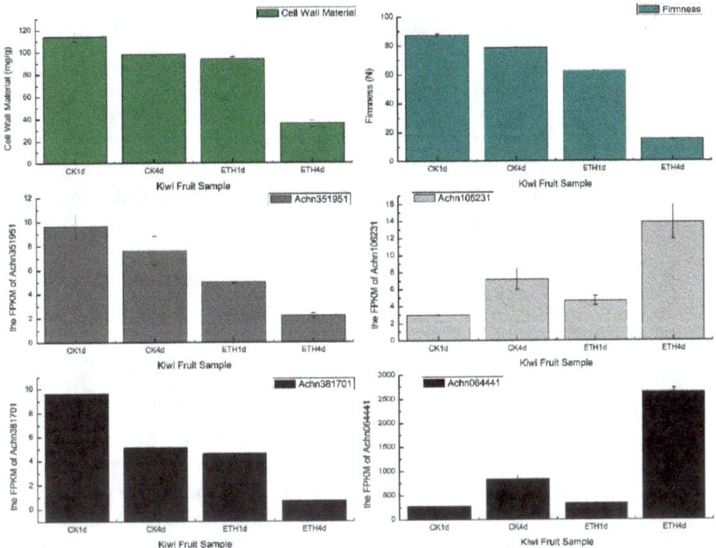

Figure 7. The cell wall material content (mg/g), firmness (N) and FPKM value of four new candidate genes (*Achn351951*, *Achn106231*, *Achn381701* and *Achn064441*) detected by C-CorA in four groups of kiwi fruit.

3.3.2. Transcriptome Analysis in Persimmon Using C-CorA

Acetaldehyde is a compound that could precipitate soluble condensed tannins into insoluble condensed tannins, which is an important process for persimmon fruit flavor [28]. The production of acetaldehyde depends on two key enzymes, pyruvate decarboxylase (PDC) and alcohol dehydrogenase (ADH) [28]. Three PDC genes (*EVM0028451*, *EVM0027273* and *EVM0022732*) and two ADH genes (*EVM0007501* and *EVM0027066*) were detected in the highest correlated module by a weighted gene co-expression network analysis (WGCNA) as reported in Kou et al [29]. When we applied the C-CorA method on the same dataset as Kou et al. [29], two more genes (PDC, *EVM0018709* and ADH, *EVM0007329*) were detected. The gene expressions and ethanol/acetaldehyde production are shown in Figure 8. The expression patterns of the two genes were consistent with ethanol and acetaldehyde content. The new genes found by C-CorA are worth further research.

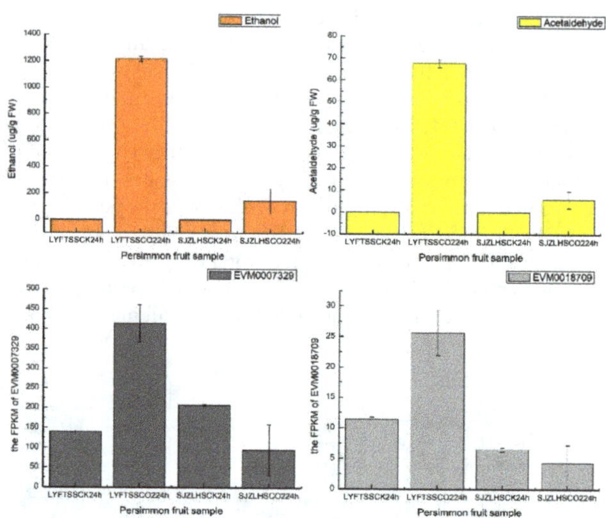

Figure 8. The ethanol production (μg/g FW) and acetaldehyde production (μg/g FW) of four persimmon fruit groups, along with the FPKM value of the two newly detected genes (*EVM0007329* and *EVM0018709*) detected by C-CorA.

4. Discussion

In this work, we used a cluster-based correlation analysis method (C-CorA) to analyze the loquat RNA-Seq and identified some additional lignification-related transcripts which were not detected in the previous study [12]. In the C-CorA workflow, the cluster step and correlation analysis step are treated as two independent modules. In the cluster step, many alternative methods could be used rather than the *k*-means method, depending on the features of the input data. An advantage of using the clustering result for the correlation coefficient calculations is that it can process various types of data in different dimensions. For the correlation analysis of RNA-Seq data, C-CorA can deal with different cases, including gene vs. gene/phenotype, multi-gene vs. gene/phenotype and multi-phenotype vs. gene. The correlation coefficient calculated by the C-CorA method can also be used in other correlation-based analyses, such as the WGCNA [3].

The *k*-means method has some limitations. The number of groups must be determined in advance. In this work, the parameter *k* was set to 4. By choosing a relative value of *k* and then combining these groups into 2-group patterns, the potential correct cluster will be included. Also, the clustering result given by the *k*-mean method is sometimes not unique. To overcome this deficiency, multiple rounds of calculations should be performed for each data set.

For RNA-Seq analysis, alignment-free methods are quickly developed. Kallisto [30] and Sailfish [31] use the unique *k*mer from each transcript to calculate the read count for the calculation of abundance. The C-CorA method can also use these *k*mer counts to do the correlation analysis. The *k*mer distribution can describe the difference between transcripts or genome sequences, meaning that the C-CorA method can be extended to other applicable sequence-based analyses. For data from hybrid population analyses, the cluster method matches the phenomena of segregation of character. The C-CorA method has the potential to be applied to these analyses.

According to previous work [32,33], CAD (cinnamyl alcohol dehydrogenase) is an important enzyme that catalyzes the final step in the biosynthesis of lignin precursors, and MYB transcription factors [9,10] have been identified that are involved in loquat fruit chilling lignification by regulating lignin biosynthetic genes. By performing a C-CorA analysis of gene expression vs. lignin content, we identified *EjCAD3* (*UN68301*) and

EjCAD4 (*UN26767*) as candidate genes. In addition, a MYB transcription factor (*UN00328*) was highly correlated to the lignin content, which suggests that it could be a positive regulator of lignin biosynthesis. Using the same approach on gene vs. gene, we found two transcription factors, BHLH (*UN35332*) and WRKY (*UN10110*), that might work as negative regulators of *EjCAD4*. For the correlation analysis of multi-gene vs. lignin content, we identified six calcium correlated genes that may cooperate with *EjCAD4* in loquat fruit lignification.

Several differentially expressed genes previously identified [12] as candidate genes, such as *EjCAD3* (*UN68301*), also showed a high correlation with the lignin content using the C-CorA method. This indicates that the C-CorA method is an effective method for RNA-Seq data analysis. Furthermore, the correlation analysis of gene vs. gene and multi-gene vs. phenotype performed by C-CorA can help to build a network of the target genes.

5. Conclusions

The cluster-based correlation analysis method described in this work, C-CorA, is a new correlation coefficient calculation algorithm. It uses the clustering result from the original data sets to do the correlation analysis. It can deal with various types of data and data in different dimensions. This method was applied to a set of RNA-Seq data of loquat fruit. It identified the fruit lignification correlated structural genes and transcription factors as candidate genes for further research. Using this analysis, we also identified two additional genes (*EjCAD4* and MYB), which were not detected in our previous work, as candidates for involvement in the lignification process. The loquat fruit case indicates that compared to other correlation methods, the C-CorA may be a better choice for fruit quality studies.

Supplementary Materials: The following are available online at https://www.mdpi.com/article/10.3390/horticulturae8020124/s1: Table S1: genes correlated with the lignin content of loquat fruit during postharvest treatments (0 oC, LTC, HT) from C-CorA, showing their correlation coefficients and annotations; Table S2: genes correlated with the EjCAD4 (UN26767) from C-CorA.; Table S3: genes act synergistically with EjCAD4 on the fruit lignin content identified by C-CorA.

Author Contributions: Conception and design: K.C., Q.W., W.L. and W.C. Algorithm design and implementation: W.L. Data analysis and interpretation: J.Q., W.L., W.C., Y.S. and M.Z. Manuscript writing: J.Q. and W.L. All authors have read and agreed to the published version of the manuscript.

Funding: This research was supported by grants from the National Natural Science Foundation of China (11771393, 11632015 and 31630067) and the Natural Science Foundation of Zhejiang Province, China (LZ14A010002). The funding bodies did not play any role in the design of the study, data collection and analysis or preparation of the manuscript.

Data Availability Statement: The raw RNA-Seq data used in this work was submitted to the NCBI Sequence Read Archive (SRA) under the accession SRP128075.

Acknowledgments: We thank Don Grierson for helpful suggestions and comments on the manuscript.

Conflicts of Interest: The authors declare no conflict of interest.

References

1. Langmead, B.; Trapnell, C.; Pop, M.; Salzberg, S.L. Ultrafast and memory-efficient alignment of short DNA sequences to the human genome. *Genome Biol.* **2009**, *10*, 1–10. [CrossRef] [PubMed]
2. Robinson, M.D.; McCarthy, D.J.; Smyth, G.K. edgeR: A Bioconductor package for differential expression analysis of digital gene expression data. *Bioinformatics* **2010**, *26*, 139–140. [CrossRef] [PubMed]
3. Langfelder, P.; Horvath, S. WGCNA: An R package for weighted correlation network analysis. *BMC Bioinform.* **2008**, *9*, 1–13. [CrossRef] [PubMed]
4. Zhang, B.; Horvath, S. A general framework for weighted gene co-expression network analysis. *Stat. Appl. Genet. Mol. Biol.* **2005**, *4*, 17. [CrossRef] [PubMed]
5. Li, H.; Sun, Y.; Zhan, M. Exploring pathways from gene co-expression to network dynamics. In *Computational Systems Biology. Methods in Molecular Biology*; Ireton, R., Montgomery, K., Bumgarner, R., Samudrala, R., McDermott, J., Eds.; Humana Press: Totowa, NJ, USA, 2009; pp. 249–267.

6.   Sax, K.; Gowen, J.W. Permanence of tree performance in a clonal variety and a critique of the theory of bud mutation. *Genetics* **1923**, *8*, 179. [CrossRef]

7.   Cai, C.; Chen, K.; Xu, W.; Zhang, W.; Li, X.; Ferguson, I. Effect of 1-MCP on postharvest quality of loquat fruit. *Postharvest Biol. Technol.* **2006**, *40*, 155–162. [CrossRef]

8.   Cai, C.; Xu, C.; Li, X.; Ferguson, I.; Chen, K. Accumulation of lignin in relation to change in activities of lignification enzymes in loquat fruit flesh after harvest. *Postharvest Biol. Technol.* **2006**, *40*, 163–169. [CrossRef]

9.   Xu, Q.; Yin, X.R.; Zeng, J.K.; Ge, H.; Song, M.; Xu, C.J.; Chen, K.S. Activator-and repressor-type MYB transcription factors are involved in chilling injury induced flesh lignification in loquat via their interactions with the phenylpropanoid pathway. *J. Exp. Bot.* **2014**, *65*, 4349–4359. [CrossRef]

10.   Wang, W.Q.; Zhang, J.; Ge, H.; Li, S.J.; Li, X.; Yin, X.R.; Chen, K.S. EjMYB8 transcriptionally regulates flesh lignification in loquat fruit. *PLoS ONE* **2016**, *11*, e0154399. [CrossRef]

11.   Zeng, J.K.; Li, X.; Xu, Q.; Chen, J.Y.; Yin, X.R.; Ferguson, I.B.; Chen, K.S. EjAP2-1, an AP 2/ERF gene, is a novel regulator of fruit lignification induced by chilling injury, via interaction with Ej MYB transcription factors. *Plant Biotechnol. J.* **2015**, *13*, 1325–1334. [CrossRef]

12.   Liu, W.; Zhang, J.; Jiao, C.; Yin, X.; Fei, Z.; Wu, Q.; Chen, K. Transcriptome analysis provides insights into the regulation of metabolic processes during postharvest cold storage of loquat (*Eriobotrya japonica*) fruit. *Hortic. Res.* **2019**, *6*, 1–11. [CrossRef] [PubMed]

13.   Shan, L.L.; Li, X.; Wang, P.; Cai, C.; Zhang, B.; De Sun, C.; Chen, K.S. Characterization of cDNAs associated with lignification and their expression profiles in loquat fruit with different lignin accumulation. *Planta* **2008**, *227*, 1243–1254. [CrossRef] [PubMed]

14.   Zhong, S.; Joung, J.G.; Zheng, Y.; Chen, Y.R.; Liu, B.; Shao, Y.; Giovannoni, J.J. High-throughput illumina strand-specific RNA sequencing library preparation. *Cold Spring Harbor Protoc.* **2011**, *2011*, 940. [CrossRef]

15.   Bolger, A.M.; Lohse, M.; Usadel, B. Trimmomatic: A flexible trimmer for Illumina sequence data. *Bioinformatics* **2014**, *30*, 2114–2120. [CrossRef] [PubMed]

16.   Quast, C.; Pruesse, E.; Yilmaz, P.; Gerken, J.; Schweer, T.; Yarza, P.; Glöckner, F.O. The SILVA ribosomal RNA gene database project: Improved data processing and web-based tools. *Nucleic Acids Res.* **2012**, *41*, D590–D596. [CrossRef] [PubMed]

17.   Grabherr, M.G.; Haas, B.J.; Yassour, M.; Levin, J.Z.; Thompson, D.A.; Amit, I.; Regev, A. Full-length transcriptome assembly from RNA-Seq data without a reference genome. *Nat. Biotechnol.* **2011**, *29*, 644–652. [CrossRef] [PubMed]

18.   Zheng, Y.; Zhao, L.; Gao, J.; Fei, Z. iAssembler: A package for *de novo* assembly of Roche-454/Sanger transcriptome sequences. *BMC Bioinform.* **2011**, *12*, 1–8. [CrossRef] [PubMed]

19.   Lee, S.; Mo, H.; Im Kim, J.; Chapple, C. Genetic engineering of *Arabidopsis* to overproduce disinapoyl esters, potential lignin modification molecules. *Biotechnol. Biofuels* **2017**, *10*, 1–13. [CrossRef] [PubMed]

20.   Tronchet, M.; Balague, C.; Kroj, T.; Jouanin, L.; Roby, D. Cinnamyl alcohol dehydrogenases-C and D, key enzymes in lignin biosynthesis, play an essential role in disease resistance in *Arabidopsis*. *Mol. Plant Pathol.* **2010**, *11*, 83–92. [CrossRef]

21.   Du, Q.; Gong, C.; Pan, W.; Zhang, D. Development and application of microsatellites in candidate genes related to wood properties in the Chinese white poplar (*Populus tomentosa* Carr.). *DNA Res.* **2013**, *20*, 31–44. [CrossRef]

22.   Knight, H.; Trewavas, A.J.; Knight, M.R. Cold calcium signaling in *Arabidopsis* involves two cellular pools and a change in calcium signature after acclimation. *Plant Cell* **1996**, *8*, 489–503. [PubMed]

23.   Lu, G.; Li, Z.; Zhang, X.; Wang, R.; Yang, S. Expression analysis of lignin-associated genes in hard end pear (*Pyrus pyrifolia* Whangkeumbae) and its response to calcium chloride treatment conditions. *J. Plant Growth Regul.* **2015**, *34*, 251–262. [CrossRef]

24.   Figueroa, C.R.; Opazo, M.C.; Vera, P.; Arriagada, O.; Díaz, M.; Moya-León, M.A. Effect of postharvest treatment of calcium and auxin on cell wall composition and expression of cell wall-modifying genes in the Chilean strawberry (*Fragaria chiloensis*) fruit. *Food Chem.* **2012**, *132*, 2014–2022. [CrossRef]

25.   Braccini, I.; Pérez, S. Molecular basis of Ca^{2+}-induced gelation in alginates and pectins: The egg-box model revisited. *Biomacromolecules* **2001**, *2*, 1089–1096. [CrossRef] [PubMed]

26.   Zhang, Q.Y.; Ge, J.; Liu, X.C.; Wang, W.Q.; Liu, X.F.; Yin, X.R. Consensus Co-expression network analysis identifies AdZAT5 regulating pectin degradation in ripening kiwifruit. *J. Adv. Res.* **2021**; *in press*. [CrossRef]

27.   Anderson, C.T. We be jammin': An update on pectin biosynthesis, trafficking and dynamics. *J. Exp. Bot.* **2015**, *67*, erv501. [CrossRef]

28.   Min, T.; Yin, X.R.; Shi, Y.N.; Luo, Z.R.; Yao, Y.C.; Grierson, D.; Ferguson, I.B.; Chen, K.S. Ethylene-responsive transcription factors interact with promoters of ADH and PDC involved in persimmon (*Diospyros kaki*) fruit de-astringency. *J. Exp. Bot.* **2012**, *63*, 6393–6405. [CrossRef]

29.   Kou, S.M.; Jin, R.; Wu, Y.Y.; Huang, J.W.; Zhang, Q.Y.; Sun, N.J.; Yang, Y.; Guan, C.F.; Wang, W.Q.; Zhu, C.Q.; et al. Transcriptome analysis revealed the roles of carbohydrate metabolism on differential acetaldehyde production capacity in persimmon fruit in response to high-CO_2 treatment. *J. Agric. Food Chem.* **2021**, *69*, 836–845. [CrossRef]

30.   Nicolas, B.; Harold, P.; Páll, M.; Lior, P. Near-optimal RNA-Seq quantification. *Nat. Biotechnol.* **2015**, *34*, 525–527.

31. Patro, R.; Mount, S.M.; Kingsford, C. Sailfish enables alignment-free isoform quantification from RNA-seq reads using lightweight algorithms. *Nat. Biotechnol.* **2014**, *32*, 462–464. [CrossRef]
32. Salentijn, E.M.; Aharoni, A.; Schaart, J.G.; Boone, M.J.; Krens, F.A. Differential gene expression analysis of strawberry cultivars that differ in fruit-firmness. *Physiol. Plant.* **2003**, *118*, 571–578. [CrossRef]
33. Xu, M.; Zhang, M.X.; Shi, Y.N.; Liu, X.F.; Li, X.; Grierson, D.; Chen, K.S. EjHAT1 participates in heat alleviation of loquat fruit lignification by suppressing the promoter activity of key lignin monomer synthesis gene EjCAD5. *J. Agric. Food Chem.* **2019**, *67*, 5204–5211. [CrossRef] [PubMed]

 horticulturae

Article

QTL Mapping of Resistance to Bacterial Wilt in Pepper Plants (*Capsicum annuum*) Using Genotyping-by-Sequencing (GBS)

Soo-Young Chae [1,†], Kwanuk Lee [2,†], Jae-Wahng Do [1], Sun-Cheul Hong [1], Kang-Hyun Lee [1], Myeong-Cheoul Cho [2], Eun-Young Yang [2,*] and Jae-Bok Yoon [1,*]

[1] Pepper and Breeding Institute, K-Seed Valley, Gimje 54324, Korea; cotez@naver.com (S.-Y.C.); wahng0@hanmail.net (J.-W.D.); imp2049@naver.com (S.-C.H.); kang3410@gmail.com (K.-H.L.)

[2] National Institute of Horticultural & Herbal Science, Rural Development Administration, Wanju 55365, Korea; kwanuk01@korea.kr (K.L.); chomc@korea.kr (M.-C.C.)

[*] Correspondence: yangyang2@korea.kr (E.-Y.Y.); yoonjb2@snu.ac.kr (J.-B.Y.); Tel.: +82-(0)63-238-6613 (E.-Y.Y.); +82-(0)63-543-0586 (J.-B.Y.); Fax: +82-(0)63-238-6605 (E.-Y.Y.); +82-(0)63-543-0580 (J.-B.Y.)

[†] These authors contributed equally to this work.

Citation: Chae, S.-Y.; Lee, K.; Do, J.-W.; Hong, S.-C.; Lee, K.-H.; Cho, M.-C.; Yang, E.-Y.; Yoon, J.-B. QTL Mapping of Resistance to Bacterial Wilt in Pepper Plants (*Capsicum annuum*) Using Genotyping-by-Sequencing (GBS). *Horticulturae* **2022**, *8*, 115. https://doi.org/10.3390/horticulturae8020115

Academic Editors: Fei Chen and Jia-Yu Xue

Received: 23 November 2021
Accepted: 25 January 2022
Published: 27 January 2022

Abstract: Bacterial wilt (BW) disease, which is caused by *Ralstonia solanacearum*, is one globally prevalent plant disease leading to significant losses of crop production and yield with the involvement of a diverse variety of monocot and dicot host plants. In particular, the BW of the soil-borne disease seriously influences solanaceous crops, including peppers (sweet and chili peppers), paprika, tomatoes, potatoes, and eggplants. Recent studies have explored genetic regions that are associated with BW resistance for pepper crops. However, owing to the complexity of BW resistance, the identification of the genomic regions controlling BW resistance is poorly understood and still remains to be unraveled in the pepper cultivars. In this study, we performed the quantitative trait loci (QTL) analysis to identify genomic loci and alleles, which play a critical role in the resistance to BW in pepper plants. The disease symptoms and resistance levels for BW were assessed by inoculation with *R. solanacearum*. Genotyping-by-sequencing (GBS) was utilized in 94 F_2 segregating populations originated from a cross between a resistant line, KC352, and a susceptible line, 14F6002-14. A total of 628,437 single-nucleotide polymorphism (SNP) was obtained, and a pepper genetic linkage map was constructed with putative 1550 SNP markers via the filtering criteria. The linkage map exhibited 16 linkage groups (LG) with a total linkage distance of 828.449 cM. Notably, QTL analysis with CIM (composite interval mapping) method uncovered *pBWR-1* QTL underlying on chromosome 01 and explained 20.13 to 25.16% by R^2 (proportion of explained phenotyphic variance by the QTL) values. These results will be valuable for developing SNP markers associated with BW-resistant QTLs as well as for developing elite BW-resistant cultivars in pepper breeding programs.

Keywords: pepper plant; *Ralstonia solanacearum*; single-nucleotide polymorphism (SNP); genotyping-by-sequencing (GBS); quantitative trait loci (QTL); pepper breeding

1. Introduction

Pepper plants (*Capsicum* spp.) derived from the regions of American tropics belong to the Capsicum genus and Solanaceae family, including peppers, paprika, tomatoes, eggplants, and potatoes. It is regarded as one of the most important vegetable crops worldwide owing to diverse positive aspects in field of cuisine, medicine and healthcare, and economy [1–3]. Pepper fruits are largely consumed as fresh and dried ingredients as well as processed foods and render a wide variety of essential bioactive elements, such as vitamins, minerals, phenolics, carotenoids, and capsaicinoids [4–8]. The consumption of pepper has been gradually increased for several decades, together with the cultivation area and production in agriculture (http://www.fao.org/faostat (accessed on 9 August 2021) [9]. On the basis of Food and Agriculture Organization (FAO) [9], the cultivation area has occupied around 4.5 million hectare, and pepper production has reached around 67 million tons, including fresh and dried peppers, in the world [10]. Moreover, in terms of the world trade value of crops, the amount of chili peppers has ranked with the second position after tomato plants among the Solanaceae family [11]. However, pepper production is naturally threatened by biotic factors, including bacteria, fungi, and viruses in the agronomic field [12]. In particular, bacterial wilt (BW) is one of most destructive diseases throughout the world and has been widely spread in pepper crops across all over the Asia [13–15]. In 2017, BW led to a significant reduction of pepper yields and productions ranging from over 20 to less than 50% at most in the world cultivation area [16].

BW disease is caused by a soil-borne bacterial pathogen, *R. solanacearum,* and it is one of the global plant diseases [17]. The BW is seriously harmful for a large amount of the solanaceous family, including the vegetable crops of chili and sweet peppers, paprika, tomatoes, potatoes, and eggplants, which cause a plant-wilting disease. It is recognized as a wide range of hosts by invading over 450 different plant species through the broad climate spectrums containing tropical, subtropical, and temperate regions [18–20]. The pathogen of *R. solanacearum* enters plants via the natural opening and the wounded layers at the emergence sites of secondary roots or at root tips, which immigrate and colonize the host root cortex [21,22]. Subsequently, the *R. solanacearum* infects the parenchyma of the plant vascular system. The success of the invasion into xylem causes plant pathogenicity, including high population of the increased bacterial cells, diverse enzymes, and viscous di-and poly-saccharides by *R. solanacearum.* As such, the xylem vessels in plant roots are filled and blocked by the pathogenicity [23,24]. Interestingly, several reports have studied that the resistant cultivars possess an ability to restore the xylem transport system directly after the bacterial attacks, whereas the susceptible cultivars are observed with the xylem blocked by the occlusion derived from bacteria [25,26], leading to the damage of the water flow system inside the plant's xylem. The malfunction results from plant yellowing, wilting, and dying depending on the severity of the disease symptoms [25].

To date, multiple management methods to govern the BW disease have been developed and applied in agriculture. Indeed, the effects of combatting the devastating BW have been shown with agronomical, physical, chemical, and cultural methods, including the crop rotation of different types, utilization of bactericides, and plant breeding programs [26]. However, the management strategies have been reported with limited and insufficient effects on the BW disease regulation owing to the wide variety of host range, diverse genetic variations of the pathogen, and long-term survival in plants [26–28].

In general, it is considered that the most effective control method is to breed elite, resistant cultivars in the pepper crops against the BW [29]. Remarkably, a wide variety of BW resistant-pepper accessions have been determined. For example, the BW resistant-pepper accessions (*Capsicum* spp.) with LS2341, PI358812, Kerting, PI322726, PI322727, PI369998, PI377688, PI322728, Jatilaba, MC4, MC5, PBC 066, PBC 437, PBC 631, and PBC 1347 display high BW resistance against a wide array of BW pathogens [30–33]. In addition to this, some researches have studied that BW resistance is involved in a quantitative inheritance and is polygenically governed by multiple genes (≥ 2 genes) in the pepper cultivar Mie-Midori [29]. A pepper cultivar, PM687, has been determined to have additive effects, which are influenced by the involvement of 2 to 5 candidate genes to regulate the BW resistance [34]. Additionally, A pepper accession called LS2341 has been identified with polygenes and linkage to a major quantitative trait loci (QTL), *Bw1,* located on chromosome 08, which possesses putative 44 candidate-resistance genes against *R. solanacearum* [35]. A recent report has uncovered a marker ID10-194305124 on the major QTL *qRRs-10.1* on chromosome 10, which consists of five candidate R genes containing putative leucine-rich repeat (LRR) receptor or NB-ARC proteins and three defense-associated genes in the resistance pepper cultivar BVRC1 [36].

QTL mapping is a basic and powerful tool for genetic investigation of quantitative traits and high-density linked markers. Although the conventional QTL mapping is a time-consuming, labor-intensive, and costly procedure [37], advances in next-generation sequencing (NGS) technologies reduce sequencing cost and contribute to the rapid identification of QTL and the assessment of genome-wide single nucleotide polymorphism (SNP) in pepper crop [38]. The utilization of NGS for SNP discovery is beneficial, as it generates a large amount of sequence data that can be used for genotype-phenotype association [37,39,40]. Genotyping-by-sequencing (GBS) technology is a simple and robust method that is practical as a high-throughput genotyping tool for a large number and huge amount of DNA as well as for complex genomes of crop species [41,42]. Furthermore, GBS is a cost-effective technique with the advantages of reduced sampling time, decreased sequencing cost, and no limited reference genome sequences [37,43]. Recent studies have reported that the GBS tool is successfully applied to the various crops, including chickpea, maize, wheat, onion, soybean, pepper, and rice, for QTL mapping, high-density genetic mapping, SNP discovery, GWAS, genomic selection (GS), and genotyping [44–52]. Since the GBS tool possesses a wide applicability of QTL identification in pepper (*C. annuum*), genetic analysis of quantitative traits and high-resolution-linked markers for BW would contribute to more accurate marker-assisted selection (MAS) in plant genetics and breeding via GBS.

In this study, we produced 94 F_2 recombinant lines derived from a cross between a resistant source of *C. annuum*, KC352, and a susceptible source of *C. annuum*, 14F6002-14, for QTL mapping of BW resistance to *R. solanacearum* isolates. Next, we constructed a genetic map with 94 F_2 recombinant offspring. High-resolution SNP markers associated with BW resistance revealed novel QTL regions on chromosome 01 via GBS. The result will be utilized for developing SNP markers involved in BW resistance and for selecting and breeding elite BW-resistant pepper plants.

2. Materials and Methods

2.1. Plant Materials and Growth Conditions

The parental lines of *C. annuum* KC352 and 14F6002-14 were provided by pepper and breeding institute (Gimje, Korea), respectively. A total of 94 F_2 individual lines were derived from a cross between KC352 and 14F6002-14. The pepper plants used for the experiment were grown at National Institute of Horticultural and Herbal Science (NIHHS, Wanju, Korea, 35°83′ N, 127°03′ E) under glasshouse conditions where temperature was controlled to 26/18 °C (day/night), and relative humidity (RH) was within 60–70%. The soil in pots was prepared as previously described in [53]. In brief, the commercial media (Bio Sangto, Seoul, Korea) was composed of coco peat (47.2%), peat moss (35%), vermiculite (10.0%), zeolite (7%), dolomite (0.6%), humectant (0.006%), and fertilizers (0.194%, 270 mg kg^{-1} N, P, and K), respectively.

2.2. DNA Extraction

A total of 96 plant samples from 9–10 leaf stage of 4-week-old plants were ground with the TissueLyser II (Qiagen, Hilden, Germany), and genomic DNA (gDNA) from the samples was extracted using cetyl trimethyl ammonium bromide (CTAB) method as previously described in [54].

2.3. Disease Assay and Resistance Index Scoring

To evaluate resistance against *R. solanacearum*, resistance screening was conducted in 45–48 F_2 offspring derived from a cross between parental lines of KC352 and 14F6002-14 with three independent replicates (Supplementary Figure S1). A total of 94 F_2 lines were tested for the resistance screening and subsequent GBS analysis. To prepare the inoculum, the *R. solanacearum* of WR-1 strain isolates (race 1, biovar 3) were cultivated on NA medium and incubated at 28 °C for bacterial cell growth for 2–3 days. The plates were collected with distilled water to harvest the cells. The concentration was determined by measuring OD at 600 nm = 0.3, and the cell density was adjusted to approximately 10^7–10^8 cfu per mL before inoculation. Two parental lines and 94 F_2 offspring were inoculated at the 6–7 leaf stage onto the plant roots with 5 mL of the bacterial suspension after wounding the plant roots by stabbing a scalpel along with two sides at a soil depth of 1–2 cm. The inoculated plants were kept under vinyl-protected conditions at 25 to 30 °C, and disease resistance was continuously observed and recorded after inoculation. *C. annuum* KC352 was utilized as the resistant control, whereas *C. annuum* 14F6002-14 was utilized as the susceptible control to compare the severity of disease symptoms in F_2-segregating populations. The disease symptoms and disease resistance index were evaluated on the basis with disease scale of 0–4 as previously described in [25], where 0 = no visible symptoms, 1 = 1 to 25% of wilted leaves, 2 = 26 to 50% of wilted leaves, 3 = 51 to 75% of wilted leaves, and 4 = 76 to 100% of wilted leaves.

2.4. Preparation of Libraries for Genotype-by-Sequencing (GBS) Analysis

A total of 96 individuals were subjected to GBS analysis. The quantity and quality of extracted gDNAs were validated using 1% agarose gel electrophoresis before running next-generation sequencing (NGS). The preparation of GBS libraries was conducted as provided by SEEDERS sequencing company (Daejeon, Korea). To construct GBS libraries, gDNAs were digested with *ApeKI* (New England Biolabs, Ipswitch, MA, USA) [41] with a minor modification. In detail, oligonucleotides for the top and bottom strands of each barcode adapter and a common adapter were separately diluted in 50 µM TE buffer and annealed in thermocycler conditions followed with 95 °C, 2 min (ramp down to 25 °C by 0.1 °C/s; 25 °C, 30 min; 4 °C hold). Barcodes and common adapters were diluted in $10\times$ adapter buffer, including 500 mM NaCl and 100 mM Tris-Cl to 10 µM, and 2.4 µL of the mixture was applied into a 96-well PCR plate. Then, 100 ng/µL of DNA samples were added to individual adapter-containing wells and digested for overnight at 75 °C with 3.6 U ApeKI (New England Biolabs, Ipswitch, MA, USA) in 20-µL volumes. Adapters were then ligated to sticky ends by adding 30 µL of a mixture containing $10\times$ ligase buffer and 200 unit of T4 DNA ligase (MG Med, Seoul, Korea) to individual wells. The samples were incubated at 22 °C for 2 h and heated to 65 °C for 20 min to remove the activity of the T4 DNA ligase. The 96 digested DNA samples possessing a different barcode adapter were combined with each 5 µL and were purified using a purification kit (QIAquick PCR Purification Kit; Qiagen, Valencia, CA, USA) following the manufacturer's instructions. Restriction fragments from each library were then amplified in 50-µL volumes containing 2 µL pooled DNA fragments, Herculase II Fusion DNA Polymerase (Agilent, Santa Clara, CA, USA), and 25 pmol with each of the primers in [41]. Polymerase chain reaction (PCR) was conducted with the following conditions: one cycle at 95 °C for 2 min, 16 cycles at 95 °C for 30 s, 62 °C for 30 s, 68 °C for 30 s, and stopped at 68 °C for 5 min. The amplified fragment size and library quality were assessed with Agilent Tape station with high-sensitivity DNA chip. Whole-genome sequences were conducted using Illumina Hiseq X ten platform (Illumina, San Diego, CA, USA).

2.5. Sequencing, Alignment, and SNP Genotyping

Raw reads were de-multiplexed, and the barcode sequences were trimmed using SEEDERS in-house python script as previously described in [49]. Reads were also trimmed using the Cutadapt (ver. 1.8.3) to eliminate the sequences of the common adapter. The de-multiplexed reads were processed by the SolexaQA package ver.1.13 [41]. Next, bad-quality bases with low Phred quality score (Q = 20 or 0.05 probability of error) were trimmed using DynamicTrim in the SolexaQA package, and the read lengths lower than 25 bp with poor-quality sequence were discarded, using Lengthsort program in the SolexaQA package [55]. The processed and cleaned reads were applied to align with *C. annuum* cv. CM334 reference genome (ver. 1.55, http://www.sgn.cornell.edu/ (accessed on 19 November 2019), and the read depth was counted by the number of aligned reads via the Burrows–Wheeler Aligner (BWA, 0.6.1-r104) program as described in [56]. The BWA was carried out with following default options: gap open penalty ($-O$) = 15, number of threads ($-t$) = 16, mismatch penalty ($-M$) = 6, maximum differences in the seed ($-k$) = 1, and gap extension penalty ($-E$) = 8, except for seed length ($-l$) = 30. The detection of raw SNPs and the consensus sequences were acquired from the resulting mapped reads with BAM format file using

SAMtool (v.0.1.16) utilities [57]. For SNP calling, the varFilter command in the SAMtool was utilized with default options as previously described in [17,49]. Finally, on the basis of ratio of SNP/InDel reads in the mapped reads, variant types of SNP were grouped with three categories: homozygous SNP/InDel for read rate $\geq$90%, heterozygous SNP/InDel for read rate $\geq$40% and $\leq$60%, and the rest defined as "etc." [49,58,59].

2.6. Linkage Map Construction

Genetic linkage maps of F_2 segregating lines were illustrated using the JoinMap ver. 4.0 (Kyazma B.V., Wageningen, The Netherlands). A total of 1550 SNPs were grouped into 16 linkage groups (LGs) with a logarithm of the odds (LOD) threshold score $\geq$5.5, and a maximum distance of 30 centiMorgans (cM) were used. The genetic map distance of the SNP markers was converted to cM using the Kosambi's mapping function [60]. To remove the skewed SNP and the segregation distortion, the chi-square test ($p < 0.001$) was applied, and the SNP markers were filtered with identical segregation or missing rate $\geq$30%. Final genetic linkage maps of KC352 and 14F6002-14 with the 1550 SNPs were drawn using MapChart ver. 2.3 software [61].

2.7. QTL Analysis and Candidate Genes Prediction

QTL analysis was performed with composite interval mapping (CIM) to map the QTLs involving pepper bacterial wilt resistance using the Windows QTL Cartographer v. 2.5 program [62]. The CIM was operated at a 1.0-cM walk speed using the model 6 parameters (standard model) and the forward and backward regression model. The LOD threshold level for significance of each QTL was determined as 1000 permutations of $p < 0.05$. To identify candidate genes, the positions of highly significant QTLs regions on the genetic map were compared with their physical positions on the *C. annuum* cv. CM334 reference genome (ver. 1.55, http://www.sgn.cornell.edu/ (accessed on 19 November 2019), and 1 Mb left and right sequences were mined for candidate genes from respective corresponding marker. Putative functions of the candidate genes were further annotated with an application of sequence alignment using CM334 reference genome (ver. 1.55, https://solgenomics.net/ (accessed on 21 December 2021), Kyoto Encyclopedia of Genes and Genomes (KEGG) (https://www.genome.jp/kegg/ (accessed on 21 December 2021), SwissProt (https://www.uniprot.org/ (accessed on 21 December 2021), Gene Ontology (GO) (https://www.geneontology.org/ (accessed on 23 December 2021), and the NCBI non-redundant protein (NR) (https://ncbi.nih.gov/blast/db/ (accessed on 23 December 2021) with default values.

2.8. Data Analysis

The Tukey's HSD/Kramer test ($p < 0.05$) and the descriptive statistics of disease index were analyzed using SPSS program (IBM SPSS v27.0, Chicago, IL, USA). The Pearson's correlation coefficients were calculated using the R statistical Software (ver. 4.0.1, https://www.r-project.org (accessed on 3 January 2022)).

3. Results

3.1. Phenotyping of the Resistance for Bacterial Wilt Disease

In order to evaluate bacterial wilt (BW) resistance, we observed phenotypes for 5–21 days after inoculation of *R. solanacearum* suspension into parental lines and 94 F_2 lines. The degree of disease severity was evaluated as disease index (DI), with disease scale ranging from 0 (no symptoms) to 4 (76 to 100% wilted leaves) (Figure 1A). Among the 94 inoculated plants, 16 plants were observed with no visible symptoms (DI = 0), representing a resistant line to BW, whereas 49 plants were observed with 76–100% wilted leaves (DI = 4), representing a susceptible line to BW (Figure 1B). In addition to this, we observed 14 plants with 1 to 25% wilted leaves (DI = 1), 7 plants with 26 to 50% wilted leaves (DI = 2), and 8 plants with 51 to 75% wilted leaves (DI = 3) (Figure 1B). Overall, the average DI value of the F_2 population was 2.638, and the wilt rate (%) was 68.085. The skewness and kurtosis value of the DI was −0.604 and −1.352, respectively, suggesting that the resistance level of the plants to *R. solanacearum* is susceptible, and the population might be a non-normal distribution rather than a normal distribution (Table 1).

(A)

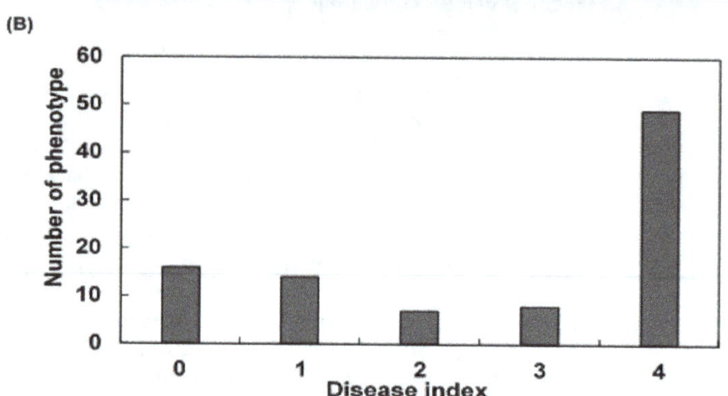

Figure 1. Distribution of the disease index after inoculation with *R. solanacearum*. (A) disease index was classified into disease scales of 0–4, where 0 = no visible symptoms, 1 = 1 to 25% wilted leaves, 2 = 26 to 50% wilted leaves, 3 = 51 to 75% wilted leaves, and 4 = 76 to 100% wilted leaves, and (B) disease index (DI) was recorded in F_2-segregating populations, including 94 genotypes from the cross between a resistant parent (KC352) and a susceptible parent (14F6002-14).

Table 1. Descriptive statistics of disease index.

Trait	F_2 Population (n = 94)								
	Max	Min	Range	Wilt Rate (%) [a]	Average [b]	Standard Error	Variation	Skewness	Kurtosis
Disease index	4	0	0–4	68.085	2.638	0.167	2.620	−0.604	−1.352

[a] Wilting rate was defined as DI that was more than 2. [b] Average was calculated on the basis of 0–4 rating scale as follows: DI = $\sum$(disease score x the number of plants corresponding to each disease score)/total number of plants.

3.2. GBS Analysis

Next, in order to conduct a genotyping-by-sequencing (GBS) analysis, all 96 plants were collected, and a construction of 96-plex GBS library was generated. As such, a total of approximately 108.0 Gbp of DNA sequences (715,257,004 reads) were obtained from single-lane sequencing using Illumina Hiseq X ten platform (Table S1). The GBS raw data ere de-multiplexed according to the 96 barcode sequences. The de-multiplexed sequences of the 96 samples were trimmed by eliminating the sequences of the barcode and adaptor and removing the low-quality information. Finally, the average number and total length of trimmed reads were 6,046,776 and 696 Mbp, respectively. In addition, the average length of trimmed reads (bp) and the total trimmed raw data were 119.99 bp and

94.45%, respectively (Table S1). The trimmed data were further mapped to the reference genome: *Capsicum. annuum* cv. CM334 ver. 1.55, sourced by Sol Genomics Network (http://www.sgn.cornell.edu/ (accessed on 19 November 2019). The average numbers of mapped reads and mapped regions were 5,216,671 and 125,718, respectively (Table S1). The average depth and length of the mapped regions were 14.75 and 233.34 bp, which covered 0.56% of the reference genome. To further mine SNPs from the sequence data, in-house GBS analysis pipeline was applied with filtering criteria. A total of 628,437 raw SNPs was identified in 94 F_2 lines (Table S2). The SNPs were filtered to identify putative markers using the criteria of 30% missing values across the genotyped individual and MAF $\geq$ 25%, which yielded a total of 146,217 SNPs. Moreover, 11,020 SNPs were filtered using both missing <30% and MAF > 25% condition. Furthermore, 4387 homozygous SNPs in KC352 were selected (Table S2). After generating a map of the genotyping and SNP selections using 500 kb as the window size, 1643 SNP markers were identified, and a total of 1639 SNPs were produced with polymorphic SNP between KC352 and 14F6002-14 as the parents (Table S2).

3.3. Construction of Linkage Mapping

In order to construct a pepper genetic linkage map, 1639 SNPs were utilized for linkage grouping using the SNP matrix from GBS analysis. As such, the linkage map consisted of 1550 SNP markers on 16 linkage groups (LG) (Table 2 and Figure 2). The linkage map covered a total length of 828.449 cM with an average distance of 0.676 cM between adjacent markers (Table 2 and Figure 2). Next, LG_04 showed the maximum lengths, which were 78.045 cM (chromosome 04) in the largest LG, whereas LG_05-2 showed the minimum lengths, which were 15.952 cM (chromosome 05) in the smallest LG (Table 2). The number of mapped SNPs per chromosome ranged from the minimum 11 in chromosome 10 to the maximum 172 in chromosome 03, with an average number of 96.875 SNP markers per LG (Table 2 and Figure 2). Moreover, the correlation coefficient between genetic and physical maps was estimated among the 16 linkage groups. Chr.07 exhibited the highest correlation coefficient of 0.876, and the average correlation coefficient was 0.670.

Table 2. Summary of the pepper genetic linkage map constructed using SNP markers derived from genotyping-by-sequencing (GBS) analysis for bacterial wilt (BW) disease resistance. The higher Pearson's correlation coefficient indicates the closer correlation.

Chromosome	Linkage Group	Number of SNP Marker	Genetic Distance (cM)	Correlation Coefficient
Chr.01	LG_01-1	123	82.134	0.755
	LG_01-2	49	38.254	0.385
Chr.02	LG_02-1	60	20.799	0.096
	LG_02-2	90	70.704	0.644
Chr.03	LG_03	172	59.561	0.790
Chr.04	LG_04	117	78.045	0.867
Chr.05	LG_05-1	91	47.582	0.840
	LG_05-2	12	15.952	0.437
Chr.06	LG_06	146	52.635	0.867
Chr.07	LG_07	136	61.809	0.876
Chr.08	LG_08	54	47.681	0.513
Chr.09	LG_09	114	74.982	0.786
Chr.10	LG_10-1	11	19.247	0.603
	LG_10-2	139	54.224	0.857
Chr.11	LG_11	123	59.579	0.867
Chr.12	LG_12	113	45.261	0.541
Total	16	1550	828.449	-
Average	-	96.875	51.778	0.670

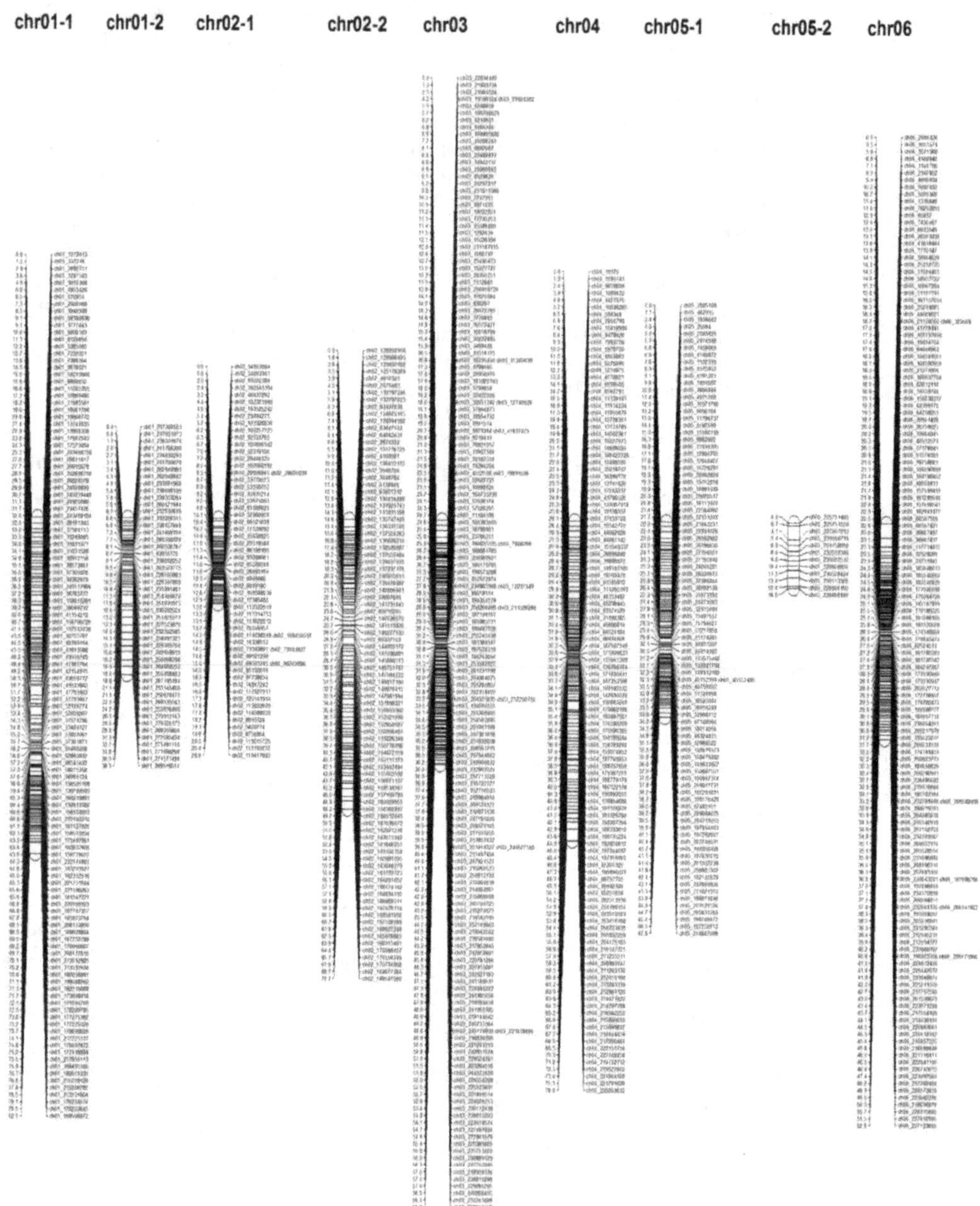

Figure 2. *Cont.*

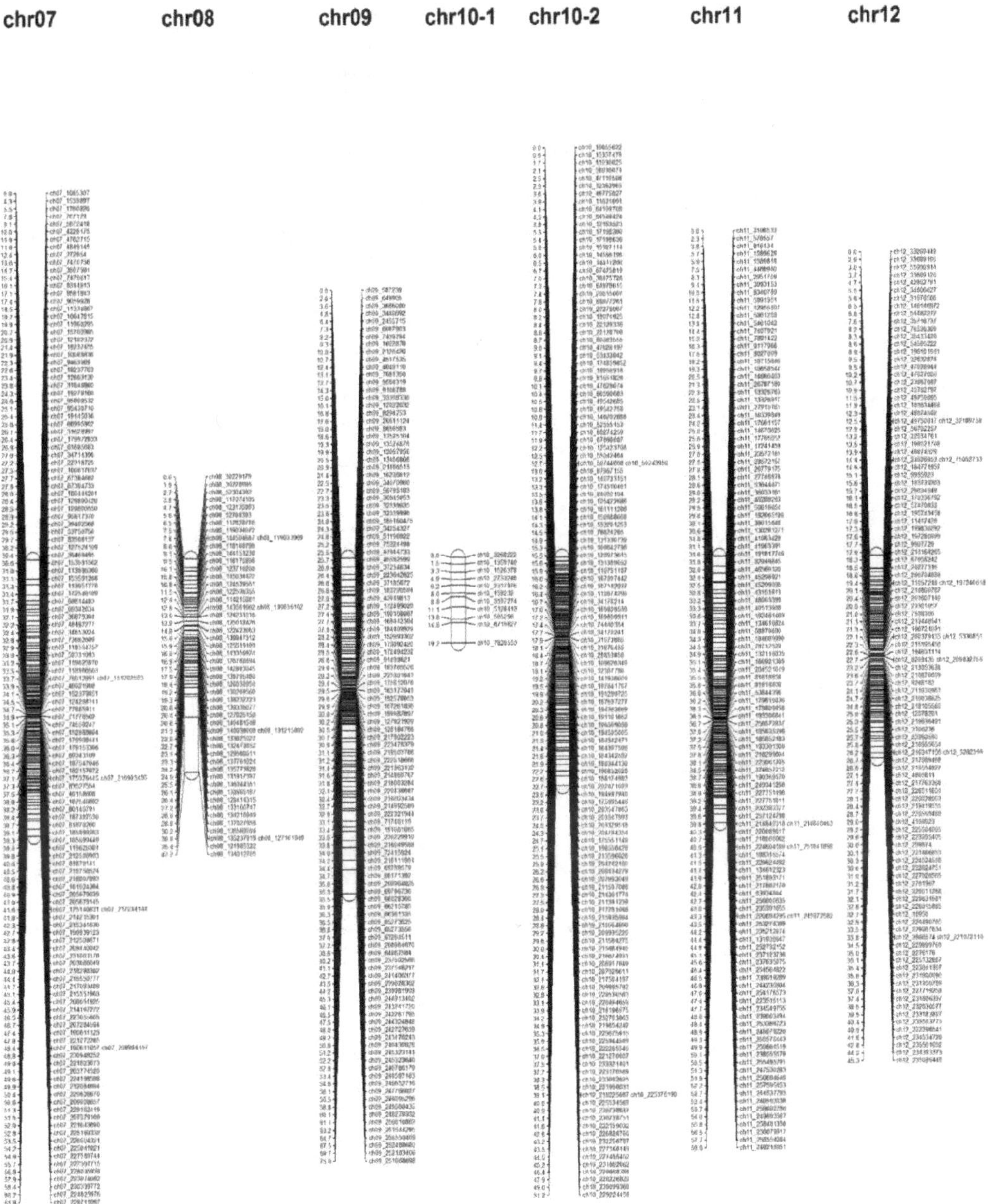

Figure 2. Distribution of single-nucleotide polymorphism (SNP) markers on 16 linkage groups of F_2 pepper population. The pepper genetic linkage map consisted of 1550 SNP markers derived from genotyping-by-sequencing (GBS) analysis. SNP names were shown on the right side of the linkage map and genetic distances (cM) between SNPs on the left.

3.4. QTL Analysis for Bacterial Wilt (BW) Resistance

In order to analyze significant quantitative trait loci (QTL) regions, the 1000 permutation was tested at $p < 0.05$, and the LOD threshold was calculated as 5.5 by QTL cartographer. Notably, the LOD values ranged from 5.69 to 7.06, which was detected in chromosome 01 via CIM (Figure 3A). On the basis of the threshold levels, we designed one QTL, *pBWR-1*, named the pepper Bacterial Wilt Resistance-1 on chromosomes 01 (Figure 3B). In addition, the significant LOD $\geq$ 5.5 regions identified at the *pBWR-1* QTL region were within LG_01-1 on chromosome 01. The *pBWR-1* QTL was located between the ch01_47793963 and ch01_161127926 markers on chromosome 01 with twelve markers (Figure 3B and Table 3). The explained phenotypic variance of the QTL was ranged from 20.13 to 25.16% (Table 3).

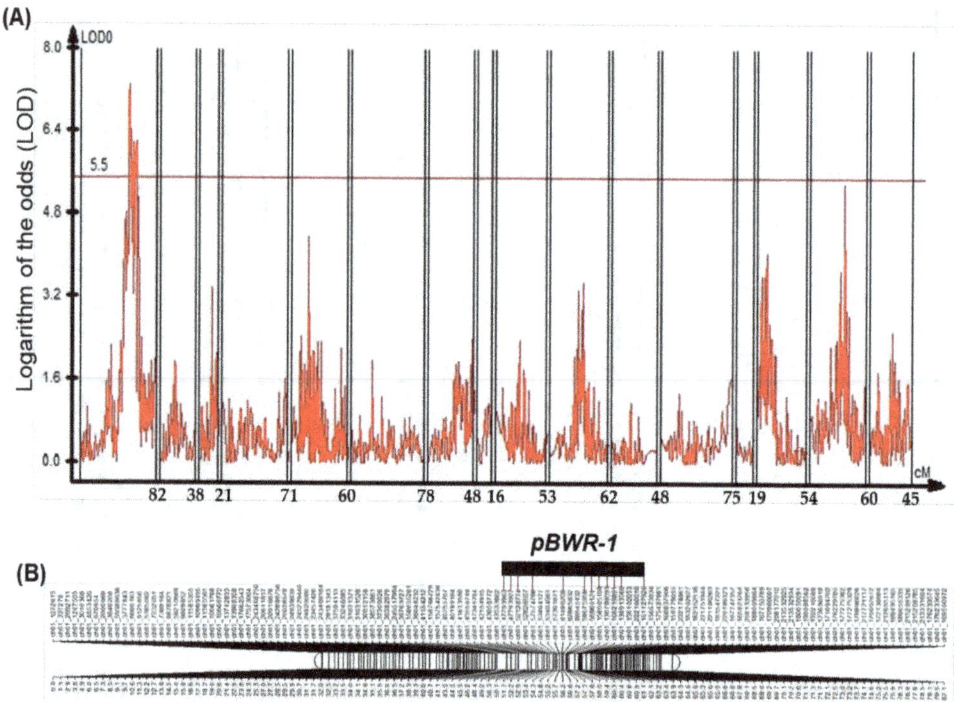

Figure 3. (**A**) Quantitative trait loci (QTL) plot. QTL associated with bacterial wilt (BW) disease resistance in the F$_2$ population derived from the parental lines of KC352 and 14F6002-14 in the linkage groups obtained via CIM. The LOD threshold is indicated by the red-colored horizontal line. (**B**) Physical map of chromosome 01 with SNP markers used for the mapping of BW resistance locus as shown in Figure 2. The QTL positions (LOD $\geq$ 5.5) and SNP markers of BW resistance are indicated with a black bar combined with red-colored vertical lines above the linkage map.

Table 3. Summary of significant QTLs (LOD $\geq$ 5.5) regions with composite interval mapping (CIM) analysis.

Chr.	Marker (Chr_pos)	QTL Position (cM)	LOD	Additive	Dominant	[a] R^2 (%)	[b] |d/a|	[c] |d/a| Value	[d] Number of Gene
	ch01_47793963	51.73	7.065493	−0.85811	0.701208	23.9114	D	0.82	1
	ch01_47793907	52.21	7.323219	−0.88307	0.690647	25.1692	PD	0.78	1
	ch01_52128774	52.94	6.962209	−0.88383	0.603339	24.0817	PD	0.68	5
	ch01_57574786	54.32	6.440147	−0.85645	0.651621	23.1222	PD	0.76	0
	ch01_61496208	56.25	6.180448	−0.89693	0.496732	24.5181	PD	0.55	2
Chr. 01-1	ch01_58075358	57.31	5.682495	−0.79976	0.700647	20.9856	D	0.88	0
	ch01_56960124	57.65	5.983697	−0.93776	0.399064	23.0298	PD	0.43	0
	ch01_156505108	58.98	6.161348	−0.88705	0.631756	24.0344	PD	0.71	11
	ch01_130789593	59.46	5.602784	−0.81563	0.56539	20.171	PD	0.69	0
	ch01_160219891	60.27	5.506165	−0.95201	0.162341	22.5886	A	0.17	2
	ch01_130913560	60.36	6.216155	−0.92824	0.283328	22.2838	PD	0.31	0
	ch01_161127926	61.1	5.692397	−0.85691	0.394691	20.1396	PD	0.46	10

[a] R^2, proportion of phenotypic variance explained by a major QTL; [b] \|d/a\|, estimation of gene action; A, (additive effect) 0–0.20; PD, (partial dominance) 0.21–0.80; [c] D, (dominance) 0.81–1.20; OD, (overdominance) >1.20. [d] Number of gene was selected with the 1 Mb to the left and right of the corresponding marker.

3.5. Prediction and Annotation of Candidate Genes

In order to identify candidate genes within the major QTL region, the number of genes was selected with the 1 Mb to the left and right of the corresponding markers, and a total of 31 candidate genes were annotated on the basis of CM334 reference genome, Swiss-Prot, and the NCBI non-redundant protein (NR) databases (Table 4). In addition, functional classification of 31 predicted genes was further analyzed along with the Kyoto Encyclopedia of Genes Genomes (KEGG) pathway and Gene Ontology (GO) term (Table 4). As such, the analysis of GO term enrichment identified one gene (CA01g20110) annotated with "defense response" (GO: 0006952), and the KEGG analysis identified one gene (CA01g20130) with "plant–pathogen interaction" (KO: 13457). Overall, four of the 31 candidate genes were assigned to defense-associated genes; CA01g18650 in ch01_156505108 marker and CA01g20130 (putative disease resistance protein RPM1), CA01g20110 (Thaumatin-like protein), and CA01g20140 (NB-ARC domain-containing protein) in ch01_161127926 marker were predicted as disease-resistance proteins. Taken together, the annotated four disease-resistance/defense-associated genes would be crucial candidate genes for *pBWR-1* QTL in the present study.

Table 4. The candidate genes within the significant QTL regions.

Gene Name	Gene Start	Gene End	CM334(v.1.55)	KEGG/GO	SwissProt	NR
CA01g12080	47862177	47863580	Putative amino acid transporter	-	Putative polyamine transporter	Putative polyamine transporter
CA01g12450	52035941	52041393	Calcium-dependent protein kinase 17	K13412 (calcium-dependent protein kinase)	Calcium-dependent protein kinase 17	Calcium-dependent protein kinase 17-like
CA01g12460	52113432	52115689	Rho GTPase-activating protein	-	Rho GTPase-activating protein 7	rho GTPase-activating protein 7
CA01g12470	52119767	52121954	Putative rho GTPase activator	-	Rho GTPase-activating protein 7	PREDICTED: rho GTPase-activating protein 7-like
CA01g12480	52128303	52129214	Rho GTPase-activating protein 7-like isoform X2 (*Solanum tuberosum*)	-	Rho GTPase-activating protein 7	PREDICTED: rho GTPase-activating protein 7-like
CA01g12490	52147172	52148002	Phospholipase A22	-	Uncharacterized protein	Putative rho GTPase-activating protein 7-like isoform X1
CA01g13370	61438907	61440480	Arsenite-resistance protein, putative	-	Serrate RNA effector molecule	Serrate RNA effector molecule
CA01g13380	61446486	61447578	Arsenite-resistance protein, putative	-	Serrate RNA effector molecule	Serrate RNA effector molecule
CA01g18610	156433050	156436475	Detected protein of unknown function	-	O-fucosyltransferase family protein	Putative siroheme synthase-like
CA01g18620	156438428	156442076	AT5g40850/MHK7_8	K02303 (uroporphyrin-III C-methyltransferase)	Siroheme synthase	PREDICTED: siroheme synthase
CA01g18630	156454093	156456205	Putative glutathione S-transferase T1	K00799 (glutathione S-transferase)	Putative glutathione S-transferase	PREDICTED: probable glutathione S-transferase
CA01g18640	156457355	156460628	Cohesin subunit rad21, putative	K06670 (RAD21; cohesin complex subunit SCC1)	Uncharacterized protein	PREDICTED: sister chromatid cohesion 1 protein 3
CA01g18650	156487695	156490046	Anthranilate phosphoribosyltransferase-like protein	-	Uncharacterized protein	Putative disease-resistance protein RPM1-like
CA01g18660	156497587	156514448	DNA-directed RNA polymerase	-	DNA-directed RNA polymerase subunit beta	PREDICTED: DNA-directed RNA polymerase I subunit 2

Table 4. *Cont.*

Gene Name	Gene Start	Gene End	CM334(v.1.55)	KEGG/GO	SwissProt	NR
CA01g18670	156515197	156518484	Putative synaptotagmin	-	Protein QUIRKY	Protein QUIRKY
CA01g18680	156539078	156540145	Detected protein of unknown function	K22733 (NIPA, SLC57A2S, magnesium transporter)	Magnesium transporter	Putative magnesium transporter NIPA8
CA01g18690	156550800	156555986	DnaJ-like protein	K03686 (DnaJ, molecular chaperone DnaJ)	Uncharacterized protein	Hypothetical protein T459_02496
CA01g18700	156563245	156580138	Eukaryotic translation initiation factor 3 subunit, putative	K03255 (TIF31, CLU1, protein TIF31)	Clustered mitochondria protein	Clustered mitochondria protein
CA01g18710	156602485	156602895	Detected protein of unknown function	K11253 (Histone H3)	Histone H3	PREDICTED: histone H3.2-like
CA01g19920	160217017	160229898	Ankyrin repeat family protein (*Populus trichocarpa*)	K24810 (IBTK, inhibitor of Bruton tyrosine kinase)	Ankyrin repeat domain-containing protein	Hypothetical protein FXO37_07210
CA01g19930	160315849	160316316	Polypeptide with a gag-like domain	-	LRRNT-2 domain-containing protein	Uncharacterized protein
CA01g20080	161050160	161051870	Transcription factor bHLH130-like isoform X1 (*Citrus sinensis*)	-	Transcription factor bHLH130	PREDICTED: transcription factor bHLH130-like
CA01g20090	161051871	161052295	DNA binding protein, putative	-	Transcription factor bHLH130	PREDICTED: transcription factor bHLH130-like
CA01g20100	161113518	161117641	Ras-related protein RGP1-like (*Solanum tuberosum*)	K07904 (RAB11A, Ras-related protein Rab-11A)	Ras-related protein RABA4c	Ras-related protein RABA4c
CA01g20110	161118723	161120113	Thaumatin-like protein SE39b	-/GO (0006952): defense response	Uncharacterized protein	PREDICTED: thaumatin-like protein 1b isoform X1
CA01g20120	161154079	161154489	Histone H3 family protein (*Populus trichocarpa*)	K11253 (H3, histone H3)	Histone H3	PREDICTED: histone H3.2-like
CA01g20130	161161494	161162582	Disease resistance protein RPM1-like (*Fragaria vesca*) subsp. vesca	K13457 (RPM1, RPS3; disease-resistance protein RPM1)	Uncharacterized protein	Hypothetical protein T459_01449

Table 4. *Cont.*

Gene Name	Gene Start	Gene End	CM334(v.1.55)	KEGG/GO	SwissProt	NR
CA01g20140	161184987	161187221	Detected protein of unknown function	-	NB-ARC domain-containing protein	PREDICTED: putative late blight-resistance protein homolog R1B-23
CA01g20150	161212363	161212704	Detected protein of unknown function	-	Uncharacterized protein	Putative cellulose synthase A catalytic subunit 3 (UDP-forming)-like
CA01g20160	161215973	161218180	Detected protein of unknown function	-	Uncharacterized protein	Uncharacterized protein LOC107839230 isoform X2
CA01g20170	161221263	161222494	Detected protein of unknown function	-	Uncharacterized protein	Uncharacterized protein LOC107839213

4. Discussion

BW is one of the most destructive pepper diseases worldwide, leading to the reduction of yield and production in pepper cultivation [16,63]. It is difficult to manage BW disease owing to a wide array of plant host range, a huge number of diverse BW isolates, and its long survivability in pepper plants [18–20]. Thus, it is indispensable to breed resistant pepper cultivars against BW. Although molecular marker-assisted selection (MAS) for BW resistance can contribute to a rapid selection of BW-resistant breeding in pepper crops, a few studies have determined QTL regions [17,36,62]. In this study, we performed QTL analysis to develop molecular markers that are associated with BW resistance in pepper (*Capsicum annuum*) by evaluating the 94 F_2 recombinant lines obtained by a cross between a resistant and a susceptible parental line. We first constructed a genetic linkage map using GBS approach and identified significant QTL regions on chromosome 01 associated with BW resistance.

4.1. Construction of Pepper Genetic Map

GBS is a genome-wide genotyping, powerful, and straightforward approach that takes advantage of enzyme-based genome analysis, thereby conferring a rapid and cost-effective analysis of the huge and complex genome in organisms. The GBS tool has been widely utilized in genotyping segregated plants via the combination of a high-throughput next-generation sequencing (NGS) technology to generate multiplexed libraries using barcoded adapters [64,65]. With the application of GBS analyses, it has been reported that a large amount of barely SNPs are produced, and $\geq$34,000 SNPs are mapped onto its reference genome sequences, and $\geq$20,000 wheat SNPs are constructed onto its reference map [65]. Moreover, the method has identified 9998 SNPs and 64,754 SNPs located on the peach and pea genomes, respectively [66,67]. Notably, recent studies have explored and evaluated 1,399,567 SNP in onion using a GBS library [50], and a total of 91,132 raw SNPs were uncovered via a QTL study involved in flowering time in perilla [49]. In addition to this, current applications of GBS analysis have exhibited a total of 22,446 SNPs for QTL mapping of the resistance against the cucumber mosaic virus in a cucumber crop as well as a total of 66,405 SNPs for *Phytophthora capsici* resistance in a pepper crop [48,59]. In the present study, a total of 628,437 raw SNPs were identified and successfully genotyped with 94 F_2 offspring using GBS. Around 108 Gbp of raw data (Tables S1 and S2) and a total 1639 SNPs were finally produced, and the SNP markers were shown with genome-wide distribution, covering the whole pepper genome (Figure 2). A total of 1550 SNP markers were ultimately constructed on a genetic linkage map, which comprised 16 LGs, including one linkage group on chromosome 03, 04, 06, 07, 08, 09, 11, and 12 as well as two linkage groups on chromosome 01, 02, 05, and 10, respectively (Figure 2 and Table 2). In general, SNP markers used for genetic mapping are based on the polymorphic markers of the parent. When the polymorphism of the parent is absent in a large region within the middle of the genome, separating two linkage groups on one chromosome can often be produced although the genetic map is physically one chromosome. In particular, the phenomena often occur when working with breeding lines. Indeed, previous reports have shown that Yellow lupin (*Lupinus luteus* L.) possesses 26 chromosomes, but 40 linkage groups were constructed for QTL mapping via NGS approaches [68], and a linkage group of LG01 was divided into two LGs on chromosome 01 in perilla via GBS [49]. Recently, it has been determined that linkage map is constructed with the resistance trait of powdery mildew from F_5 pepper population via GBS. The LG07 is separated into two linkage groups on chromosome 07 [69].

4.2. Genetic Inheritance of pBWR-1 QTL on Chromosome 01

As mentioned above, it has been reported that the genetic analysis and identification of resistance genes play a crucial role in the field of crop breeding against BW [24,30,31,34–36,38,63]. Nonetheless, the genetic inheritance is poorly understood in the involvement of BW resistance in pepper crops, and the mechanism of genetic inheritance is still unclear since different pepper sources result in different values of BW resistance. In our results, we evaluated 49 F_2 offspring as the DI value 4 (the most severe symptoms, 76 to 100% wilted leaves), whereas 14 plants were evaluated as DI value 1 (no visible symptoms) with the comparison of the parental lines after BW inoculation (Figure 1). We observed susceptible plants 3.5 times more than resistant plants in the F_2 population, implying that BW resistance might be a partially recessive trait in the pepper lines used for our experiment. In similar line with our result, the genetic inheritance of BW resistance has been unraveled, and the resistance homozygous recessive (rr) allele was identified using F_2 populations derived from a cross between resistant Anugraha and susceptible Pusa Jwala of near-isogenic lines (NILs) in pepper crops [70]. On the contrary, it has been determined that the inheritance action of BW resistance is involved in an incomplete dominance with more than two BW-resistance genes using the progeny derived from a cross between the *capsicum* Mie-Midori (resistant line) and the *capsicum* AC2258 (susceptible line) [29]. It has been also studied that the BW resistance is governed by two to five genes using the progeny derived from a cross between the PM687 (resistant line) and the Yolo (susceptible line) [34]. In addition to this, researches have demonstrated that the disease severity in F_1 hybrids is close to or lower than the generation-means of mid-parent values [29,49], and the progeny derived from BVRC 1 (resistant line) and BVRC25 (susceptible line) [36] as well as from MC4 (resistant line) and Subicho (susceptible line) exhibited the association of more than two resistance genes against *R. solanacearum*, indicating that the BW resistance would be involved in a partial dominance effect [71]. Although the complex mechanism still needs to be elucidated, the contradictory findings might result from diverse factors, including the different pepper sources of breeding lines, the different inoculation methods, the bacterial isolates, the different criteria using DI calculation, and the different environmental growth factors. It is, therefore, of our interest to further study the complex action of genetic inheritance in the pepper source used for the experiment with the comparison of other BW-resistant lines.

4.3. Detection of Major QTL Controlling Resistance to R. solanacearum

Multiple management strategies have been actively developed and applied to control the BW disease. However, the effects have been limited and insufficient for the control of destructive BW disease owing to the different pepper sources, the bacterial isolates, and different inoculation methods as mentioned above [26–28,63]. Besides, high disease-resistance phenotype is not always associated with a good performance of horticultural traits, such as good fruit shape, fruit size, fruit yield, and fruit quality [71]. Thus, it is crucial to understand the genetic basis for resistance to BW disease to utilize pepper breeding programs using MAS, thereby ultimately integrating BW resistance with desirable traits during a breeding process [47,48,71]. In previous studies, the pepper accessions of LS2341 and BVRC1 have been determined with a *Bw1* QTL and a major *qRRs-10.1* QTL underlying on chromosome 08 and 10, respectively [35,36]. Initially, Mimura et al. (2009) reported that the CAMS451 marker of *Bw1* QTL in linkage group 11 (LG11) was located on chromosome 01 with the comparison of a LG01 on chromosome 01 of SNU3 map, which was integrated by a genetic linkage map of an interspecific cross between *C. annuum* and *C. chinense* [35]. However, the group reported the linkage group was shifted to chromosome 08 from 01 again [72]. The current studies suggest that the LG 11 is possessed by chromosome 08 rather than 01 in *C. annuum*. Mathew (2020) recently reported that the pepper CAMS451 marker of *Bw1 QTL* lies on chromosome 08 (position: 122704651-124710667) and has annotated 44 defense-associated genes from 1 Mb upstream and downstream of the marker [73]. In addition to this, another research on the BW resistance against *R. solanacearum* demonstrated that a major *qRRs-10.1* QTL region is located on chromosome 10 (position: 56910000-69110000, 111090000-183670000) in *C. annuum*. Interestingly, 54 genes were annotated, and five putative R genes lie on the regions of 193.4–196.3 Mb, which are nearly closed to the markers of ID10-194305124 and ID10-196208712 [36]. In contrast, in the present study, we identified the major *pBWR-1* QTL region on chromosome 01 via a CIM method using the GBS analysis on the basis of the threshold levels (Figure 3), which exhibited the remarkable LODs from 5.69 to 7.06 in LG_01-1 (Table 3). Importantly, our finding shows that the major *pBWR-1* QTL region underlies on chromosome 01 (position: 47793907–61496208, 130789593–161127926) with 12 markers, which encode 31 candidate genes in the region (Tables 3 and 4). These discrepancies of previous and our current result on the genome loci would result from a variety of materials and methods as aforementioned reasons.

4.4. pBWR-1 Candidate Genes for Resistance to R. solanacearum

We annotated the 31 candidate genes on the basis of sequence alignment using CM334 reference genome, KEGG, Swiss-Prot, GO, and NR databases (Table 4). Among them, four candidate genes were annotated as defense-associated genes: one gene, CA01g18650 (putative disease-resistance protein RPM1), in ch01_156505108 marker and three genes, such as nearly tandem-arrayed genes CA01g20110 (GO: 0006952, Thaumatin-like protein), CA01g20130 (putative disease-resistance protein RPM1, KO: 13457), and CA01g20140 (NB-ARC domain-containing protein), in ch01_161127926 marker. In comparison with previous publications, the identified candidate genes might not be similar to Mathew's (2020) results on chromosome 08, whereas the genes encoding LRR proteins and NB-ARC domain-containing protein on chromosome 10 in Du et al. (2019) are shared with our results, implying that these genes are possibly indispensable for BW resistance.

Previous studies have reported that pathogen defense-associated R genes are tandemly located in chromosomes [74–76]. For example, eight genes encoding an amino terminal coiled-coil domain (CC), a central nucleotide binding (NB) site, leucine-rich repeat (LRR) domain are tandemly arrayed in the *Pvr4* locus of CM334 genome in *C. annuum*. Fourteen genes encoding NB-LRR tandemly lie on the *Tsw* locus of PI159236 genome in *C. chinense*. Moreover, among of three tandem-arrayed R genes within the *qRRS-10.1* QTL, two genes, including CA10g13010 and CA10g13020, were annotated as *Bs2*, which is classified into the NB-LRR family, suggesting that NB-LRR proteins play a crucial role in disease resistance against pathogens [75]. It has been studied that elongation factor tu receptor (EFR) from *Arabidopsis* and *Bs2* from pepper are expressed in tomato plant for controlling BW and bacterial spot (BS), respectively [77]. Intriguingly, the EFR was determined as a critical component in plant defense of PAMP-triggered immunity (PTI) via the interaction between conserved PAMPs and bacterial pathogens [78]. Furthermore, Du et al. (2019) reported that CA10g12520 within the *qRRS-10.1* QTL encodes PR-1 gene, indicating the participation in the interaction of plant and pathogen via PTI. In similar line with previous results, we identified CA01g18650 and CA01g20130 encoding putative disease-resistance protein RPM1. The bacterial resistance to *Pseudomonas syringae pv. maculicola 1* (RPM1) encodes a CC-NB-LRR family protein, which is a peripheral plasma membrane protein [79,80]. The RPM1 recognizes the effector proteins of *avrB* or *avrRpm1* of *Pseudomonas syringae* in *Arabidopsis*, resulting in the rapid generation of a hypersensitive response (HR) [79–81]. Moreover, identified CA01g20140 harbors NB-ARC domain-containing protein. It has been shown that effector-triggered immunity (ETI) is associated with nucleotide-binding leucine-rich repeat (NLR) proteins [82]. A variety of plant NLRs possess an NB-ARC domain (nucleotide-binding adaptor shared by Apaf-1, R proteins, and CED4), which is able to interact with NLR domain-containing proteins, indicating that an NB-ARC domain is important for pathogen defense [83,84]. Importantly, we also identified CA01g20110 encoding Thaumatin-like protein (GO: 0006952). Previous researches have reported that Thaumatin-like proteins (TLPs) are classified into PR (pathogenesis-related protein)-5 class protein, and TLP genes are upregulated in peanut (*Arachis hypogaea* L) with the treatment of the leaf spot pathogen, *Phaeoisariopsis personata* [85], as well as in wheat (*Triticum aestivum*) by leaf rust fungus, *Puccinia triticina* [86]. In addition, constitutive expression of *Arabidopsis* Thaumatin-like protein 1 (ATLP1) in potato are observed with reduced lesions and percent reductions in response to *Alternaria solani* and *Phytophthora infestans* [87], implying that TLPs are essential in diverse biotic response. Although we cannot completely rule out the effect of differential expression of other candidate genes in the list (Table 4) and minor QTL effects that were not detected on the BW resistance in the current study, it is our endeavor for future research to focus on the understanding of genetic mechanisms, such as inheritance factors, as well as on functional analysis of the identified candidate genes in the resistance to *R. solanacearum*.

5. Conclusions

In the present study, the disease symptoms and resistance for BW were evaluated by the inoculation with *R. solanacearum* using 94 F_2-segregating populations. Using GBS, 628,437 SNPs were identified, and the filtered 1550 SNP were subsequently constructed into the genetic linkage map displaying 16 LG. QTL analysis revealed that *pBWR-1* QTL was located on chromosome 01, and the number of 31 candidate genes were identified in the significant QTL regions. Importantly, the identified four genes encode defense resistance-associated proteins, such as CC-NB-LRR family protein, NB-ARC domain-containing protein, and Thaumatin-like proteins. Our finding will contribute to deep insights into the information for developing SNP markers associated with BW-resistant QTL as well as for developing BW-resistant cultivars in pepper breeding programs.

Supplementary Materials: The following are available online at https://www.mdpi.com/article/10.3390/horticulturae8020115/s1, Supplementary Figure S1: Distribution of the disease index after inoculation with *Ralstonia solanacearum*; Table S1: Genotyping-by-sequencing (GBS) statistics for 96 samples; Table S2: Single-nucleotide polymorphism (SNP) filtering criteria.

Author Contributions: Conceptualization, J.-B.Y.; methodology, S.-Y.C., J.-W.D., S.-C.H. and K.-H.L.; investigation, S.-Y.C., J.-W.D., S.-C.H. and K.-H.L.; writing—original draft preparation, S.-Y.C. and K.L.; writing—review and editing, K.L., M.-C.C. and E.-Y.Y.; visualization, K.L. and E.-Y.Y.; supervision, M.-C.C. and J.-B.Y. All authors have read and agreed to the published version of the manuscript.

Funding: This study was supported by a grant from Golden Seed Project (Project No.: 213006-05-05-SB630 "Development of molecular markers for Bacterial wilt resistance and high-pungency in chili pepper").

Institutional Review Board Statement: Not applicable.

Informed Consent Statement: Not applicable.

Data Availability Statement: The datasets presented in this study are available upon request to the corresponding author.

Acknowledgments: The authors thank Sung-Hwan Jo at SEEDERS Inc., Republic of Korea, for assistance via intellectual discussions.

Conflicts of Interest: The authors declare no conflict of interest.

References

1. Bhutia, K.; Khanna, V.; Meetei, T.; Bhutia, N. Effects of climate change on growth and development of chilli. *Agrotechnology* **2018**, *7*, 2. [CrossRef]
2. Prohens, J.; Nuez, F. *Handbook of Plant Breeding. Vegetables II: Fabaceae, Liliaceae, Solanaceae and Umbelliferae*; Springer: New York, NY, USA, 2008; Volume 3, pp. 30–40.
3. Fraenkel, L.; Bogardus, S.T.; Concato, J.; Wittink, D.R. Treatment options in knee osteoarthritis: The patient's perspective. *Arch. Intern. Med.* **2004**, *164*, 1299–1304. [CrossRef] [PubMed]
4. Sun, T.; Xu, Z.; Wu, C.T.; Janes, M.; Prinyawiwatkul, W.; No, H. Antioxidant activities of different colored sweet bell peppers (*Capsicum annuum* L.). *J. Food Sci.* **2007**, *72*, S98–S102. [CrossRef] [PubMed]
5. Blum, E.; Mazourek, M.; O'connell, M.; Curry, J.; Thorup, T.; Liu, K.; Jahn, M.; Paran, I. Molecular mapping of capsaicinoid biosynthesis genes and quantitative trait loci analysis for capsaicinoid content in *Capsicum. Theor. Appl. Genet.* **2003**, *108*, 79–86. [CrossRef] [PubMed]
6. Stewart Jr, C.; Mazourek, M.; Stellari, G.M.; O'Connell, M.; Jahn, M. Genetic control of pungency in C. chinense via the Pun1 locus. *J. Exp. Bot.* **2007**, *58*, 979–991. [CrossRef]
7. Aza-González, C.; Núñez-Palenius, H.G.; Ochoa-Alejo, N. Molecular biology of capsaicinoid biosynthesis in chili pepper (*Capsicum* spp.). *Plant Cell Rep.* **2011**, *30*, 695–706. [CrossRef]
8. Luo, X.-J.; Peng, J.; Li, Y.-J. Recent advances in the study on capsaicinoids and capsinoids. *Eur. J. Pharmacol.* **2011**, *650*, 1–7. [CrossRef]
9. Faostat. 2021. Available online: http://www.fao.org (accessed on 9 August 2021).

10. Rajametov, S.N.; Lee, K.; Jeong, H.-B.; Cho, M.-C.; Nam, C.-W.; Yang, E.-Y. The Effect of Night Low Temperature on Agronomical Traits of Thirty-Nine Pepper Accessions (*Capsicum annuum* L.). *Agronomy* **2021**, *11*, 1986. [CrossRef]

11. Comtrade UN. UN Comtrade Database. Available online: http://comtrade.un.org (accessed on 15 October 2020).

12. APS. Common Names of Plant Disease. Available online: https://www.apsnet.org/edcenter/resources/commonnames/Pages/default.aspx (accessed on 24 October 2020).

13. Jeong, Y.; Kim, J.; Kang, Y.; Lee, S.; Hwang, I. Genetic diversity and distribution of Korean isolates of *Ralstonia solanacearum*. *Plant Dis.* **2007**, *91*, 1277–1287. [CrossRef]

14. Lee, Y.K.; Kang, H.W. Physiological, biochemical and genetic characteristics of *Ralstonia solanacearum* strains isolated from pepper plants in Korea. *Res. Plant Dis.* **2013**, *19*, 265–272. [CrossRef]

15. Jiang, G.; Peyraud, R.; Remigi, P.; Guidot, A.; Ding, W.; Genin, S.; Peeters, N. Modeling and experimental determination of infection bottleneck and within-host dynamics of a soil-borne bacterial plant pathogen. *bioRxiv* **2016**, 061408. [CrossRef]

16. Jiang, G.; Wei, Z.; Xu, J.; Chen, H.; Zhang, Y.; She, X.; Macho, A.P.; Ding, W.; Liao, B. Bacterial wilt in China: History, current status, and future perspectives. *Front. Plant Sci.* **2017**, *8*, 1549. [CrossRef]

17. Kang, Y.J.; Ahn, Y.-K.; Kim, K.-T.; Jun, T.-H. Resequencing of *Capsicum annuum* parental lines (YCM334 and Taean) for the genetic analysis of bacterial wilt resistance. *BMC Plant Biol.* **2016**, *16*, 235. [CrossRef]

18. Hayward, A. Biology and epidemiology of bacterial wilt caused by *Pseudomonas solanacearum*. *Annu. Rev. Phytopathol.* **1991**, *29*, 65–87. [CrossRef]

19. Denny, T. Plant pathogenic *Ralstonia* species. In *Plant-Associated Bacteria*; Springer: Berlin/Heidelberg, Germany, 2007; pp. 573–644.

20. Hayward, A. The hosts of *Pseudomonas solanacearum*. In *Bacterial Wilt: The Disease and Its Causative Agent, Pseudomonas solanacearum*; CAB International Press: Wallingford, UK, 1994; pp. 9–24.

21. Denny, T.P. *Ralstonia solanacearum*—A plant pathogen in touch with its host. *Trends Microbiol.* **2000**, *8*, 486–489. [CrossRef]

22. Vasse, J.; Frey, P.; Trigalet, A. Microscopic studies of intercellular infection and protoxylem invasion of tomato roots by *Pseudomonas solanacearum*. *Mol. Plant Microbe Interact.* **1995**, *8*, 241–251. [CrossRef]

23. Rahman, M.; Abdullah, H.; Vanhaecke, M. Histopathology of susceptible and resistant *Capsicum annuum* cultivars infected with Ralstonia solanacearum. *J. Phytopathol.* **1999**, *147*, 129–140. [CrossRef]

24. Wang, J.-F.; Olivier, J.; Thoquet, P.; Mangin, B.; Sauviac, L.; Grimsley, N.H. Resistance of tomato line Hawaii7996 to *Ralstonia solanacearum* Pss4 in Taiwan is controlled mainly by a major strain-specific locus. *Mol. Plant Microbe Interact.* **2000**, *13*, 6–13. [CrossRef]

25. Winstead, N. Inoculation techniques for evluating resistance to *Pseudomonas solanacearum*. *Phytopathology* **1952**, *42*, 623–634.

26. Mamphogoro, T.; Babalola, O.; Aiyegoro, O. Sustainable management strategies for bacterial wilt of sweet peppers (*Capsicum annuum*) and other Solanaceous crops. *J. Appl. Microbiol.* **2020**, *129*, 496–508. [CrossRef]

27. Buddenhagen, I. Bacterial wilt of certain seed-bearing *Musa* spp. caused by tomato strain of *Pseudomonas solanacearum*. *Phytopathology* **1962**, *52*, 286.

28. Hayward, A. Characteristics of Pseudomonas solanacearum. *J. Appl. Bacteriol.* **1964**, *27*, 265–277. [CrossRef]

29. Matsunaga, H.; Sato, T.; Monma, S. In Inheritance of bacterial wilt resistance in the sweet pepper cv. Mie-Midori. In Proceedings of the 10th Eucarpia Meeting on Genetics and Breeding of Capsicum and Eggplant, Avignon, France, 7–11 September 1998; p. 172.

30. Lopes, C.A.; Boiteux, L.S. Biovar-specific and broad-spectrum sources of resistance to bacterial wilt (*Ralstonia solanacearum*) in Capsicum. *Embrapa Hortaliças-Artig. Periódico Indexado (ALICE)* **2004**, *4*, 350–355. [CrossRef]

31. Mimura, Y.; Yoshikawa, M.; Hirai, M. Pepper accession LS2341 is highly resistant to *Ralstonia solanacearum* strains from Japan. *HortScience* **2009**, *44*, 2038–2040. [CrossRef]

32. Kim, B.; Cheung, J.; Cha, Y.; Hwang, H. Resistance to bacterial wilt of introduced peppers. *Korean J. Plant Pathol.* **1998**, *14*, 217–219.

33. Tung, P.X.; Rasco, E.T.; Vander Zaag, P.; Schmiediche, P. Resistance to *Pseudomonas solanacearum* in the potato: I. Effects of sources of resistance and adaptation. *Euphytica* **1990**, *45*, 203–210. [CrossRef]

34. Lafortune, D.; Béramis, M.; Daubèze, A.-M.; Boissot, N.; Palloix, A. Partial resistance of pepper to bacterial wilt is oligogenic and stable under tropical conditions. *Plant Dis.* **2005**, *89*, 501–506. [CrossRef]

35. Mimura, Y.; Kageyama, T.; Minamiyama, Y.; Hirai, M. QTL analysis for resistance to *Ralstonia solanacearum* in *Capsicum* accession 'LS2341'. *J. Jpn. Soc. Hortic. Sci.* **2009**, *78*, 307–313. [CrossRef]

36. Du, H.; Wen, C.; Zhang, X.; Xu, X.; Yang, J.; Chen, B.; Geng, S. Identification of a major QTL (qRRs-10.1) that confers resistance to *Ralstonia solanacearum* in pepper (*Capsicum annuum*) using SLAF-BSA and QTL mapping. *Int. J. Mol. Sci.* **2019**, *20*, 5887. [CrossRef]

37. Sonah, H.; Bastien, M.; Iquira, E.; Tardivel, A.; Légaré, G.; Boyle, B.; Normandeau, É.; Laroche, J.; Larose, S.; Jean, M. An improved genotyping by sequencing (GBS) approach offering increased versatility and efficiency of SNP discovery and genotyping. *PLoS ONE* **2013**, *8*, e54603. [CrossRef]

38. Takagi, H.; Abe, A.; Yoshida, K.; Kosugi, S.; Natsume, S.; Mitsuoka, C.; Uemura, A.; Utsushi, H.; Tamiru, M.; Takuno, S. QTL-seq: Rapid mapping of quantitative trait loci in rice by whole genome resequencing of DNA from two bulked populations. *Plant J.* **2013**, *74*, 174–183. [CrossRef]

39. Varshney, R.K.; Nayak, S.N.; May, G.D.; Jackson, S.A. Next-generation sequencing technologies and their implications for crop genetics and breeding. *Trends Biotecnol.* **2009**, *27*, 522–530. [CrossRef]

40. Hayward, A.; Mason, A.; Dalton-Morgan, J.; Zander, M.; Edwards, D.; Batley, J. SNP discovery and applications in *Brassica napus*. *J. Plant Biotechnol.* **2012**, *39*, 49–61. [CrossRef]

41. Elshire, R.; Glaubitz, J.; Sun, Q.; Poland, J.A.; Kawamoto, K.; Buckler, E.S.; Mitchell, S.E. A robust, simple genotyping-by-sequencing (GBS) approach for high diversity species. *PLoS ONE* **2011**, *6*, e19379. [CrossRef]

42. Baldwin, S.; Pither-Joyce, M.; Wright, K.; Chen, L.; McCallum, J. Development of robust genomic simple sequence repeat markers for estimation of genetic diversity within and among bulb onion (*Allium cepa* L.) populations. *Mol. Breed.* **2012**, *30*, 1401–1411. [CrossRef]

43. Davey, J.W.; Hohenlohe, P.A.; Etter, P.D.; Boone, J.Q.; Catchen, J.M.; Blaxter, M.L. Genome-wide genetic marker discovery and genotyping using next-generation sequencing. *Nat. Rev. Genet.* **2011**, *12*, 499–510. [CrossRef]

44. Romay, M.C.; Millard, M.J.; Glaubitz, J.C.; Peiffer, J.A.; Swarts, K.L.; Casstevens, T.M.; Elshire, R.J.; Acharya, C.B.; Mitchell, S.E.; Flint-Garcia, S.A. Comprehensive genotyping of the USA national maize inbred seed bank. *Genome Biol.* **2013**, *14*, R55. [CrossRef]

45. Li, H.; Vikram, P.; Singh, R.P.; Kilian, A.; Carling, J.; Song, J.; Burgueno-Ferreira, J.A.; Bhavani, S.; Huerta-Espino, J.; Payne, T. A high density GBS map of bread wheat and its application for dissecting complex disease resistance traits. *BMC Genom.* **2015**, *16*, 216. [CrossRef]

46. Iquira, E.; Humira, S.; François, B. Association mapping of QTLs for sclerotinia stem rot resistance in a collection of soybean plant introductions using a genotyping by sequencing (GBS) approach. *BMC Plant Biol.* **2015**, *15*, 5. [CrossRef]

47. Jaganathan, D.; Thudi, M.; Kale, S.; Azam, S.; Roorkiwal, M.; Gaur, P.M.; Kishor, P.K.; Nguyen, H.; Sutton, T.; Varshney, R.K. Genotyping-by-sequencing based intra-specific genetic map refines a "QTL-hotspot" region for drought tolerance in chickpea. *Mol. Genet. Genom.* **2015**, *290*, 559–571. [CrossRef]

48. Siddique, M.I.; Lee, H.-Y.; Ro, N.-Y.; Han, K.; Venkatesh, J.; Solomon, A.M.; Patil, A.S.; Changkwian, A.; Kwon, J.-K.; Kang, B.-C. Identifying candidate genes for *Phytophthora capsici* resistance in pepper (*Capsicum annuum*) via genotyping-by-sequencing-based QTL mapping and genome-wide association study. *Sci. Rep.* **2019**, *9*, 9962. [CrossRef]

49. Kang, Y.-J.; Lee, B.-M.; Nam, M.; Oh, K.-W.; Lee, M.-H.; Kim, T.-H.; Jo, S.-H.; Lee, J.-H. Identification of quantitative trait loci associated with flowering time in perilla using genotyping-by-sequencing. *Mol. Biol. Rep.* **2019**, *46*, 4397–4407. [CrossRef]

50. Jo, J.; Purushotham, P.M.; Han, K.; Lee, H.-R.; Nah, G.; Kang, B.-C. Development of a genetic map for onion (*Allium cepa* L.) using reference-free genotyping-by-sequencing and SNP assays. *Front. Plant Sci.* **2017**, *8*, 1606. [CrossRef]

51. Reyes, V.P.; Angeles-Shim, R.B.; Mendioro, M.S.; Manuel, M.; Carmina, C.; Lapis, R.S.; Shim, J.; Sunohara, H.; Nishiuchi, S.; Kikuta, M. Marker-Assisted Introgression and Stacking of Major QTLs Controlling Grain Number (Gn1a) and Number of Primary Branching (WFP) to NERICA Cultivars. *Plants* **2021**, *10*, 844. [CrossRef]

52. Kitony, J.K.; Sunohara, H.; Tasaki, M.; Mori, J.-I.; Shimazu, A.; Reyes, V.P.; Yasui, H.; Yamagata, Y.; Yoshimura, A.; Yamasaki, M. Development of an Aus-Derived Nested Association Mapping (Aus-NAM) Population in Rice. *Plants* **2021**, *10*, 1255. [CrossRef]

53. Rajametov, S.N.; Lee, K.; Jeong, H.B.; Cho, M.C.; Nam, C.W.; Yang, E.Y. Physiological Traits of Thirty-Five Tomato Accessions (*Solanum lycopersicum* L.) in Response to Low Temperature. *Agriculture* **2021**, *11*, 792. [CrossRef]

54. Allen, G.; Flores-Vergara, M.; Krasynanski, S.; Kumar, S.; Thompson, W. A modified protocol for rapid DNA isolation from plant tissues using cetyltrimethylammonium bromide. *Nat. Protoc.* **2006**, *1*, 2320–2325. [CrossRef]

55. Cox, M.P.; Peterson, D.A.; Biggs, P.J. SolexaQA: At-a-glance quality assessment of Illumina second-generation sequencing data. *BMC Bioinform.* **2010**, *11*, 485. [CrossRef]

56. Li, H.; Durbin, R. Fast and accurate short read alignment with Burrows–Wheeler transform. *Bioinformatics* **2009**, *25*, 1754–1760. [CrossRef]

57. Li, H.; Handsaker, B.; Wysoker, A.; Fennell, T.; Ruan, J.; Homer, N.; Marth, G.; Abecasis, G.; Durbin, R. The sequence alignment/map format and SAMtools. *Bioinformatics* **2009**, *25*, 2078–2079. [CrossRef]

58. Kim, J.-E.; Oh, S.-K.; Lee, J.-H.; Lee, B.-M.; Jo, S.-H. Genome-wide SNP calling using next generation sequencing data in tomato. *Mol. Cells* **2014**, *37*, 36. [CrossRef]

59. Eun, M.H.; Han, J.-H.; Yoon, J.B.; Lee, J. QTL mapping of resistance to the Cucumber mosaic virus P1 strain in pepper using a genotyping-by-sequencing analysis. *Hortic. Environ. Biotechnol.* **2016**, *57*, 589–597. [CrossRef]

60. Kosambi, D. The estimation of map distance. *Ann. Eugen.* **1944**, *12*, 505–525.

61. Voorrips, R. MapChart: Software for the graphical presentation of linkage maps and QTLs. *J. Hered.* **2002**, *93*, 77–78. [CrossRef]

62. Wang, S.; Basten, C.; Zeng, Z. *Windows QTL Cartographer*; Bioinformatics Research Center, North Carolina State University: Raleigh, NC, USA, 2007.

63. Tran, N.H.; Kim, B.S. Sources of resistance to bacterial wilt found in Vietnam collections of pepper (*Capsicum annuum*) and their nuclear fertility restorer genotypes for cytoplasmic male sterility. *Plant. Pathol. J.* **2012**, *28*, 418–422. [CrossRef]

64. Deschamps, S.; Llaca, V.; May, G.D. Genotyping-by-sequencing in plants. *Biology* **2012**, *1*, 460–483. [CrossRef]

65. Poland, J.A.; Rife, T.W. Genotyping-by-sequencing for plant breeding and genetics. *Plant Genome* **2012**, *5*. [CrossRef]

66. Bielenberg, D.G.; Rauh, B.; Fan, S.; Gasic, K.; Abbott, A.G.; Reighard, G.L.; Okie, W.R.; Wells, C.E. Genotyping by sequencing for SNP-based linkage map construction and QTL analysis of chilling requirement and bloom date in peach [*Prunus persica* (L.) Batsch]. *PLoS ONE* **2015**, *10*, e0139406.

67. Boutet, G.; Carvalho, S.A.; Falque, M.; Peterlongo, P.; Lhuillier, E.; Bouchez, O.; Lavaud, C.; Pilet-Nayel, M.-L.; Rivière, N.; Baranger, A. SNP discovery and genetic mapping using genotyping by sequencing of whole genome genomic DNA from a pea RIL population. *BMC Genom.* **2016**, *17*, 121. [CrossRef]

68. Iqbal, M.M.; Huynh, M.; Udall, J.A.; Kilian, A.; Adhikari, K.N.; Berger, J.D.; Erskine, W.; Nelson, M.N. The first genetic map for yellow lupin enables genetic dissection of adaptation traits in an orphan grain legume crop. *BMC Genet.* **2019**, *20*, 68. [CrossRef]

69. Manivannan, A.; Choi, S.; Jun, T.-H.; Yang, E.-Y.; Kim, J.-H.; Lee, E.-S.; Lee, H.-E.; Kim, D.-S.; Ahn, Y.-K. Genotyping by Sequencing-Based Discovery of SNP Markers and Construction of Linkage Map from F5 Population of Pepper with Contrasting Powdery Mildew Resistance Trait. *BioMed Res. Int.* **2021**, *2021*, 6673010. [CrossRef]

70. Thakur, P.P.; Mathew, D.; Nazeem, P.; Abida, P.; Indira, P.; Girija, D.; Shylaja, M.; Valsala, P. Identification of allele specific AFLP markers linked with bacterial wilt [*Ralstonia solanacearum* (Smith) Yabuuchi et al.] resistance in hot peppers (*Capsicum annuum* L.). *Physiol. Mol. Plant Pathol.* **2014**, *87*, 19–24. [CrossRef]

71. Kwon, J.-S.; Nam, J.-Y.; Yeom, S.-I.; Kang, W.-H. Leaf-to-whole plant spread bioassay for pepper and *Ralstonia solanacearum* interaction determines inheritance of resistance to bacterial wilt for further breeding. *Int. J. Mol. Sci.* **2021**, *22*, 2279. [CrossRef]

72. Mimura, Y.; Inoue, T.; Minamiyama, Y.; Kubo, N. An SSR-based genetic map of pepper (*Capsicum annuum* L.) serves as an anchor for the alignment of major pepper maps. *Breed. Sci.* **2012**, *62*, 93–98. [CrossRef]

73. Mathew, D. Analysis of QTL Bw1 and marker CAMS451 associated with the bacterial wilt resistance in hot pepper (*Capsicum annuum* L.). *Plant Gene* **2020**, *24*, 100260. [CrossRef]

74. Kim, S.B.; Kang, W.H.; Huy, H.N.; Yeom, S.I.; An, J.T.; Kim, S.; Kang, M.Y.; Kim, H.J.; Jo, Y.D.; Ha, Y. Divergent evolution of multiple virus-resistance genes from a progenitor in *Capsicum* spp. *New Phytol.* **2017**, *213*, 886–899. [CrossRef]

75. Tai, T.H.; Dahlbeck, D.; Clark, E.T.; Gajiwala, P.; Pasion, R.; Whalen, M.C.; Stall, R.E.; Staskawicz, B.J. Expression of the Bs2 pepper gene confers resistance to bacterial spot disease in tomato. *Proc. Natl. Acad. Sci. USA* **1999**, *96*, 14153–14158. [CrossRef]

76. Kearney, B.; Staskawicz, B.J. Widespread distribution and fitness contribution of *Xanthomonas campestris* avirulence gene *avrBs2*. *Nature* **1990**, *346*, 385–386. [CrossRef]

77. Kunwar, S.; Iriarte, F.; Fan, Q.; Evaristo da Silva, E.; Ritchie, L.; Nguyen, N.S.; Freeman, J.H.; Stall, R.E.; Jones, J.B.; Minsavage, G.V. Transgenic expression of *EFR* and *Bs2* genes for field management of bacterial wilt and bacterial spot of tomato. *Phytopathology* **2018**, *108*, 1402–1411. [CrossRef]

78. Boutrot, F.; Zipfel, C. Function, discovery, and exploitation of plant pattern recognition receptors for broad-spectrum disease resistance. *Annu. Rev. Phytopathol.* **2017**, *55*, 257–286. [CrossRef]

79. Grant, M.R.; Godiard, L.; Straube, E.; Ashfield, T.; Lewald, J.; Sattler, A.; Innes, R.W.; Dangl, J.L. Structure of the *Arabidopsis* RPM1 gene enabling dual specificity disease resistance. *Science* **1995**, *269*, 843–846. [CrossRef]

80. Boyes, D.C.; Nam, J.; Dangl, J.L. The *Arabidopsis thaliana* RPM1 disease resistance gene product is a peripheral plasma membrane protein that is degraded coincident with the hypersensitive response. *Proc. Natl. Acad. Sci. USA* **1998**, *95*, 15849–15854. [CrossRef]

81. Al-Daoude, A.; de Torres Zabala, M.; Ko, J.-H.; Grant, M. RIN13 is a positive regulator of the plant disease resistance protein RPM1. *Plant Cell* **2005**, *17*, 1016–1028. [CrossRef]

82. Van Ooijen, G.; Mayr, G.; Kasiem, M.M.; Albrecht, M.; Cornelissen, B.J.; Takken, F.L. Structure–function analysis of the NB-ARC domain of plant disease resistance proteins. *J. Exp. Bot.* **2008**, *59*, 1383–1397. [CrossRef]

83. Qi, D.; Innes, R.W. Recent advances in plant NLR structure, function, localization, and signaling. *Front. Immunol.* **2013**, *4*, 348. [CrossRef]

84. de Araújo, A.C.; Fonseca, F.C.D.A.; Cotta, M.G.; Alves, G.S.C.; Miller, R.N.G. Plant NLR receptor proteins and their potential in the development of durable genetic resistance to biotic stresses. *Biotechnol. Res. Innov.* **2019**, *3*, 80–94. [CrossRef]

85. Singh, N.K.; Kumar, K.R.R.; Kumar, D.; Shukla, P.; Kirti, P. Characterization of a pathogen induced thaumatin-like protein gene AdTLP from *Arachis diogoi*, a wild peanut. *PLoS ONE* **2013**, *8*, e83963. [CrossRef]

86. Zhang, J.; Wang, F.; Liang, F.; Zhang, Y.; Ma, L.; Wang, H.; Liu, D. Functional analysis of a pathogenesis-related thaumatin-like protein gene TaLr35PR5 from wheat induced by leaf rust fungus. *BMC Plant Biol.* **2018**, *18*, 76. [CrossRef]

87. Ali, G.S.; Hu, X.; Reddy, A. Overexpression of the *Arabidopsis* thaumatin-like protein 1 in transgenic potato plants enhances resistance against early and late blights. *BioRxiv* **2019**, 621649. [CrossRef]

 *horticulturae*

Article

GDS: A Genomic Database for Strawberries (*Fragaria* spp.)

Yuhan Zhou [1], Yushan Qiao [1], Zhiyou Ni [1], Jianke Du [1], Jinsong Xiong [1], Zongming Cheng [1,*] and Fei Chen [2,*]

[1] College of Horticulture, Nanjing Agricultural University, Nanjing 210095, China; 2020104011@stu.njau.edu.cn (Y.Z.); qiaoyushan@njau.edu.cn (Y.Q.); 2020204004@stu.njau.edu.cn (Z.N.); 2017204014@njau.edu.cn (J.D.); jsxiong@njau.edu.cn (J.X.)

[2] College of Tropical Crops, Hainan University, Haikou 570228, China

[*] Correspondence: zmc@njau.edu.cn (Z.C.); feichen@hainanu.edu.cn (F.C.)

Abstract: Strawberry species (*Fragaria* spp.) are known as the "queen of fruits" and are cultivated around the world. Over the past few years, eight strawberry genome sequences have been released. The reuse of these large amount of genomic data, and the more large-scale comparative analyses are very challenging to both plant biologists and strawberry breeders. To promote the reuse and exploration of strawberry genomic data and enable extensive analyses using various bioinformatics tools, we have developed the Genome Database for Strawberry (GDS). This platform integrates the genome collection, storage, integration, analysis, and dissemination of large amounts of data for researchers engaged in the study of strawberry. We collected and formatted the eight published strawberry genomes. We constructed the GDS based on Linux, Apache, PHP and MySQL. Different bioinformatic software were integrated. The GDS contains data from eight strawberry species, as well as multiple tools such as BLAST, JBrowse, synteny analysis, and gene search. It has a designed interface and user-friendly tools that perform a variety of query tasks with a few simple operations. In the future, we hope that the GDS will serve as a community resource for the study of strawberries.

Keywords: strawberry; bioinformatics; database; genome

Citation: Zhou, Y.; Qiao, Y.; Ni, Z.; Du, J.; Xiong, J.; Cheng, Z.; Chen, F. GDS: A Genomic Database for Strawberries (*Fragaria* spp.). *Horticulturae* **2022**, *8*, 41. https://doi.org/10.3390/horticulturae8010041

Academic Editor: Young-Doo Park

Received: 28 November 2021
Accepted: 28 December 2021
Published: 31 December 2021

Publisher's Note: MDPI stays neutral with regard to jurisdictional claims in published maps and institutional affiliations.

1. Introduction

Strawberries (*Fragaria* spp.), comprising of approximately 25 species [1], are plants from the Rosaceae. Their ploidy types range from diploid to decaploid [2,3], while wild members of the genus distributed throughout the northern hemisphere and parts of western South America [4]. The main cultivated and commercial strawberry species is the octoploid *Fragaria* ×*ananassa* (2n = 8x = 56) [5–8]. The first strawberry genome sequence from woodland strawberry (*Fragaria vesca*) was released in 2010 [9]. Since then, more and more strawberry species have been sequenced and annotated. In 2013, the cultivated strawberry (*Fragaria* ×*ananassa*) genome was sequenced using the Illumina and Roche 454 sequencing platforms [10], and was re-sequenced using a combination of short- and long-read approaches, producing a higher-quality assembly [11]. Strawberry genomics research not only promotes our understanding of the origin and evolution of strawberries but also has benefits for strawberry breeding [12].

Given these recent advances in strawberry genomics, it is necessary to establish a free online resource center for the integration of strawberry genome data. Therefore, we integrated the genomes and other related data of eight strawberry species (*Fragaria* ×*ananassa*, *Fragaria iinumae*, *Fragaria nilgerrensis*, *Fragaria nipponica*, *Fragaria nubicola*, *Fragaria orientalis*, *Fragaria vesca*, and *Fragaria viridis*) referring to the databases of other species, such as Arabidopsis, kiwifruit, and walnut [13–19]. We excavated, analyzed, and appropriately clustered these data into the online platform Genome Database for Strawberry (GDS). The GDS provides a user-friendly web interface; it also integrates a series of practical bioinformatics tools that enable researchers to search, browse, or retrieve specific information.

Genomics, transcriptomics and proteomic technology has developed rapidly. The GDS developed here will greatly benefit future application of high-throughput and -omics technologies. In addition, our achievement provides a directly resource for strawberry breeders and research communities, which will further facilitate the development of new strawberry cultivars with improved flavor. Nowadays, the phylogenomic relationships among the strawberry genomes is unclear. The current debate on the evolutionary of strawberries is one of the most important issues in the world. Our database could promote research on strawberry evolution.

2. Materials and Methods

2.1. Web Server and Code

The GDS is based on the web server software Apache (v2.4.41) on Linux operating system. PHP (v7.4) and MySQL (v8.0) were used for back-end code and HTML5, CSS3, and JavaScript for front-end codes. All codes have been submitted to Github (https://github.com/, accessed on 25 December 2021) and can be accessed for free by entering "Han-Oscar/GDS-code". Data were deposited into the mysql database in batches and displayed on the website upon searching using Navicat (version 15) software.

2.2. Formatting the Genomic Data of Strawberries

The GDS cover the genomic sequences from eight strawberry species from GDR (Genome Database for Rosaceae) and Kazusa (Strawberry GARDEN). *Fragaria nipponica*, *Fragaria viridis*, and *Fragaria orientalis* each has one version of the genome data, *Fragaria ×ananassa*, *Fragaria nilgerrensis*, *Fragaria nubicola*, and *Fragaria iinumae* has two, and *Fragaria vesca* has three. Only the latest version of genome data was used for downstream analysis. One can then download the genome sequence and protein sequence and gene annotations of the eight strawberry species which were analyzed and classified well. Specifically, the gene and protein IDs of *Fragaria xananassa* and *Fragaria vesca* remain the same, and those of *Fragaria iinumae*, *Fragaria nilgerrensis*, *Fragaria nipponica*, *Fragaria nubicola*, and *Fragaria orientalis* should start with "FII_", "FNil_", "FNI_", "FNub_", and "FOR_" before the data can be analyzed by different bioinformatics software.

3. Results

3.1. Overview of the GDS

We created a user-friendly website for the GDS to make it easier for the scientific community to use. The domain name of the GDS is http://eplant.njau.edu.cn/strawberry (accessed on 25 December 2021), and it currently has two terabytes of server space. We implemented the GDS in Apache httpd, HTML5, PHP, and MySQL. The GDS web pages were created using HTML and Bootstrap, and were connected to the database through Apache, PHP, and MySQL to allow for the query of gene-related information by users.

3.2. The Homepage of GDS

The interface of the GDS included five parts (Figure 1): the "navigation bar", the "species gallery", the "tool sets", the "brief introduction", and the "live visitor statistics". At the top of the homepage, the navigation bar (Figure 1a) consists of six labels: Logo, Species, Tools, Download, Community, and Help. Below the navigation bar are quick links to the eight strawberry species. A suite of bioinformatics tools is on the right (Figure 1b,c). Below the species gallery and tool set is a brief introduction to the GDS (Figure 1d). Finally, live statistics (Figure 1e) tools were implemented to collect the number and location of visits.

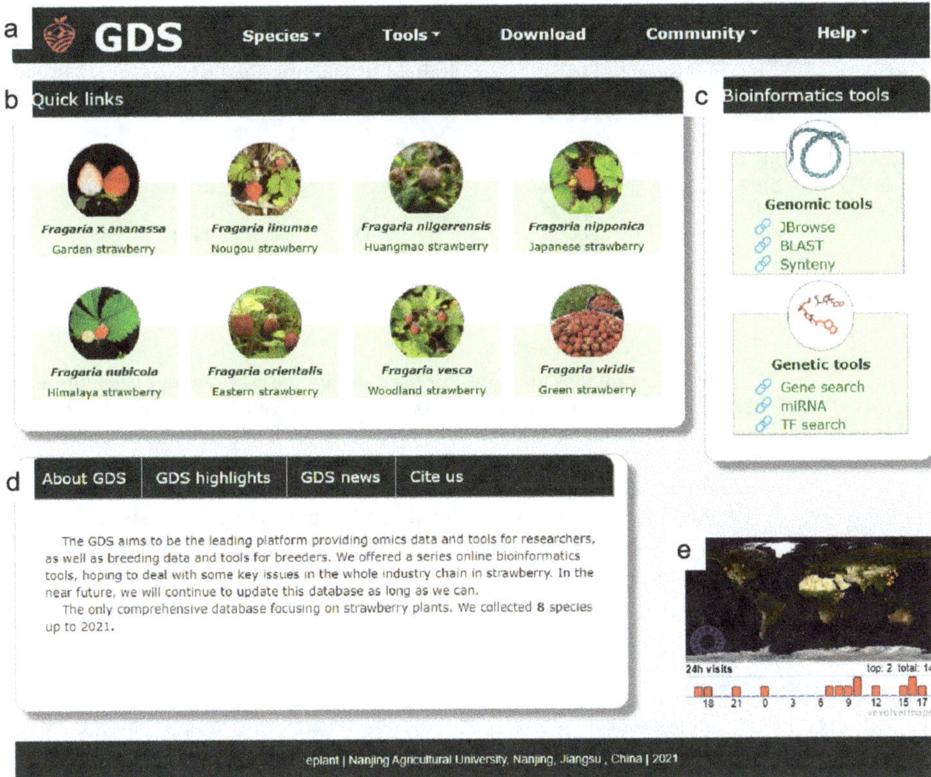

Figure 1. The GDS homepage. (**a**) Navigation bar; (**b**) species gallery; (**c**) tool set; (**d**) brief introduction, and (**e**) live visitor statistics.

3.3. Introduction to the Strawberry Species and Genomes

In the individual species module, we provide visitors with the eight species of strawberry. The first part of the module includes the Latin, English, and Chinese names of the species. The second part provides taxonomic information. In the third part, we summarize detailed and accurate information for the species. The fourth part lists genome assembly details such as genome size, contig N50, and sequencing technology, and the final part provides references to the relevant genome report.

3.4. Data Sets

The reference genome sequence and general feature format (GFF), coding sequence (CDS), protein sequence (PEP) files, and expression data are included in the GDS [20]. A summary of the genomic data currently available in the GDS is presented in Table 1. The versions of the genomes from top to bottom in the table are v1.0a2, r1.1, v1.0, r1.1, r1.1, r1.1, v4.0a2, and v1.0 [21–25].

Table 1. Statistics of the genome features for eight strawberry species.

Species	Assembly Size (Mb)	Ploidy	Scaffold N50 (kb)	Contig N50 (kb)	BUSCO V5 (%)
Fragaria × ananassa	805.5	8x = 56	5980.469	79.973	99.6
Fragaria iinumae	199. 6	2x = 14	4.112	0.824	98.4
Fragaria nipponica	206.4	2x = 14	1.952	0.617	46.7
Fragaria nubicola	203.7	2x = 14	1.982	0.618	92.0
Fragaria orientalis	214.2	4x = 28	1.913	0.480	23.9
Fragaria vesca	220.8	2x = 14	36,100	7900	98.2
Fragaria viridis	214.9	2x = 14	29,200	3500	94.2
Fragaria nilgerrensis	270.3	2x = 14	38,300	8510	93.8

3.5. Completeness of the Genomes

BUSCO provides measures for the quantitative assessment of genome assembly, gene set, and transcriptome completeness based on evolutionarily informed expectations of gene content from near-universal single-copy orthologs selected from OrthoDB [26]. On the species introduction page, we have integrated BUSCO5 results of the eight species, including complete and single-copy, complete and duplicated, fragmented, and missing orthologs. The results are presented as a bar graph (Figure 2), showing that *F. ananassa* currently has the most complete gene assembly of the eight strawberry species.

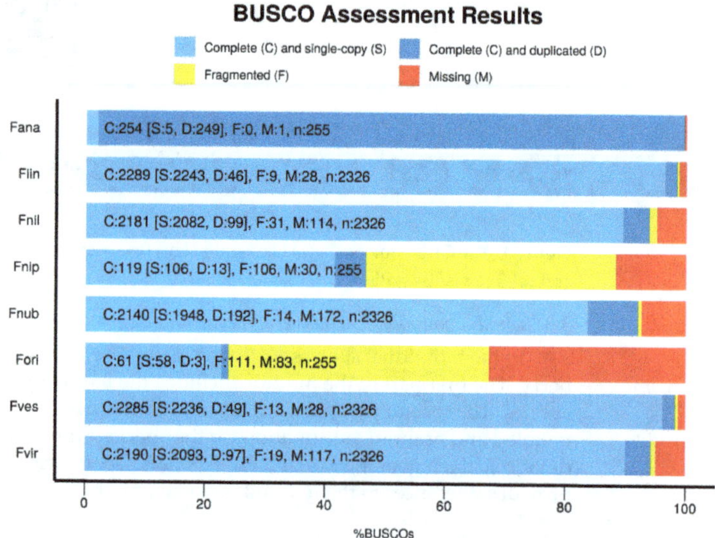

Figure 2. Genomic completeness of eight strawberry species evaluated using BUSCO V5 software. Fana, *Fragaria × ananassa*; Fiin, *Fragaria iinumae*; Fnil, *Fragaria nilgerrensis*; Fnip, *Fragaria nipponica*; Fnub, *Fragaria nubicola*; Fori, *Fragaria orientalis*; Fves, *Fragaria vesca*; and Fvir, *Fragaria viridis*.

3.6. Phylogenomic Relationships among the Strawberry Genomes

OrthoFinder [27,28] is a fast, accurate, and comprehensive platform for comparative genomics. It identifies orthogroups and orthologs, infers rooted gene trees for all orthogroups, and identifies all of the gene duplication events in the gene trees. To analyze the relationships among eight strawberry genomes, we constructed a phylogenetic tree using OrthoFinder (V2.5.2) software and included two additional species, *Rosa chinensis*, and *Arabidopsis thaliana* (Figure 3a).

Figure 3. Phylogenomic analysis of eight strawberry species. (**a**) A phylogenomic species tree of eight strawberry species; (**b**) *F. ×ananassa*; (**c**) *F. iinumae*; (**d**) *F. nilgerrensis*; (**e**) *F. nipponica*; (**f**) *F. nubicola*; (**g**) *F. orientalis*; (**h**) *F. vesca*, and (**i**) *F. viridis*. The picture (**c**) is from https://en.wikipedia.org/wiki/File:Fragaria_iinumae_(fruits).jpg (accessed on 10 November 2021); picture (**i**) is from https://en.wikipedia.org/wiki/File:%D0%9A%D0%BB%D1%83%D0%B1%D0%BD%D0%B8%D0%BA%D0%B0_(Fragaria_viridis).jpeg (accessed on 10 November 2021); picture (**f**) is from http://www.fpcn.net/uploads/allimg/131107/2-13110G30231V8.JPG (accessed on 10 November 2021).

Fragaria virginiana and *Fragaria chiloensis* are the genomes of the progenitor species of *Fragaria ×ananassa*. However, the dispute over its diploid ancestor has lasted for more than half a century and is still unresolved. In 2019, Edger et al. speculated that it has four different diploid ancestors, *F. vesca*, *F. iinumae*, *F. viridis* and *F. nipponica* [11]. Unexpectedly, just a few months later, Liston and others completed a reanalysis of the same set of data, but they came to a completely different conclusion. They believed that the octoploid strawberry has only two existing ancestors, *F. vesca* and *F. iinumae* [29]. Edger et al. insisted on the previous conclusion [30]. The structure of our phylogenetic tree [31] clearly indicates that *Fragaria vesca* is closest to *Fragaria ×ananassa*. However, because of the low sequencing and assembled technology or gene introgression, there is no direct evidence indicating the origin of the cultivated strawberry.

The pictures below the tree (Figure 3b–i) shows *F. ×ananassa*, *F. iinumae*, *F. nilgerrensis*, *F. nipponica*, *F. nubicola*, *F. orientalis*, *F. vesca*, and *F. viridis*. Photographs b, d, e, g and h were provided by our colleague, Dr. Qiao and the others were obtained from Wikipedia or Baidu.

3.7. Genomic Comparison of Gene Orthogroups

To provide an overview of the comparison among these strawberry genomes, we compared the number of gene orthogroups identified by Orthofinder in the strawberry genomes. We uploaded the orthogroup data to an online Venn diagram tool (https://www.

vandepeerlab.org/?q=tools/venn-diagrams, accessed on 10 November 2021) to generate a Venn diagram showing the shared and unique gene orthogroups in *F. iinumae*, *F. nilgerrensis*, *F. nipponica*, *F. vesca*, and *F. viridis*. The gene orthogroup numbers are shown in each segment of the diagram (Figure 4); 13,766 gene orthogroups were shared among the five species, and 13,380 gene families appeared to be unique to *F. nipponica*.

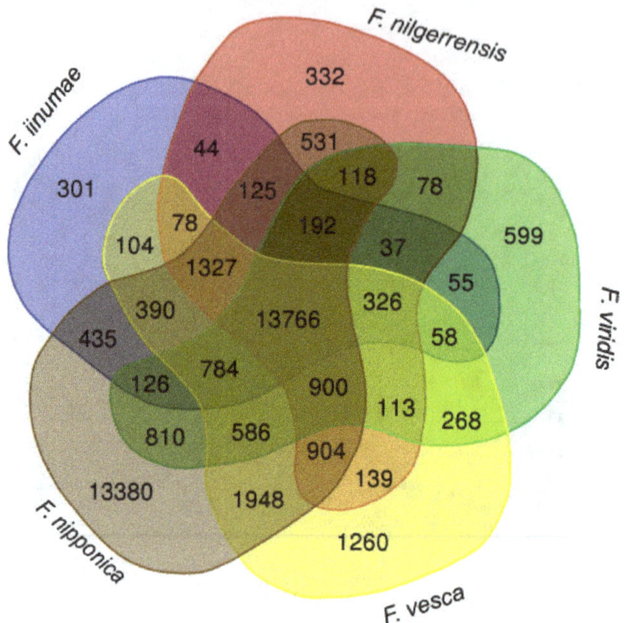

Figure 4. Venn diagram of gene orthogroups in five diploid and wild *Fragaria* species. Comparison of the number of shared gene families among five diploid strawberries, *F. iinumae*, *F. nilgerrensis*, *F. nipponica*, *F. vesca*, and *F. viridis*.

The number of predicted genes was quite higher in F. nipponica, compared with the other four species, and the number of specific genes for F. nipponica was also extremely high (13,380). The reason why so many genes were predicted in *F. nipponica* is because there were more than 80,000 proteins annotated, only using Illumina sequencing technology, in 2014. In the future, with the improvement of the technology of sequencing, this problem will disappear.

3.8. Gene Annotations

There are tens of thousands of genes and proteins in the eight strawberry species, and these sequences contain large amounts of valuable species information for which researchers are searching. Consequently, we have integrated millions of data into the GDS Gene Search tool for obtaining detailed information on target genes. The following are the types of detailed gene information that our tool integrates.

1. Gene family annotation. The ancestral genes of strawberries have undergone genomic duplication and mutation during their long evolutionary history [32], resulting in a series of related genes with similar conserved sequence motifs. The Pfam [33] (http://pfam.xfam.org/, accessed on 10 November 2021) database is a large collection of protein families. Each family is represented by multiple sequence alignments and a hidden Markov model (HMMs) [34]. We have analyzed the proteins of the eight strawberry species using the Pfam 34.0 database and hmmscan (version 3.3) software.

2. KEGG (Kyoto Encyclopedia of Genes and Genomes) annotation. KEGG is a resource for understanding the functions and utilities of biological systems, such as the cell,

organism, and ecosystem. It contains molecular-level information, especially large-scale datasets generated by genome sequencing and other high-throughput technologies [35]. KofamKOALA is a web server that assigns KEGG Orthologs (KOs) to protein sequences by homology search against a database of profile hidden Markov models (KOfam) with precomputed adaptive thresholds. KofamKOALA was installed using Ruby (v2.4 and above, v2.7 was used in this study), HMMER (v3.1 and above, v3.3 was used here), and Parallel (the latest version). The GDS uses KofamKOALA [36] software to make KEGG predictions that contain the KO IDs and more exhaustive information from the official website (https://www.kegg.jp/, accessed on 10 November 2021). We use KofamKOALA (v1.2), which relies on a file named exec_annotation, to analyze the protein files of eight kinds of strawberries.

3. GO annotation. GO [37] is a database established by the Gene Ontology Consortium. It aims to establish a database that is applicable to various species and that limits and describes the functions of genes and proteins. The updated semantic vocabulary standard is applicable to all species. By establishing a set of controlled vocabulary terms with a dynamic form, GO annotations can describe the roles of genes and proteins in cells and organisms. InterPro [38,39] was developed based on Java and aggregates data resources from multiple functional annotation databases such as Pfam, Panther, SMART, SUPERFAMILY, and tmhmm. It predicts the biological functions of proteins by classifying their sequences into protein families and predicting protein domains. InterProscan (v5.5) was used to annotate proteins from the eight strawberry species. A comparison library is available upon downloading the latest version of InterProscan. Instructions on InterProscan can be obtained by entering "./interproscan.sh" in the terminal. The final data can be obtained from the MySQL database.

4. Signal peptide prediction [40]. Signal peptides are short (5–30 amino acid) peptide chains that guide the transfer of newly synthesized proteins to the secretory pathway. The SignalP [41] software tool predicts whether there is a potential signal peptide cleavage site and identifies its location in a given amino acid sequence. Users may enter the "singalP" folder of the download interface to download data. SignalP (5.0 version) is used here with command "signalp -batch 30,000 -org euk -fasta proteins" for the analysis of proteins from the eight strawberry species.

To date, eight nuclear genomes, 436,160 protein sequences, 3,107,804 GO annotations, 27,481 signal peptides, and 1918 transcription factors [42] (Table 2) have been downloaded, analyzed, and organized in the GDS MySQL database.

Table 2. Statistics of whole datasets in GDS.

Data Type	Count
Nuclear genome	8
Choroplast genome	7
Coding sequence	455,467
Protein	436,160
GO term	3,107,804
KEGG	309,589
Gene family	243,687
Signal peptide	27,481
TF	1918

3.9. Sequence Searches Using Basic Local Alignment Search Tool (BLAST)

Sequence similarity comparison is a widely used basic bioinformatics tool for the identification of possible homology between sequences and potential similarities in gene function [43]. Hence, it is necessary for most users to find regions of similarity between biological sequences in gene information databases. GDS employs the free, open-source, and powerful Sequenceserver software for BLAST searches [44]. This interface of Sequenceserver is simple, user-friendly, and powerful (Figure 5). SequenceServer has a simple interface, it performs BLAST and visually inspect BLAST results for biological interpretation. It uses simple algorithms to prevent potential errors during analysis and provides flexible text-based, visual outputs to support researchers' work efficiency. SequenceServer is a BLAST+ server for personal use with a clear and thoughtful design. It contains genomic sequences, CDS, and protein sequences of strawberries, and uses jstree to optimize BLAST to offer clear visualization of complicated results.

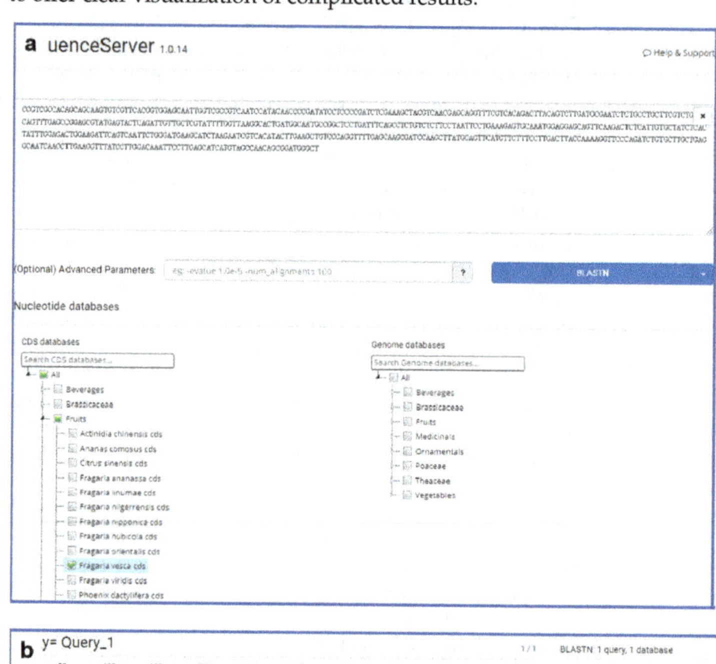

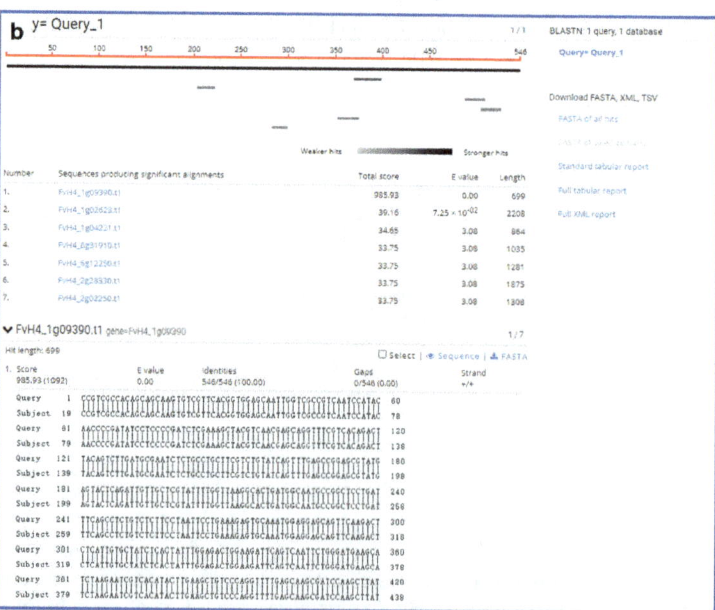

Figure 5. The BLAST tool integrated into GDS. (**a**) A user can enter genomic, PEP, or CDS sequences into the text box, then select the species name below; (**b**) the resulting alignment scores are ranked from high to low, and the details of each sequence alignment are given below the list of scores.

3.10. Genomic Visualization Using JBrowse

A genome browser is a software tool that can be deployed on the server side so that users can access online platforms. JBrowse is a fully featured genome browser that can visualize various types of genome-located data, located in a variety of different data stores, and interfacing to other client and server applications. We used JBrowse [45] built using HTML5 and JavaScript. It integrates and visualizes various existing genome data, including eight nuclear genomes and seven chloroplast genomes [46] so that users can visually browse and analyze the genome and various types of annotation data with strong scalability (Figure 6). In addition, the genome browser can support other types of data, such as repetitive sequences.

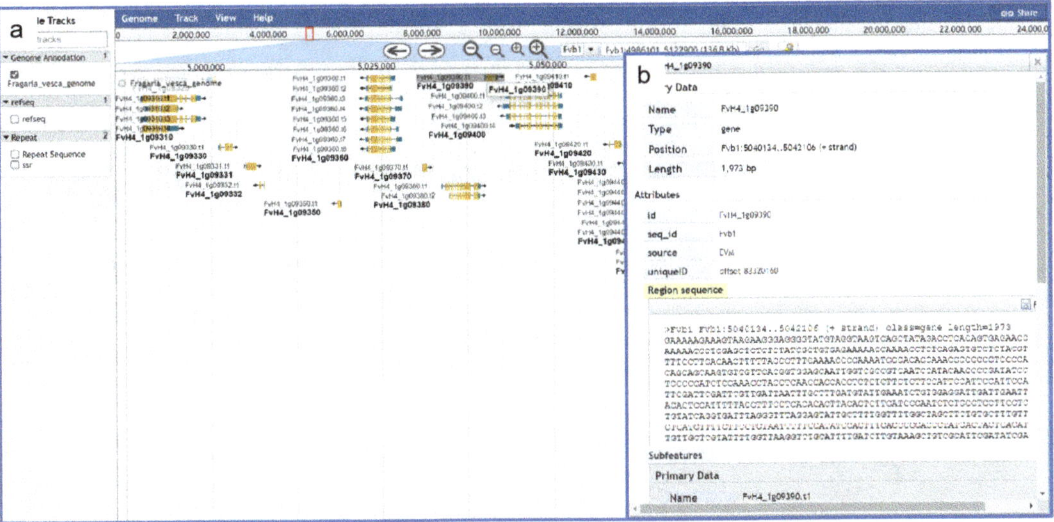

Figure 6. The JBrowse tool integrated into GDS for visualization of strawberry genomic details. (**a**) Gene visualization interface; (**b**) detailed data on individual genes.

3.11. Tracing Whole-Genome Duplication Using Synteny Browse Search

Given the close phylogenetic relationships among strawberry species, there are likely to be many homologous gene blocks in their genomes. The Python version of MCScan [47] was used to identify homologous gene blocks in the genomes of the strawberry species. We selected four species (Figure 7), including cultivated strawberry, to use in searches of homologous genes, as well as upstream and downstream genes. Scientists can look for syntenic genes of *F. vesca* by entering a gene identifier, finding the homologous gene(s), and using them as input for a subsequent gene search.

3.12. microRNA Search

microRNAs (miRNAs) are a class of non-coding single-stranded RNA molecules with a length of approximately 22 nucleotides that are encoded by endogenous genes. The Rfam database [48,49] is a collection of RNA families, each represented by multiple sequence alignments, consensus secondary structures, and covariance models. The GDS use cmscan [50] from the Infernal (V1.1.4) software package to predict the miRNAs in the six species of strawberry with high-quality assemblies.

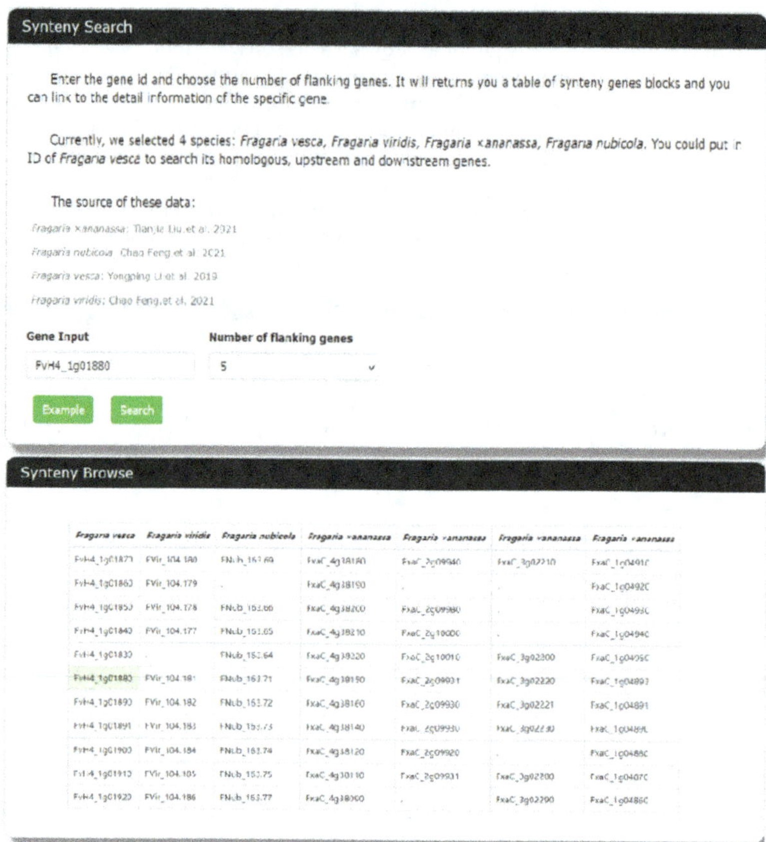

Figure 7. The synteny search tool in GDS is designed for whole-genome duplication analyses. Researchers can use Synteny Browse Search to look for syntenic genes by entering a gene identifier and selecting a number of flanking genes to be presented.

3.13. Transcription Factor Search

Transcription factors play an important role in all biological processes, from seed germination to senescence. Therefore, it is critical for researchers to gain a good understanding of the relationship between the structures and functions of various transcription-factor families. iTAK [51] is a program that can identify plant transcription factors (TFs), transcriptional regulators (TRs), and protein kinases (PKs) based on protein or nucleotide sequences. It then classifies individual TFs, TRs, and PKs into different gene families. iTAK (v1.7) is used here to identify and analyze transcription factors from the six highly assembled strawberry species.

3.14. Gene Search

From the search results of BLAST and JBrowse, scientists can enter a gene identifier to search for information about the gene version (Figure 8a), protein and CDS sequence (Figure 8b), KEGG annotation, gene family, signal peptides (Figure 8c), and GO annotation (Figure 8d). The results include links to the corresponding annotation databases for more information, as well as gene expression data (Figure 8e) from mature pollen.

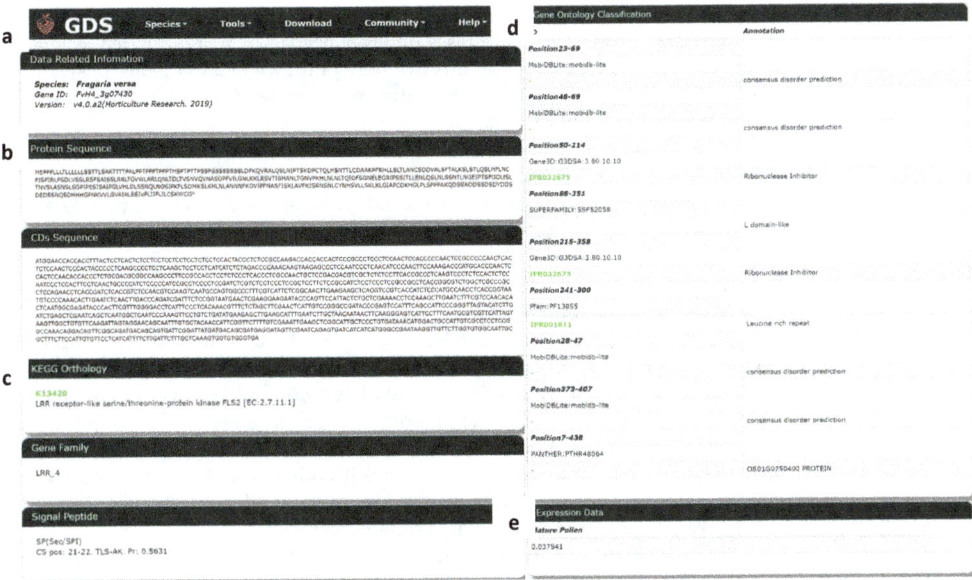

Figure 8. Gene search tool in GDS. (**a**) Related information; (**b**) protein and CDS sequence; (**c**) KEGG annotation, gene family, and signal peptides; (**d**) GO annotation, and (**e**) expression.

3.15. Download

The Download module provides access to the genome assembly, CDS and PEP sequences, annotation data, and miRNA downloads in FASTA and GFF3 formats. Chloroplast genomes of seven strawberry species and related publications are also available. We implemented an FTP site to store and share the data, which users can download at a rapid speed.

3.16. Community of Strawberry Researchers

In the Community module, we provided links to 39 horticultural conferences and 11 relevant publications on strawberry genome research. We also included an FAQ section to explain how to use the database and a contact list for the researchers who established it.

4. Discussion

The rapid development of genome sequencing technology has enabled the sequencing, assembly, and annotation of many plant genomes, providing genetic information on plant growth, development, and evolution. Genome sequencing and analysis technologies have not only deepened our understanding of plant species but also accelerated gene functional studies and molecular breeding. There are many strawberry species and varieties with complicated genomes, and their genome data are refined and updated very often. For example, the *Fragaria vesca* genome (214.4 Mb) was the first sequenced strawberry genome [10]. However, its quality was not ideal. Later on, Edger and associates from the University of California sequenced the genome of woodland strawberry Hawaii-4 using single-molecule real-time (SMRT) sequencing [22] and constructed a more complete genome map (V4.0). The SMRT sequencing can produce much longer contigs, greatly facilitating genome assembly and annotation. Specifically, the length of contig N50 of V4.0 reached 7.9 Mb, 300 times longer than those of V1.0, and >99.8% of the contigs were successfully mapped to the seven chromosomes. This new strawberry genome map offers more accurate sequences and detailed location information. The polyploidization of strawberries, which contain diploid, tetraploid (*Fragaria orientalis*) [5], and octoploid (*Fragaria ×ananassa*) varieties, has made the sharing, analysis, and integration of their genomic data a difficult task. A more convenient online database with multiple integrated and classified strawberry genomes is urgently needed. It will facilitate the gene-functional studies, thereby promoting the improvement of the yield and quality of strawberries [52]. To our knowledge, GDS is currently the only up-to-date database for strawberries that integrates multiple bioinformatics tools.

The storage and analysis of strawberry genome data are also hot research topics, and databases such as GDR (Genome Database for Rosaceae) and Kazusa (Strawberry garden) were created for these purposes. Although GDR and Kazusa have developed databases for strawberries, these databases have a number of problems that require urgent solutions. First, most of the genomic data are unprocessed and scattered. The data lack functional annotations, are not clustered in gene families, and are not preformatted for searching. After downloading data from these public databases, one must process the data oneself, which is a challenge for those with less expertise in bioinformatics. More importantly, some websites are difficult to access in China, and data downloads are also greatly restricted sometimes. To date, there is no specific, widely available database for strawberry research. Here, we use the latest version of the software for the analysis of strawberry proteins that are not all available on other websites. In addition, our laboratory specializes in strawberry research and has extensive collaboration with other strawberry research groups. In the future, newly released strawberry sequencing data will be updated in the database and made available to all researchers in a timely manner.

GDS stores the genome sequences of eight strawberry species and related gene annotations. The advanced and popular BLAST and JBrowse tools have been implemented, as well as a syntenic block search tool and an miRNA finder. This database serves as a central portal for the strawberry research community, enabling researchers to download genomes, protein sets, transcription data, and recently published articles on the strawberry genome. The GDS will be constantly updated when new genomes, transcriptomes, and other types of genetic datasets are published. In the future, we will develop and establish more gene online analysis tools to facilitate strawberry researchers in conducting online analysis. We will do our best to develop and deploy new omics tools in the GDS to provide a better user experience. Furthermore, GDS will contain studies and statistics on strawberries' breeding. In summary, this new database incorporates published strawberry plant genomes, multiple analysis tools, new features for strawberry plant genomic data analysis, gene function characterization, synteny and miRNA search, and publication, which is easily accessed and can potentially benefit the strawberry plant research community.

Author Contributions: Z.C. and F.C. designed and led this project. Y.Z. constructed the GDS. Y.Q., J.D., Z.N. and J.X. analyzed the data. Y.Z. wrote the draft manuscript. All authors have read and agreed to the published version of the manuscript.

Funding: Fei Chen acknowledges funding from the Fundamental Research Fund for the Central University (KYXJ202004) and a starting fund from Nanjing Agricultural University (804012). Zongming Cheng and Jinsong Xiong acknowledge funding from Priority Academic Program Development of Jiangsu Higher Education Institutions (PAPD). This work was supported by the National Natural Science Foundation of China (Grant no. 32072540, 31872056).

Institutional Review Board Statement: Not applicable.

Informed Consent Statement: Not applicable.

Data Availability Statement: All code about the database is available on Github (https://github.com/Han-Oscar/GDS-code, accessed on 10 November 2021).

Acknowledgments: We thank Yushan Qiao, Zhiyou Ni, Jianke Du for valuable comments and suggestions on our database. We thank Xiaogang Lei and Xiaojiang Li for providing technical assistance to our database development and Fei Chen, Zongming Cheng for assistance with the correction of the English language in the manuscript.

Conflicts of Interest: The authors declare that they have no conflict of interest.

References

1. Kim, E.H. A New Species of Fragaria (Rosaceae) from Oregon. *J. Bot. Res. Inst. Tex.* **2012**, *6*, 9–15.
2. Kim, E.H.; Preeda, N.; Tomohiro, Y. Decaploidy in *Fragaria iturupensis* (Rosaceae). *Am. J. Bot.* **2009**, *96*, 713–719. [CrossRef]
3. Van de Peer, Y.; Mizrachi, E.; Marchal, K. The evolutionary significance of polyploidy. *Nat. Rev. Genet.* **2017**, *18*, 411–424. [CrossRef] [PubMed]
4. Lei, J.J.; Xue, L.; Guo, R.X.; Dai, H.P. The *Fragaria* species native to China and their geographical distribution. *Acta Hortic.* **2017**, *1156*, 37–46. [CrossRef]
5. Detlef, U.; Klaus, O. Diversity of volatile patterns in sixteen *Fragaria vesca* L. accessions in comparison to cultivars of *Fragaria* ×*ananassa*. *J. Appl. Bot. Food Qual.* **2013**, *86*, 37–46. [CrossRef]
6. Hendrix, B.; Stewart, J.M. Estimation of the nuclear DNA content of gossypium species. *Ann. Bot.* **2005**, *95*, 789–797. [CrossRef]
7. Tennessen, J.A.; Govindarajulu, R.; Ashman, T.L.; Liston, A. Evolutionary origins and dynamics of octoploid strawberry subgenomes revealed by dense targeted capture linkage maps. *Genome Biol. Evol.* **2014**, *6*, 3295–3313. [CrossRef]
8. Cappelletti, R.; Sabbadini, S.; Mezzetti, B. Strawberry *(Fragaria* ×*ananassa)*. *Methods Mol. Biol.* **2015**, *1224*, 217–227. [CrossRef]
9. Shulaev, V.; Sargent, D.J.; Crowhurst, R.N.; Mockler, T.C.; Folkerts, O.; Delcher, A.L.; Jaiswal, P.; Mockaitis, K.; Liston, A.; Mane, S.P.; et al. The genome of woodland strawberry *(Fragaria vesca)*. *Nat. Genet.* **2011**, *43*, 109–116. [CrossRef]
10. Hirakawa, H.; Shirasawa, K.; Kosugi, S.; Tashiro, K.; Nakayama, S.; Yamada, M.; Kohara, M.; Watanabe, A.; Kishida, Y.; Fujishiro, T.; et al. Dissection of the Octoploid Strawberry Genome by Deep Sequencing of the Genomes of *Fragaria* Species. *DNA Res.* **2014**, *21*, 169–181. [CrossRef]
11. Edger, P.P.; Poorten, T.J.; VanBuren, R.; Hardigan, M.A.; Colle, M.; McKain, M.R.; Smith, R.D.; Teresi, S.J.; Nelson, A.D.L.; Wai, C.M.; et al. Origin and evolution of the octoploid strawberry genome. *Nat. Genet.* **2019**, *51*, 541–547. [CrossRef]
12. Chen, F.; Song, Y.; Li, X.; Chen, J.; Mo, L.; Zhang, X.; Lin, Z.; Zhang, L. Genome sequences of horticultural plants: Past, present, and future. *Hortic. Res.* **2019**, *6*, 112. [CrossRef]
13. Xiaoming, S.; Fulei, N.; Wei, C.; Xiao, M.; Ke, G.; Qihang, Y.; Jinpeng, W.; Nan, L.; Pengchuan, S.; Qiaoying, P.; et al. Coriander Genomics Database: A genomic, transcriptomic, and metabolic database for coriander. *Hortic. Res.* **2020**, *7*, 55. [CrossRef]
14. Tam, P.S.; Peter, L.; Scott, C.E. GigaDB: Announcing the GigaScience database. *Gigascience* **2012**, *1*, 11. [CrossRef]
15. Junyang, Y.; Jiacheng, L.; Wei, T.; Ya, Q.W.; Xiaofeng, T.; Wei, L.; Ying, Y.; Lihuan, W.; Shengxiong, H.; Congbing, F.; et al. Kiwifruit Genome Database (KGD): A comprehensive resource for kiwifruit genomics. *Hortic. Res.* **2020**, *7*, 117. [CrossRef]
16. Xiao, Q.; Li, Z.; Qu, M.; Xu, W.; Su, Z.; Yang, J. LjaFGD: *Lonicera japonica* functional genomics database. *J. Integr. Plant Biol.* **2021**, *63*, 1422–1436. [CrossRef]
17. Wenlei, G.; Junhao, C.; Jian, L.; Jianqin, H.; Zhengjia, W.; Kean-Jin, L. Portal of *Juglandaceae*: A comprehensive platform for *Juglandaceae* study. *Hortic. Res.* **2020**, *7*, 35. [CrossRef]
18. Lamesch, P.; Berardini, T.Z.; Li, D.; Swarbreck, D.; Wilks, C.; Sasidharan, R.; Muller, R.; Dreher, K.; Alexander, D.L.; Garcia-Hernandez, M.; et al. The *Arabidopsis* Information Resource (TAIR): Improved gene annotation and new tools. *Nucleic Acids Res.* **2012**, *40*, D1202–D1212. [CrossRef]
19. Chen, F.; Dong, W.; Zhang, J.; Guo, X.; Chen, J.; Wang, Z.; Lin, Z.; Tang, H.; Zhang, L. The Sequenced Angiosperm Genomes and Genome Databases. *Front. Plant Sci.* **2018**, *9*, 418. [CrossRef]
20. Bo, L.; Victor, R.; Ron, M.S.; James, A.T.; Colin, N.D. RNA-Seq gene expression estimation with read mapping uncertainty. *Bioinformatics* **2010**, *26*, 493–500. [CrossRef]
21. Liu, T.; Li, M.; Liu, Z.; Ai, X.; Li, Y. Reannotation of the cultivated strawberry genome and establishment of a strawberry genome database. *Hortic. Res.* **2021**, *8*, 41. [CrossRef]
22. Edger, P.P.; VanBuren, R.; Colle, M.; Poorten, T.J.; Wai, C.M.; Niederhuth, C.E.; Alger, E.I.; Ou, S.; Acharya, C.B.; Wang, J.; et al. Single-molecule sequencing and optical mapping yields an improved genome of woodland strawberry *(Fragaria vesca)* with chromosome-scale contiguity. *Gigascience* **2018**, *7*, 1–7. [CrossRef]
23. Zhang, J.; Lei, Y.; Wang, B.; Li, S.; Yu, S.; Wang, Y.; Li, H.; Liu, Y.; Ma, Y.; Dai, H.; et al. The high-quality genome of diploid strawberry *(Fragaria nilgerrensis)* provides new insights into anthocyanin accumulation. *Plant Biotechnol. J.* **2020**, *18*, 1908–1924. [CrossRef]
24. Feng, C.; Wang, J.; Harris, A.J.; Folta, K.M.; Zhao, M.; Kang, M. Tracing the Diploid Ancestry of the Cultivated Octoploid Strawberry. *Mol. Biol. Evol.* **2021**, *38*, 478–485. [CrossRef]
25. Li, Y.; Pi, M.; Gao, Q.; Liu, Z.; Kang, C. Updated annotation of the wild strawberry *Fragaria vesca* V4 genome. *Hortic. Res.* **2019**, *6*, 61. [CrossRef]
26. Seppey, M.; Manni, M.; Zdobnov, E.M. BUSCO: Assessing Genome Assembly and Annotation Completeness. *Methods Mol. Biol. (Clifton N.J.)* **2019**, *1962*, 227–245. [CrossRef]
27. David, M.E.; Steven, K. OrthoFinder: Phylogenetic orthology inference for comparative genomics. *Genome Biol.* **2019**, *20*, 238. [CrossRef]
28. David, M.E.; Steven, K. OrthoFinder: Solving fundamental biases in whole genome comparisons dramatically improves orthogroup inference accuracy. *Genome Biol.* **2015**, *16*, 157. [CrossRef]
29. Liston, A.; Wei, N.; Tennessen, J.A.; Junmin, L.; Ming, D.; Tia-Lynn, A. Revisiting the origin of octoploid strawberry. *Nat. Genet.* **2020**, *52*, 2–4. [CrossRef]

30. Edger, P.P.; McKain, M.R.; Yocca, A.E.; Knapp, S.J.; Qiao, Q.; Zhang, T. Reply to: Revisiting the origin of octoploid strawberry. *Nat. Genet.* **2020**, *52*, 5–7. [CrossRef]

31. Daniel, P.; James, J.L.; Richard, E.H. Phylogenetic Relationships Among Species of *Fragaria* (Rosaceae) Inferred from Non-coding Nuclear and Chloroplast DNA Sequences. *Syst. Bot.* **2000**, *25*, 337–348. [CrossRef]

32. Chen, P.; Liu, Q.Z. Genome-wide characterization of the WRKY gene family in cultivated strawberry (*Fragaria × ananassa* Duch.) and the importance of several group III members in continuous cropping. *Sci. Rep.* **2019**, *9*, 8423. [CrossRef] [PubMed]

33. Sara, E.; Jaina, M.; Alex, B.; Sean, R.E.; Aurélien, L.; Simon, C.P.; Matloob, Q.; Lorna, J.R.; Gustavo, A.S.; Alfredo, S.; et al. The Pfam protein families database in 2019. *Nucleic Acids Res.* **2019**, *47*, D427–D432. [CrossRef]

34. Finn, R.D.; Clements, J.; Eddy, S.R. HMMER web server: Interactive sequence similarity searching. *Nucleic Acids Res.* **2011**, *39*, 29–37. [CrossRef]

35. Kanehisa, M.; Sato, Y. KEGG Mapper for inferring cellular functions from protein sequences. *Protein Sci.* **2020**, *29*, 28–35. [CrossRef]

36. Aramaki, T.; Blanc-Mathieu, R.; Endo, H.; Ohkubo, K.; Kanehisa, M.; Goto, S.; Ogata, H. KofamKOALA: KEGG Ortholog assignment based on profile HMM and adaptive score threshold. *Bioinformatics* **2020**, *36*, 2251–2252. [CrossRef]

37. The Gene Ontology Consortium. The Gene Ontology Resource: 20 years and still GOing strong. *Nucleic Acids Res.* **2019**, *47*, D330–D338. [CrossRef]

38. Robert, D.F.; Teresa, K.A.; Patricia, C.B.; Alex, B.; Peer, B.; Alan, J.B.; Hsin-Yu, C.; Zsuzsanna, D.; Sara, E.; Matthew, F.; et al. InterPro in 2017—beyond protein family and domain annotations. *Nucleic Acids Res.* **2017**, *45*, D190–D199. [CrossRef]

39. Blum, M.; Chang, H.; Chuguransky, S.; Grego, T.; Kandasaamy, S.; Mitchell, A.; Nuka, G.; PaysanLafosse, T.; Qureshi, M.; Raj, S.; et al. The InterPro protein families and domains database: 20 years on. *Nucleic Acids Res.* **2020**, *49*, D344–D354. [CrossRef]

40. Henrik, N.; Konstantinos, D.T.; Søren, B.; Gunnar, H. A Brief History of Protein Sorting Prediction. *Protein J.* **2019**, *38*, 200–216. [CrossRef]

41. José, J.A.A.; Konstantinos, D.T.; Casper, K.S.; Thomas, N.P.; Ole, W.; Søren, B.; Gunnar, V.H.; Henrik, N. SignalP 5.0 improves signal peptide predictions using deep neural networks. *Nat. Biotechnol.* **2019**, *37*, 420–423. [CrossRef]

42. Pérez-Rodríguez, P.; Riaño-Pachón, D.M.; Corrêa, L.G.G.; Rensing, S.A.; Kersten, B.; Mueller-Roeber, B. PlnTFDB: Updated content and new features of the plant transcription factor database. *Nucleic Acids Res.* **2010**, *38*, D227–D234. [CrossRef]

43. Christiam, C.; George, C.; Vahram, A.; Ning, M.; Jason, P.; Kevin, B.; Thomas, L.M. BLAST+: Architecture and applications. *BioMed Cent.* **2009**, *10*, 421. [CrossRef]

44. Priyam, A.; Woodcroft, B.J.; Rai, V.; Moghul, I.; Munagala, A.; Ter, F.; Chowdhary, H.; Pieniak, I.; Maynard, L.J.; Gibbins, M.A.; et al. Sequenceserver: A Modern Graphical User Interface for Custom BLAST Databases. *Mol. Biol. Evol.* **2019**, *36*, 2922–2924. [CrossRef]

45. Robert, B.; Eric, Y.; Colin, M.D.; Richard, D.H.; Monica, M.; Gregg, H.; David, M.G.; Christine, G.E.; Suzanna, E.L.; Lincoln, S.; et al. JBrowse: A dynamic web platform for genome visualization and analysis. *Genome Biol.* **2016**, *17*, 66. [CrossRef]

46. Wambui, N.; Aaron, L.; Richard, C.; Tia-Lynn, A.; Nahla, B. Insights into phylogeny, sex function and age of *Fragaria* based on whole chloroplast genome sequencing. *Mol. Phylogenet. Evol.* **2013**, *66*, 17–29. [CrossRef]

47. Wang, Y.; Tang, H.; Debarry, J.D.; Tan, X.; Li, J.; Wang, X.; Lee, T.; Jin, H.; Marler, B.; Guo, H.; et al. MCScanX: A toolkit for detection and evolutionary analysis of gene synteny and collinearity. *Nucleic Acids Res.* **2012**, *40*, 49. [CrossRef]

48. Kalvari, I.; Nawrocki, E.P.; Argasinska, J.; Quinones-Olvera, N.; Finn, R.D.; Bateman, A.; Petrov, A.I. Non-Coding RNA Analysis Using the Rfam Database. *Curr. Protoc. Bioinform.* **2018**, *62*, 51. [CrossRef]

49. Kalvari, I.; Nawrocki, E.P.; OntiverosPalacios, N.; Argasinska, J.; Lamkiewicz, K.; Marz, M.; GriffithsJones, S.; ToffanoNioche, C.; Gautheret, D.; Weinberg, Z.; et al. Rfam 14: Expanded coverage of metagenomic, viral and microRNA families. *Nucleic Acids Res.* **2020**, *49*, 192–200. [CrossRef]

50. Nawrocki, E.P.; Kolbe, D.L.; Eddy, S.R. Infernal 1.0: Inference of RNA alignments. *Bioinformatics* **2009**, *25*, 1335–1337. [CrossRef]

51. Zheng, Y.; Jiao, C.; Sun, H.; Rosli, H.G.; Pombo, M.A.; Zhang, P.; Banf, M.; Dai, X.; Martin, G.B.; Giovannoni, J.J.; et al. iTAK: A Program for Genome-wide Prediction and Classification of Plant Transcription Factors, Transcriptional Regulators, and Protein Kinases. *Mol. Plant* **2016**, *9*, 1667–1670. [CrossRef]

52. Sook, J.; Taein, L.; Chun-Huai, C.; Katheryn, B.; Ping, Z.; Jing, Y.; Jodi, H.; Stephen, P.F.; Ksenija, G.; Kristin, S.; et al. 15 years of GDR: New data and functionality in the Genome Database for Rosaceae. *Nucleic Acids Res.* **2019**, *47*, 1137–1145. [CrossRef]

Article

Frequent Gene Duplication/Loss Shapes Distinct Evolutionary Patterns of NLR Genes in Arecaceae Species

Xiao-Tong Li [1,†], Guang-Can Zhou [2,†], Xing-Yu Feng [1], Zhen Zeng [1], Yang Liu [1,*] and Zhu-Qing Shao [1,*]

1 School of Life Sciences, Nanjing University, Nanjing 210023, China; lixt@cemps.ac.cn (X.-T.L.); fxy15928934559@163.com (X.-Y.F.); zengzhen561@163.com (Z.Z.)
2 College of Agricultural and Biological Engineering (College of Tree Peony), Heze University, Heze 274015, China; zhouguangcan@hezeu.edu.cn
* Correspondence: m18845043187@163.com (Y.L.); zhuqingshao@nju.edu.cn (Z.-Q.S.)
† These authors contributed equally to this work.

Abstract: Nucleotide-binding leucine-rich repeat (NLR) genes play a key role in plant immune responses and have co-evolved with pathogens since the origin of green plants. Comparative genomic studies on the evolution of NLR genes have been carried out in several angiosperm lineages. However, most of these lineages come from the dicot clade. In this study, comparative analysis was performed on NLR genes from five Arecaceae species to trace the dynamic evolutionary pattern of the gene family during species speciation in this monocot lineage. The results showed that NLR genes from the genomes of *Elaeis guineensis* (262), *Phoenix dactylifera* (85), *Daemonorops jenkinsiana* (536), *Cocos nucifera* (135) and *Calamus simplicifolius* (399) are highly variable. Frequent domain loss and alien domain integration have occurred to shape the NLR protein structures. Phylogenetic analysis revealed that NLR genes from the five genomes were derived from dozens of ancestral genes. *D. jenkinsiana* and *E. guineensis* genomes have experienced "consistent expansion" of the ancestral NLR lineages, whereas a pattern of "first expansion and then contraction" of NLR genes was observed for *P. dactylifera*, *C. nucifera* and *C. simplicifolius*. The results suggest that rapid and dynamic gene content and structure variation have shaped the NLR profiles of Arecaceae species.

Keywords: Arecaceae; NLR gene; plant disease resistance; phylogeny; evolutionary pattern

Citation: Li, X.-T.; Zhou, G.-C.; Feng, X.-Y.; Zeng, Z.; Liu, Y.; Shao, Z.-Q. Frequent Gene Duplication/Loss Shapes Distinct Evolutionary Patterns of NLR Genes in Arecaceae Species. *Horticulturae* **2021**, *7*, 539. https://doi.org/10.3390/horticulturae7120539

Academic Editor: Young-Doo Park

Received: 4 November 2021
Accepted: 30 November 2021
Published: 2 December 2021

Publisher's Note: MDPI stays neutral with regard to jurisdictional claims in published maps and institutional affiliations.

1. Introduction

The innate immune system can protect plants from the threats of foreign pathogens [1]. One of the core parts of the plant immune system is a set of genes, termed plant disease resistance genes (*R* genes), which recognize pathogen-derived virulence proteins (called "effectors") to activate downstream defense responses [1]. Upon the recognition of the invasion of pathogens, R proteins can activate a hypersensitivity reaction and a series of immune responses, and finally cause the cell death of the infected cells, to restrain the proliferation and spread of pathogens [1]. Nucleotide-binding leucine-rich repeat (NLR) genes are the largest type of all the different *R* genes, accounting for over 60% of the *R* genes functionally characterized to date [2]. A typical NLR protein contains a variable domain at the N-terminus, a highly conserved NBS domain in the middle, and a diverse leucine-rich repeat (LRR) domain at the C-terminus [3]. As the N-terminal domains in angiosperms are usually annotated as CC, TIR, or RPW8 domain, angiosperm NLR genes were classified into three subclasses: CC-NBS-LRR (CNL), TIR-NBS-LRR (TNL), and RPW8-NBS-LRR (RNL) [4,5]. Functionally, CNL and TNL proteins act as "sensor NLRs" that recognize specific pathogen effectors to trigger downstream immune responses, while RNL proteins serve as downstream signal transduction molecules ("helper NLR") of CNL and TNL proteins [6,7].

NLR genes constitute a large gene family in plant genomes, usually comprising hundreds of members [8], and they show very fast evolutionary modes in response to the fast-evolving pathogens [4]. With more and more plant genomes being sequenced, genome-wide evolutionary analyses and comparative genomic studies of NLR genes have been performed in many species and taxa, and different taxa exhibited distinct evolutionary patterns. For example, frequent gene losses and limited gene duplications resulted in a small number of NLR genes in the Cucurbitaceae species [9]. A similar pattern of NLR gene contraction, caused by gene losses or frequent gene deletions, was also reported for Poaceae species [10,11]. In contrast, NLR genes in Fabaceae and Rosaceae species exhibited a "consistent expansion" evolutionary pattern [5,12], while the five Brassicaceae species exhibited a "first expansion and then contraction" of NLR genes [13]. Moreover, species belonging to the same family may also show distinct patterns of NLR gene evolution [14,15]. For example, in four orchid species, *Phalaenopsis equestris* and *Dendrobium catenatum* exhibited an "early contraction to recent expansion" evolutionary pattern, while *Gastrodia elata* and *Apostasia shenzhenica* showed a "contraction" evolutionary pattern [14].

The distinct evolutionary patterns of these angiosperm lineages provide valuable resources to the understanding of the fast evolutionary modes of *R* genes due to threats from different pathogens. However, most of these investigated angiosperm lineages are from the dicot clade, while only two monocot lineages have been surveyed [10,11,14]. Because the monocot and dicot clades are different in NLR subclass composition [8], investigating more monocot lineages would provide new insights into the NLR gene dynamics among monocot evolution.

The Arecaceae consists of 183 genera and 2450 species, which are distributed throughout the tropical and subtropical areas in Africa, the Americas, Asia, Madagascar, and the Pacific, and widely grown as ornamentals in botanical gardens (Flora of China, www.iplant.cn/foc/, accessed on 2 August 2021). Recently, the genome of five horticultural plants from the Arecaceae family of the Arecales, including *Elaeis guineensis* (2n = 32), *Phoenix dactylifera* (2n = 36), *Daemonorops jenkinsiana* (2n = 24), *Cocos nucifera* (2n = 32) and *Calamus simplicifolius* (2n = 26), were sequenced and made available [16–19]. Among them, oil palm (*E. guineensis*) is a source of vegetable oil and has very important economic value [20], date palm (*P. dactylifera*) is the most popular fruit in the Middle East and North Africa, and *C. nucifera* is widely distributed on Earth and has considerable food and medicinal value [21]. These horticultural plants are faced with infection by various pathogens during their lifespan. However, the composition and evolutionary pattern of NLR genes in the Arecaceae family have rarely been investigated [22,23]. Deciphering the evolutionary pattern of NLR genes among the five Arecaceae species would provide an additional example of dynamic NLR gene evolution across species speciation in the monocot lineage. Additionally, the obtained NLR information may serve as a primary resource for the disease resistance breeding of the Arecaceae species.

2. Materials and Methods

2.1. Identification and Classification of the NLR Genes

The five whole genomes of the *E. guineensis*, *P. dactylifera*, *D. jenkinsiana*, *C. nucifera* and *C. simplicifolius* were used in this study. Genomic sequences and annotation files were obtained from the GigaScience database. NLR genes of the five genomes were retrieved from the ANNA database (https://biobigdata.nju.edu.cn/ANNA/, accessed on 10 August 2021). All the identified NLR genes were subjected to NCBI's conserved domain database (https://www.ncbi.nlm.nih.gov/Structure/cdd/wrpsb.cgi, accessed on 30 August 2021) using the default settings to determine whether they encoded CC, RPW8, LRR and other integrated domains (E value: 10^{-4}). The domains that commonly encoded by the NLR genes, such as NBS, LRR, TIR, RPW8, CC, AAA+ and DUF1863 were removed from the integrated domain list.

2.2. Cluster Arrangement of the Identified NLR Genes

Gene clustering was determined according to the criterion used for *Medicago truncatula* [24]: if two neighboring NLR genes were located within 250 kb on a chromosome, these two genes were regarded as members of the same gene cluster. Based on this criterion, the NLR genes in the five Arecaceae genomes were assigned to clustered loci and singleton loci.

2.3. Sequence Alignment and Phylogenetic Analysis of NLR Genes

The amino acid sequences of the NBS domain were extracted from the identified NLR genes and used for multiple alignments using ClustalW integrated in MEGA 7.0 with default settings [25]. Sequences that were too short (<190 amino acids, less than two-thirds of a regular NBS domain) or too divergent were removed to prevent interference with the alignments and subsequent phylogenetic analysis. The resulting alignments were manually corrected and improved using MEGA 7.0. The phylogenetic tree was constructed using IQ-TREE (version 1.6.12) with the maximum likelihood method, following the selection of best-fit model by ModelFinder [26,27]. Branch support values were assessed using SH-aLRT and UFBoot2 tests with 1000 replications [28].

2.4. Gene Loss/Duplication Analysis of the NLR Genes

In order to identify the gene duplication/loss events during the speciation of the five Arecaceae species, the NLR gene phylogenetic tree was reconciled with the species tree using Notung-2.9 software [29]. The types of NLR gene duplication within a genome were determined using the MCScanX package [30] based on a pair-wise all-against-all blast of protein sequences.

3. Results

3.1. Comparative Analysis of NLR Gene Composition in the Genomes of Five Arecaceae Species

A total of 399, 536, 85, 262, and 135 NLR genes from the genomes of *C. simplicifolius*, *D. jenkinsiana*, *P. dactylifera*, *E. guineensis* and *C. nucifera*, respectively, were retrieved from the ANNA database. Among them, *D. jenkinsiana* possessed the largest number of NLR genes and was as 1.34-, 6.31-, 2.05- and 3.97-times bigger than that of *C. simplicifolius*, *P. dactylifera*, *E. guineensis* and *C. nucifera*, respectively. All NLR genes were divided into the CNL and RNL subclasses based on the classification provided in ANNA, with no TNL genes found. Among the two NLR subclasses, CNL genes overwhelmingly outnumbered RNL genes, with 99.50%, 99.25%, 100%, 99.62% and 98.52% of NLRs in *C. simplicifolius*, *D. jenkinsiana*, *P. dactylifera*, *E. guineensis* and *C. nucifera*, respectively, being CNLs. There were only two, four, one, and two RNL genes in *C. simplicifolius*, *D. jenkinsiana*, *E. guineensis* and *C. nucifera*, and no RNL genes were identified in *P. dactylifera*. Domain composition analysis revealed that less than half of NLR genes in *C. simplicifolius*, *D. jenkinsiana* and *C. nucifera* encode intact NLR proteins possessing all three domains (CC/RPW8-NBS-LRR), with the rest of NLR, either lacking the CC/RPW8 domain at the N-terminus, the LRR domain at the C-terminus, or domains at both termini (Table 1). The proportion of intact NLR genes in the *E. guineensis* genome is much higher, with 144 (143 CNL and 1 RNL) of the 261 genes having all three domains. Some NLR genes were classified as "other" in CNL due to their atypical structural domain compositions (Tables 1 and S1). For example, the *D. jenkinsiana* genome encodes one "other" gene in CNL($N_{CC}LN_{CC}L$), and the *C. nucifera* genome encodes two "other" genes, including $N_{CC}LN_{CC}$ and CNCNL.

Table 1. The number of identified NLR genes in the five Arecaceae genomes.

Domain Compositions	*C. simplicifolius*	*D. jenkinsiana*	*P. dactylifera*	*E. guineensis*	*C. nucifera*
CNL subclass	**397 (99.50%)**	**532 (99.25%)**	**85 (100%)**	**261 (99.62%)**	**133 (98.52%)**
CNL (Intact)	112	171	26	143	63
CN	142	202	22	41	28
NL	60	57	26	59	28
N	83	101	11	17	12
Other	0	1 (NLNL)	0	1 (CNCNL)	2 (NLN, CNCNL)
RNL subclass	**2 (0.50%)**	**4 (0.75%)**	**0**	**1 (0.38%)**	**2 (1.48%)**
RNL (Intact)	0	0	0	1	1
RN	0	1	0	0	0
NL	2	3	0	0	0
N	0	0	0	0	1
Other	0	0	0	0	0
Total number	**399**	**536**	**85**	**262**	**135**

Integration of alien domains in addition to the three typical domains was detected for NLR genes from the five genomes, including eight, 15, five, three and five distinct integrated domains (IDs) found in 13, 14, six, seven and seven NLR genes of the *C. simplicifolius*, *D. jenkinsiana*, *P. dactylifera*, *E. guineensis* and *C. nucifera* genomes, respectively (Table S2). All the NLR-ID genes belong to the CNL subclass. The numbers of NLR genes with fused IDs (NLR-ID gene) in the five genomes show a significant positive correlation with total NLR gene numbers (Figure 1a). An average of 4.15% of NLR genes in each genome possess the NLR-ID structure. The comparison of the ID diversity in the five species shows that a total of 32 non-redundant IDs were present in the five genomes (Table S2). Some of these IDs have been detected in proteins with immune function, including the v-SNARE domain and the PKinase domain. Plant SNARE domain-containing proteins are targets of filamentous fungi effectors and are monitored by NLRs for programmed cell death [31]. PKinases are known to function in the immune pathways in both plants and mammals and are also often found in the receptor-like PKinases that transduce PAMP-triggered immunity [32]. Among the 32 different types of IDs, one of them was found in NLR genes from three species and two were found in NLR genes from two species (Table S2). In contrast, the majority of IDs were found only in one genome, suggesting frequent occurrence of species-specific domain fusions (Figure 1b).

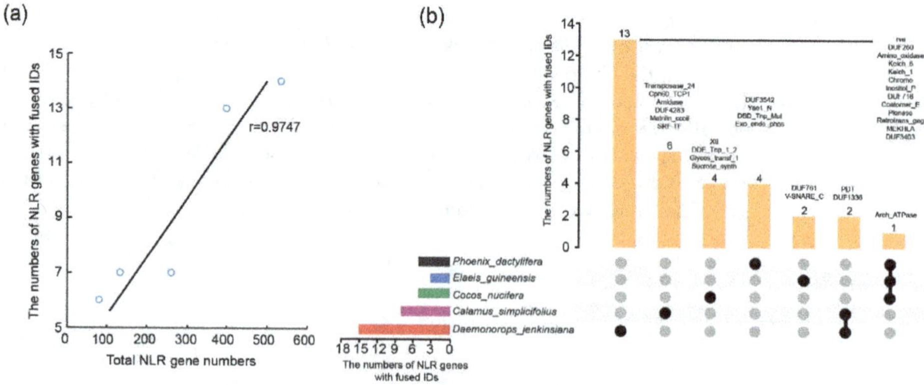

Figure 1. Exogenous fusion domains of NLR genes in *D. jenkinsiana*, *C. nucifera*, *C. simplicifolius*, *E. guineensis* and *P. dactylifera*. (**a**) Spearman correlation between the total number of species' NLR genes and the number of NLR genes fused IDs. r represents the correlation coefficient, $p < 0.05$. (**b**) Extraneous domains of NLR gene-specific fusions or convergent fusions among different species in the Arecaceae species. Black circles indicate exogenous fusion domain presence, and gray circles indicate exogenous fusion domains non- presence.

3.2. Organization of NLR Genes in Arecaceae Genomes

Clustering organization of NLR genes has been proposed as an important mechanism of generating NLR diversity and functional members. Our results show that the majority of NLR genes were organized into clusters rather than singletons in *C. nucifera*, *D. jenkinsiana*, and *E. guineensis* genomes, with 54.8%, 70.7% and 77.7% NLR genes detected in clusters, respectively (Table 2). However, there were more singleton genes in the other two Arecaceae genomes, with only 15 (17.6%) and 176 (44.1%) NLR genes organized into clusters in the *P. dactylifera* and *C. simplicifolius* genomes, respectively. Among the five Arecaceae genomes, the clustered loci in *E. guineensis* genome contained the most genes (3.37 genes/locus) on average. The largest gene clusters of *C. nucifera*, *D. jenkinsiana*, *E. guineensis*, *C. simplicifolius*, and *P. dactylifera* were in locus 47 (eight genes), locus 34 (10 genes) and 126 (10 genes), locus 80 (11 genes), locus 237 (five genes), and locus 47 (three genes), respectively (Table 2).

Table 2. The organization of NLR genes in five Arecaceae species.

Gene and Loci	*C. nucifera*	*D. jenkinsiana*	*E. guineensis*	*C. simplicifolius*	*P. dactylifera*
No. of chromosome-anchored NBS loci and genes	89 (135)	284 (536)	114 (262)	296 (399)	77 (85)
No. of singleton loci (no. of NBS genes)	61 (61)	157 (157)	60 (60)	223 (223)	70 (70)
No. of clustered loci (no. of NBS genes)	28 (74)	127 (379)	54 (202)	73 (176)	7 (15)
Clustered NBS genes/singleton NBS genes	1.21	2.41	3.37	0.34	0.21
Average (median) no. of NBS genes in clusters	2.6 (2)	3.0 (3)	3.7 (3)	1 (2)	2.1 (2)
No. of clusters with 10 or more NBS genes	0	2	2	0	0
No. of NBS genes in the largest cluster	8 (locus 47)	10 (loci 34, 126)	11 (locus 80)	5 (locus 237)	3 (locus 47)

NLR genes may undergo duplication via different mechanisms [33]. We surveyed the duplication patterns of NLR genes from the five Arecaceae species by using MCScanX software. The results showed that amplification of NLR genes in the five genomes was dominated by different types. The majority of NLR genes in *C. nucifera* (57.0%) and *E. guineensis* (70.1%) were generated by tandem/proximal duplications, whereas most NLR genes in the remaining three genomes were characterized as dispersed duplication. Only a small proportion of NLR genes in *C. nucifera*, *D. jenkinsiana* and *E. guineensis* were generated by whole genome duplications (WGD)/segmental duplication, whereas no WGD/segmental duplicated NLR genes were found in *C. simplicifolius* or *P. dactylifera* (Figure 2). However, the proportion of segmental duplicated genes might have been underestimated because the syntenic relationship of NLR genes would be disrupted during long-term evolution.

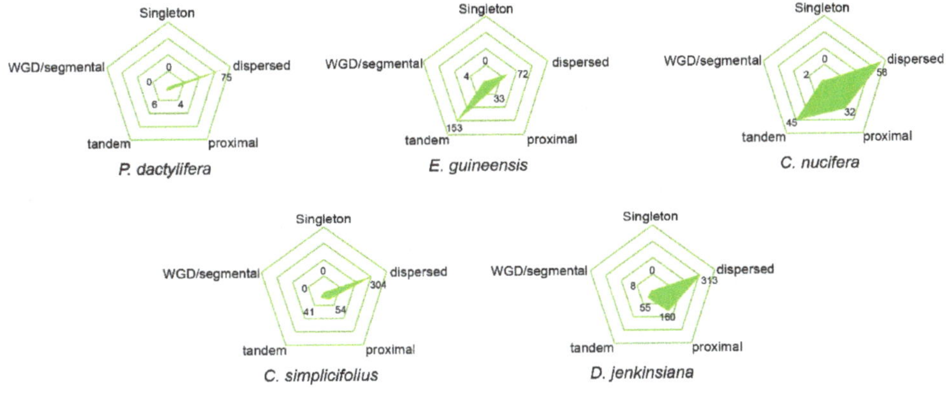

Figure 2. Type of gene duplication in *C. simplicifolius*, *D. jenkinsiana*, *P. dactylifera*, *E. guineensis* and *C. nucifera*, respectively.

3.3. Phylogenetic Analysis of the NLR Genes

To trace the evolutionary history of the NLR genes in Arecaceae, a phylogenetic tree was constructed based on the amino acid sequences of the NBS domain, using three *Amborella* TNL proteins as the outgroups (Figure 3 and Data S1). The results show that NLR genes from the five species form two monophyletic clades with high support values (>90%) (Figure S1). The two deeply diverged clades correspond to the RNL and CNL subclasses, respectively (Figure 3a), supporting the ancient separation of the two NLR clades (Figure S1; Data S1). Compared to the CNL clade, the branch lengths of the RNL clade were relatively short (Figure 3a), suggesting that RNL genes had a slow evolutionary rate. Within the CNL clade, clustering of NLR genes from a single species was frequently observed (Figure 3a), which is caused by species-specific gene duplications.

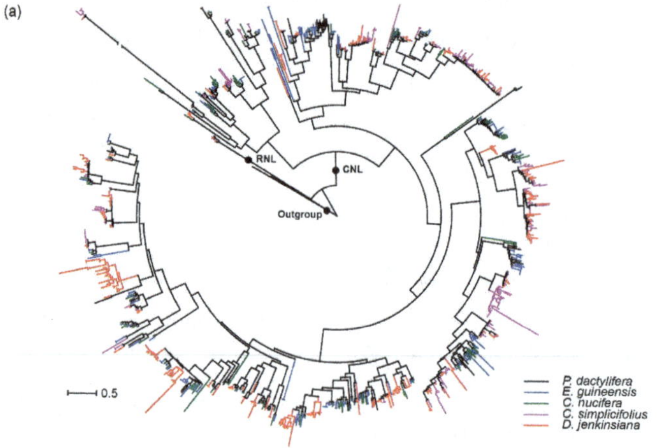

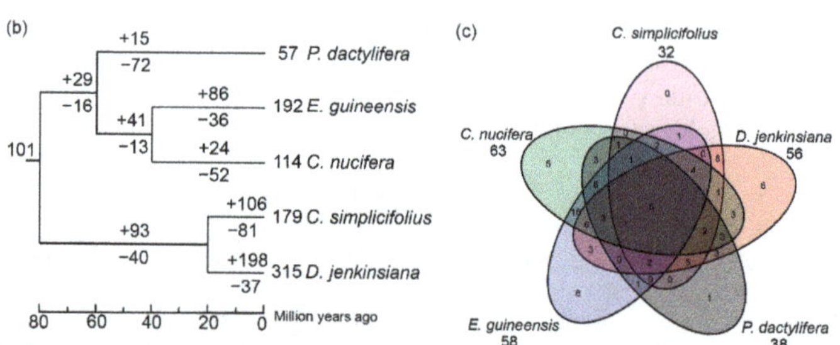

Figure 3. Phylogenetic tree of CNL and RNL genes based on conserved NBS domain sequences. (**a**) A total of 1408 CNL and 9 RNL sequences were used: 536 sequences from *D. jenkinsiana* (shown in red), 135 from *C. nucifera* (green), 399 from *C. simplicifolius* (purple), 262 from *E. guineensis* (blue), and 85 from *P. dactylifera* (black). Four TNL sequences from Arabidopsis thaliana were used as an outgroup. (**b**) Number variations of NLR genes at different stages of Arecaceae species evolution. Differential gene losses and gains are indicated by numbers with − or + on each branch. (**c**) Arecaceae ancestral lineage genes were inherited by *C. simplicifolius*, *D. jenkinsiana*, *P. dactylifera*, *E. guineensis* and *C. nucifera*, respectively.

To gain insight into the evolution of the NLR genes during the speciation of the five Arecaceae species, Notung software was used to reconcile gene duplication/loss events of the NLR genes at each node of the phylogenetic tree [34]. The result reveals that at least 101 ancestral NLR lineages (2 RNL and 99 CNL) may have existed in the common ancestor of the five Arecaceae species (Figure S1 and Data S1). We term these ancestral NLR lineages as "Arecaceae NLR lineage". The common ancestor of *P. dactylifera*, *E. guineensis* and *C. nucifera* (Pd-Eg-Cn) inherited 85 Arecaceae NLR lineages but lost 16. These Arecaceae NLR lineages further expanded to 114 NLR lineages before the divergence of the *P. dactylifera*, *E. guineensis* and *C. nucifera* (Pd-Eg-Cn). The common ancestor of *C. simplicifolius* and *D. jenkinsiana* (Cs-Dj) inherited 61 Arecaceae NLR lineages and duplicated to 154 Cs-Dj NLR lineages (Figure 3b). Subsequent gene loss and gain during the divergence of Pd-Eg-Cn and Cs-Dj lineages further shaped the NLR content in the five Arecaceae species, resulting in the current distinct NLR profile in these species. By assigning the NLR genes in each species to the 114 ancestral Arecaceae NLR lineages, the results showed that 32, 56, 38, 58 and 63 Arecaceae NLR lineages were inherited by *C. simplicifolius*, *D. jenkinsiana*, *P. dactylifera*, *E. guineensis* and *C. nucifera*, respectively (Figure 3c). Among these ancestral NLR lineages, only five of them were reserved in all five genomes, and 20 lineages were species-specific (*D. jenkinsiana*, eight; *P. dactylifera*, one; *C. simplicifolius*, zero; *E. guineensis*, six; *C. nucifera*, five). The remaining 89 lineages were differentially preserved in two to four species (Figure 3c).

3.4. Tracing the Trajectory of NLR Gene Evolution in Different Arecaceae Species

Based on the traced loss and gain of NLR genes at each species divergence node, the evolutionary trajectory of NLR genes for each of the five species can be traced. After separation from the common ancestor of the five Arecaceae species (Cs-Dj-Pd-Eg-Cn node), 40 gene-loss and 93 gain events were detected at the Cs-Dj node (Figure 3b), suggesting a pattern of NLR expansion during this period. The Cs-Dj node then diverged to generate *C. simplicifolius* and *D. jenkinsiana*. In *C. simplicifolius*, 81 NLR genes inherited from the *Cs-Dj* were lost, while the remaining NLR genes underwent intensive gene duplication, leading to the generation of 106 new members (Figure 3b). Again, a pattern of NLR expansion occurred during the period from *Cs-Dj* to *C. simplicifolius*. Taken together, a "consistent expansion" of NLR genes occurred during the evolution of *C. simplicifolius* from the common ancestor of the five Arecaceae species (Figure 4d). Using this strategy, a similar pattern of "consistent expansion" was also observed for NLR genes in *D. jenkinsiana* and *E. guineensis* (Figure 4b,e). However, a different pattern of "first expansion and then contraction" was observed for the other species, *P. dactylifera* and *C. nucifera* (Figure 4a,c).

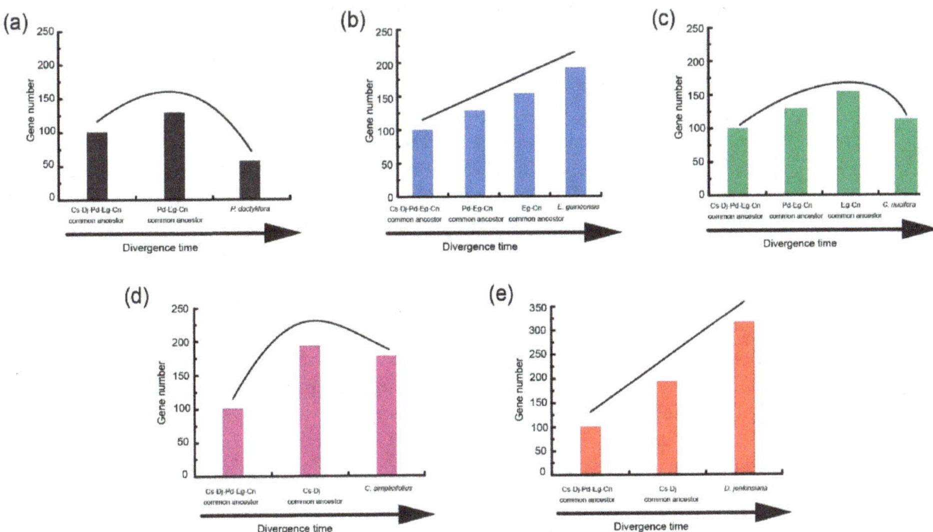

Figure 4. Dynamic evolutionary patterns of NLR genes among different evolutionary states of five Arecaceae species. (**a**) *P. dactylifera*. (**b**) *E. guineensis*. (**c**) *C. nucifera*. (**d**) *C. simplicifolius*. (**e**) *D. jenkinsiana*.

4. Discussion

The NLR genes constitute one of the largest gene families in angiosperms, with an average of about 300 genes per genome [8]. Genome-wide identification and comparative analysis of NLR genes has greatly accelerated the mining of functional NLR genes from various crops and ecological or economic important plants in recent years. The lack of NLR information has greatly hampered identification of functional disease resistance genes by using a genome-guided method for the Arecaceae species. Taking advantage of the recently released genomes from the five Arecaceae species, the NLR profile across the five species was compared in several aspects of this study, which may serve as a primary resource for molecular breeding of the Arecaceae species.

Different proportions of NLR genes in the Arecaceae species form clusters of varied size. The members within each of these NLR gene clusters provides large candidates for positive selection to act on. Additionally, high polymorphism can be maintained between NLR genes within the gene cluster through recombination, facilitating the generation of new NLR genes [35,36]. Comparative analysis of NLR genes in the five Arecaceae species showed that two of the Arecaceae genomes have more than 300 NLR genes, whereas the other three have fewer than 300 NLRs, suggesting that species-specific NLR gain and loss have occurred. NLR content dynamics is an important mechanism for plants coping with the varied environments and ecological adaption [8,37]. The species-specific NLR contraction or expansion for the five Arecaceae species suggest they may have faced different selection pressure from environmental microbes after having separated from the common ancestor, although the exact trajectory of environmental microbe diversity dynamic could hardly be traced. Reconstruction of the ancestral states of NLR genes at several divergence nodes of Arecaceae revealed 101 ancestral NLR lineages in the common ancestor, including two RNL lineages and 99 CNL lineages. The number of recovered ancestral NLR lineage in the Arecaceae family is much larger than that in the Orchidaceae family (29), but lower than that of the Poaceae (456) in the monocot clade [4,14]. The high difference of ancestral NLR lineage numbers in the three monocot families suggests that dramatic NLR contraction and amplification consistently occurred in the monocot lineage. Compared with several investigated dicot families, the ancestral NLR lineage in the Arecaceae family is fewer than the 119 ancestral NLR lineages in Fabaceae, and 166 ancestral NLR lineages in Solanaceae, and far fewer than the 228 ancestral NLR lineages found in Brassicaceae [13]. The large ancestral NLR lineage numbers in these dicot families may have benefitted from having an additional NLR subclass, TNL, in their genomes.

The current NLR profile in a genome is contributed by differential inheritance and amplification of ancestral NLR lineages [4]. Previous studies in Cucurbitaceae and Poaceae revealed that species in the two families experienced a similar pattern of NLR "contraction" [9–11], whereas Fabaceae and Rosaceae species exhibited "consistent expansion" of NLR genes after the families' radiation [5,12]. A distinct "first expansion and then contraction" pattern of NLR genes was observed in the five Brassicaceae [13]. However, species in the Arecaceae family exhibited two different modes of NLR gene evolution after diverging from the common ancestor. *D. jenkinsiana* and *E. guineensis* have experienced "consistent expansion" of the ancestral NLR lineages, whereas NLR genes in *P. dactylifera*, *C. nucifera*. and *C. simplicifolius* show "first expansion and then contraction". Such a pattern of species belonging to the same family having distinct NLR gene evolution patterns has also been observed in another monocot lineage, Orchidaceae, and two dicot families [15–17]. The results provide additional evidence to support that rapid NLR gene content variation could occur to facilitate plant adaption to changed environments.

The high diversity of NLR could be detected not only for gene content, but also for gene structure. For example, the three NLR subclasses have distinct N-terminal domains to support distinct functional mechanisms, either by making holes in the cell membrane, or by action as an enzyme [38]. Loss of characteristic domains is detected for many NLR genes in the Arecaceae species. This pattern has also been observed in several angiosperms by previous studies. For example, only a small proportion of NLR genes with intact structures were reported in *C. annuum* (23.2%), *S. lycopersicum* (42.7%), *S. tuberosum* (28.2%), *P. trichocarpa* (46.2%), *M. truncatula* (39.1%), *Lotus japonicus* (31.0%), and *Oryza sativa* (30.6%) genomes [5,39–41]. The remaining large proportion of NLR genes further expanded the diversity of NLR genes through the loss of the N-terminal, C-terminal, or both domains (Figure S2). Notably, several studies have reported that NLR genes with atypical structures also function in plant immunity [42–45], suggesting the loss of the N-terminal or C-terminal domains may also be a mechanism to generated NLR functional diversity. It is worthy of note that a deeply diverged CNL lineage showed the feature of widespread loss of the N-terminal CC domain (Figure S2). The long-term maintenance and expansion of this CNL lineage suggest that the loss of the CC domain did not completely abolish the function of genes on this lineage. Considering the CC domain had been shown to be indispensable for multimerization and forming pores on the plant cell membrane of CNL proteins, the CC-lacking structure of this CNL lineage may suggest a different functional mechanism.

Different from the domain loss found in many NLR genes, we also detected fusion of alien domains for a small proportion of Arecaceae NLR genes to form the NLR-ID structure. This provides another way to expand the structure and functional diversity. In recent years, studies have increasingly found that alien domains can be fused to plant NLR proteins to act as target proteins for pathogens' effector factors. The research on RGA4/RGA5 and Pik-1/Pik-2 of the NLR with rice blast resistance provides the first experimental evidence for this model. Both RGA5 and Pik-1 genes are fused with an HMA domain, which serves as a decoy to interact directly with pathogenic effectors to stimulate the disease resistance of RGA4 and Pik-2 [46,47]. *Arabidopsis* NLR genes RPS4/RRS1 provides another example to support the important functions of alien domains. The RRS1 protein fused with WRKY domain can directly interact with the pathogenic effectors. AvrRPS4 interacts to stimulate the disease resistance activity of RPS4 [48]. In this study, different NLR alien domains were found in five species. These domains may be the "baits" proteins of pathogenic effectors in plant cells. Among them, v-SNARE and the PKinase domains have been detected in many proteins that play a role in plant disease resistance [32,33], but SRF-TF, DUF761and DUF4283, etc. have no direct evidence of being related to plant immunity. The discovery of these alien domains is helpful to explore more potential plant immune-related proteins.

5. Conclusions

This study compared the NLR gene profiles in five Arecaceae species and revealed a high NLR content and structure variation, which serve as a resource for the evolution of functional NLR genes. Phylogenetic analysis revealed that NLR genes from the five genomes were derived from 101 ancestral genes, and distinct "consistent expansion" and "first expansion and then contraction" patterns were observed for NLR genes from the five species. The results show that dynamic gene content and structure variation have shaped the NLR profiles of different Arecaceae species. The obtained NLR profiles serve as a valuable resource for molecular breeding of these species and for further exploration of the NLR evolutionary pattern.

Supplementary Materials: The following are available online at https://www.mdpi.com/article/10.3390/horticulturae7120539/s1, Figure S1. 101 NLR gene families divided based on NLR phylogeny from *C. simplicifolius, D. jenkinsiana, P. dactylifera, E. guineensis* and *C. nucifera*. Figure S2. Phylogenetic distribution of NLRgenes with different domain compositions in five Arecaceae species. Table S1. The gene list and classification of NLR genes identified from *C. simplicifolius, D. jenkinsiana, P. dactylifera, E. guineensis* and *C. nucifera*. Table S2. Exogenous fusion domains of NLR identified from five Arecaceae species. Data S1. An NEXUS format phylogenetic tree of NLRs in five Arecaceae species with branch support values.

Author Contributions: Z.-Q.S. and Y.L. conceived and designed the study. X.-T.L., G.-C.Z., X.-Y.F., Z.Z. and Y.L. obtained and analyzed the data. X.-T.L. and G.-C.Z. wrote the manuscript. Z.-Q.S. and Y.L. revised the manuscript. All authors have read and agreed to the published version of the manuscript.

Funding: This research received no external funding.

Institutional Review Board Statement: Not applicable.

Informed Consent Statement: Not applicable.

Conflicts of Interest: The authors have no conflict of interest to declare.

References

1. Wang, W.; Feng, B.; Zhou, J.M.; Tang, D. Plant immune signaling: Advancing on two frontiers. *J. Integr. Plant Biol.* **2020**, *62*, 2–24. [CrossRef]
2. Kourelis, J.; van der Hoorn, R.A.L. Defended to the Nines: 25 Years of Resistance Gene Cloning Identifies Nine Mechanisms for R Protein Function. *Plant Cell* **2018**, *30*, 285–299. [CrossRef]
3. Meyers, B.C.; Kozik, A.; Griego, A.; Kuang, H.; Michelmore, R.W. Genome-wide analysis of NBS-LRR-encoding genes in Arabidopsis. *Plant Cell* **2003**, *15*, 809–834. [CrossRef] [PubMed]
4. Shao, Z.Q.; Xue, J.Y.; Wu, P.; Zhang, Y.M.; Wu, Y.; Hang, Y.Y.; Wang, B.; Chen, J.Q. Large-Scale Analyses of Angiosperm Nucleotide-Binding Site-Leucine-Rich Repeat Genes Reveal Three Anciently Diverged Classes with Distinct Evolutionary Patterns. *Plant Physiol.* **2016**, *170*, 2095–2109. [CrossRef]
5. Shao, Z.Q.; Zhang, Y.M.; Hang, Y.Y.; Xue, J.Y.; Zhou, G.C.; Wu, P.; Wu, X.Y.; Wu, X.Z.; Wang, Q.; Wang, B.; et al. Long-term evolution of nucleotide-binding site-leucine-rich repeat genes: Understanding gained from and beyond the legume family. *Plant Physiol.* **2014**, *166*, 217–234. [CrossRef]
6. Saile, S.C.; Jacob, P.; Castel, B.; Jubic, L.M.; Salas-Gonzales, I.; Backer, M.; Jones, J.D.G.; Dangl, J.L.; El Kasmi, F. Two unequally redundant "helper" immune receptor families mediate Arabidopsis thaliana intracellular "sensor" immune receptor functions. *PLoS Biol.* **2020**, *18*, e3000783. [CrossRef] [PubMed]
7. Castel, B.; Ngou, P.M.; Cevik, V.; Redkar, A.; Kim, D.S.; Yang, Y.; Ding, P.; Jones, J.D.G. Diverse NLR immune receptors activate defence via the RPW8-NLR NRG1. *New Phytol.* **2019**, *222*, 966–980. [CrossRef]
8. Liu, Y.; Zeng, Z.; Zhang, Y.M.; Li, Q.; Jiang, X.M.; Jiang, Z.; Tang, J.H.; Chen, D.; Wang, Q.; Chen, J.Q.; et al. An angiosperm NLR Atlas reveals that NLR gene reduction is associated with ecological specialization and signal transduction component deletion. *Mol. Plant* **2021**, *14*, 17. [CrossRef]
9. Wan, H.; Yuan, W.; Bo, K.; Shen, J.; Pang, X.; Chen, J. Genome-wide analysis of NBS-encoding disease resistance genes in Cucumis sativus and phylogenetic study of NBS-encoding genes in Cucurbitaceae crops. *BMC Genom.* **2013**, *14*, 109. [CrossRef] [PubMed]
10. Luo, S.; Zhang, Y.; Hu, Q.; Chen, J.; Li, K.; Lu, C.; Liu, H.; Wang, W.; Kuang, H. Dynamic nucleotide-binding site and leucine-rich repeat-encoding genes in the grass family. *Plant Physiol.* **2012**, *159*, 197–210. [CrossRef] [PubMed]
11. Li, J.; Ding, J.; Zhang, W.; Zhang, Y.; Tang, P.; Chen, J.Q.; Tian, D.; Yang, S. Unique evolutionary pattern of numbers of gramineous NBS-LRR genes. *Mol. Genet. Genom. MGG* **2010**, *283*, 427–438. [CrossRef]
12. Jia, Y.; Yuan, Y.; Zhang, Y.; Yang, S.; Zhang, X. Extreme expansion of NBS-encoding genes in Rosaceae. *BMC Genet.* **2015**, *16*, 48. [CrossRef] [PubMed]
13. Zhang, Y.M.; Shao, Z.Q.; Wang, Q.; Hang, Y.Y.; Xue, J.Y.; Wang, B.; Chen, J.Q. Uncovering the dynamic evolution of nucleotide-binding site-leucine-rich repeat (NBS-LRR) genes in Brassicaceae. *J. Integr. Plant Biol.* **2016**, *58*, 165–177. [CrossRef] [PubMed]
14. Xue, J.Y.; Zhao, T.; Liu, Y.; Liu, Y.; Zhang, Y.X.; Zhang, G.Q.; Chen, H.; Zhou, G.C.; Zhang, S.Z.; Shao, Z.Q. Genome- Wide Analysis of the Nucleotide Binding Site Leucine-Rich Repeat Genes of Four Orchids Revealed Extremely Low Numbers of Disease Resistance Genes. *Front. Genet* **2019**, *10*, 1286. [CrossRef]
15. Zhou, G.C.; Li, W.; Zhang, Y.M.; Liu, Y.; Zhang, M.; Meng, G.Q.; Li, M.; Wang, Y.L. Distinct Evolutionary Patterns of NBS-Encoding Genes in Three Soapberry Family (Sapindaceae) Species. *Front. Genet* **2020**, *11*, 737. [CrossRef] [PubMed]
16. Singh, R.; Ong-Abdullah, M.; Low, E.T.; Manaf, M.A.; Rosli, R.; Nookiah, R.; Ooi, L.C.; Ooi, S.E.; Chan, K.L.; Halim, M.A.; et al. Oil palm genome sequence reveals divergence of interfertile species in Old and New worlds. *Nature* **2013**, *500*, 335–339. [CrossRef] [PubMed]
17. Hazzouri, K.M.; Gros-Balthazard, M.; Flowers, J.M.; Copetti, D.; Lemansour, A.; Lebrun, M.; Masmoudi, K.; Ferrand, S.; Dhar, M.I.; Fresquez, Z.A.; et al. Genome-wide association mapping of date palm fruit traits. *Nat. Commun.* **2019**, *10*, 4680. [CrossRef]
18. Lantican, D.V.; Strickler, S.R.; Canama, A.O.; Gardoce, R.R.; Mueller, L.A.; Galvez, H.F. De Novo Genome Sequence Assembly of Dwarf Coconut (Cocos nucifera L. 'Catigan Green Dwarf') Provides Insights into Genomic Variation Between Coconut Types and Related Palm Species. *G3-Genes Genom. Genet.* **2019**, *9*, 2377–2393. [CrossRef] [PubMed]
19. Zhao, H.S.; Wang, S.B.; Wang, J.L.; Chen, C.H.; Hao, S.J.; Chen, L.F.; Fei, B.H.; Han, K.; Li, R.S.; Shi, C.C.; et al. The chromosome-level genome assemblies of two rattans (Calamus simplicifolius and Daemonorops jenkinsiana). *Gigascience* **2018**, *7*, giy097. [CrossRef]
20. Duarte Ferreira Ribeiro, C.; Barbosa Schappo, F.; da Silva Sales, I.; Santos Assuncao, L.; Murowaniecki Otero, D.; Teixeira Magalhaes-Guedes, K.; Aparecida Souza Machado, B.; Mara Block, J.; Izabel Druzian, J.; Larroza Nunes, I. Novel bioactive nanoparticles from crude palm oil and its fractions as foodstuff ingredients. *Food Chem.* **2021**, *373*, 131252. [CrossRef] [PubMed]
21. Lima, E.B.; Sousa, C.N.; Meneses, L.N.; Ximenes, N.C.; Santos Junior, M.A.; Vasconcelos, G.S.; Lima, N.B.; Patrocinio, M.C.; Macedo, D.; Vasconcelos, S.M. Cocos nucifera (L.) (Arecaceae): A phytochemical and pharmacological review. *Braz. J. Med. Biol. Res.* **2015**, *48*, 953–964. [CrossRef]

22. Durand-Gasselin, T.; Asmady, H.; Flori, A.; Jacquemard, J.C.; Hayun, Z.; Breton, F.; de Franqueville, H. Possible sources of genetic resistance in oil palm (Elaeis guineensis Jacq.) to basal stem rot caused by Ganoderma boninense–prospects for future breeding. *Mycopathologia* **2005**, *159*, 93–100. [CrossRef]

23. Hanold, D.; Randles, J.W. Coconut Cadang-Cadang Disease and Its Viroid Agent. *Plant Dis.* **1991**, *75*, 330–335. [CrossRef]

24. Ameline-Torregrosa, C.; Wang, B.-B.; O'Bleness, M.S.; Deshpande, S.; Zhu, H.; Roe, B.; Young, N.D.; Cannon, S.B. Identification and characterization of nucleotide-binding site-leucine-rich repeat genes in the model plant Medicago truncatula. *Plant Physiol.* **2008**, *146*, 5–21. [CrossRef]

25. Kumar, S.; Stecher, G.; Tamura, K. MEGA7: Molecular Evolutionary Genetics Analysis Version 7.0 for Bigger Datasets. *Mol. Biol. Evol.* **2016**, *33*, 1870–1874. [CrossRef]

26. Nguyen, L.T.; Schmidt, H.A.; von Haeseler, A.; Minh, B.Q. IQ-TREE: A fast and effective stochastic algorithm for estimating maximum-likelihood phylogenies. *Mol. Biol. Evol.* **2015**, *32*, 268–274. [CrossRef]

27. Kalyaanamoorthy, S.; Minh, B.Q.; Wong, T.K.F.; von Haeseler, A.; Jermiin, L.S. ModelFinder: Fast model selection for accurate phylogenetic estimates. *Nat. Methods* **2017**, *14*, 587–589. [CrossRef]

28. Minh, B.Q.; Nguyen, M.A.; von Haeseler, A. Ultrafast approximation for phylogenetic bootstrap. *Mol. Biol. Evol.* **2013**, *30*, 1188–1195. [CrossRef]

29. Chen, K.; Durand, D.; Farach-Colton, M. NOTUNG: A program for dating gene duplications and optimizing gene family trees. *J. Comput. Biol. A J. Comput. Mol. Cell Biol.* **2000**, *7*, 429–447. [CrossRef] [PubMed]

30. Wang, Y.; Tang, H.; Debarry, J.D.; Tan, X.; Li, J.; Wang, X.; Lee, T.H.; Jin, H.; Marler, B.; Guo, H.; et al. MCScanX: A toolkit for detection and evolutionary analysis of gene synteny and collinearity. *Nucleic Acids Res.* **2012**, *40*, e49. [CrossRef] [PubMed]

31. Heller, J.; Clave, C.; Gladieux, P.; Saupe, S.J.; Glass, N.L. NLR surveillance of essential SEC-9 SNARE proteins induces programmed cell death upon allorecognition in filamentous fungi. *Proc. Natl. Acad. Sci. USA* **2018**, *115*, E2292–E2301. [CrossRef] [PubMed]

32. Dardick, C.; Schwessinger, B.; Ronald, P. Non-arginine-aspartate (non-RD) kinases are associated with innate immune receptors that recognize conserved microbial signatures. *Curr. Opin. Plant. Biol.* **2012**, *15*, 358–366. [CrossRef]

33. Leister, D. Tandem and segmental gene duplication and recombination in the evolution of plant disease resistance gene. *Trends Genet* **2004**, *20*, 116–122. [CrossRef] [PubMed]

34. Stolzer, M.; Lai, H.; Xu, M.; Sathaye, D.; Vernot, B.; Durand, D. Inferring duplications, losses, transfers and incomplete lineage sorting with nonbinary species trees. *Bioinformatics* **2012**, *28*, i409–i415. [CrossRef]

35. Bergelson, J.; Kreitman, M.; Stahl, E.A.; Tian, D. Evolutionary dynamics of plant R-genes. *Science* **2001**, *292*, 2281–2285. [CrossRef]

36. Botella, M.A.; Parker, J.E.; Frost, L.N.; Bittner-Eddy, P.D.; Beynon, J.L.; Daniels, M.J.; Holub, E.B.; Jones, J.D. Three genes of the Arabidopsis RPP1 complex resistance locus recognize distinct Peronospora parasitica avirulence determinants. *Plant Cell* **1998**, *10*, 1847–1860. [CrossRef] [PubMed]

37. Baggs, E.L.; Monroe, J.G.; Thanki, A.S.; O'Grady, R.; Schudoma, C.; Haerty, W.; Krasileva, K.V. Convergent Loss of an EDS1/PAD4 Signaling Pathway in Several Plant Lineages Reveals Coevolved Components of Plant Immunity and Drought Response. *Plant Cell* **2020**, *32*, 2158–2177. [CrossRef]

38. Wang, J.; Hu, M.; Wang, J.; Qi, J.; Han, Z.; Wang, G.; Qi, Y.; Wang, H.W.; Zhou, J.M.; Chai, J. Reconstitution and structure of a plant NLR resistosome conferring immunity. *Science* **2019**, *364*, eaav5870. [CrossRef]

39. Qian, L.H.; Zhou, G.C.; Sun, X.Q.; Lei, Z.; Zhang, Y.M.; Xue, J.Y.; Han, Y.Y. Distinct Patterns of Gene Gain and Loss: Diverse Evolutionary Modes of NBS-Encoding Genes in Three Solanaceae Crop Species. *G3-Genes Genomes Genetics* **2017**, *7*, 1577–1585. [CrossRef] [PubMed]

40. Zhang, X.; Feng, Y.; Cheng, H.; Tian, D.; Yang, S.; Chen, J.Q. Relative evolutionary rates of NBS-encoding genes revealed by soybean segmental duplication. *Mol. Genet. Genom.* **2011**, *285*, 79–90. [CrossRef] [PubMed]

41. Yang, S.; Gu, T.; Pan, C.; Feng, Z.; Ding, J.; Hang, Y.; Chen, J.Q.; Tian, D. Genetic variation of NBS-LRR class resistance genes in rice lines. *Theor Appl Genet.* **2008**, *116*, 165–177. [CrossRef]

42. Kato, H.; Saito, T.; Ito, H.; Komeda, Y.; Kato, A. Overexpression of the TIR-X gene results in a dwarf phenotype and activation of defense-related gene expression in Arabidopsis thaliana. *J. Plant Physiol.* **2014**, *171*, 382–388. [CrossRef] [PubMed]

43. Nandety, R.S.; Caplan, J.L.; Cavanaugh, K.A.; Perroud, B.; Wroblewski, T.; Michelmore, R.W.; Meyers, B.C. The role of TIR-NBS and TIR-X proteins in plant basal defense responses. *Plant Physiol.* **2013**, *162*, 1459–1472. [CrossRef]

44. Cai, H.; Wang, W.; Rui, L.; Han, L.; Luo, M.; Liu, N.; Tang, D. The TIR-NBS protein TN13 associates with the CC-NBS-LRR resistance protein RPS5 and contributes to RPS5-triggered immunity in Arabidopsis. *Plant J. Cell Mol. Biol.* **2021**, *107*, 775–786. [CrossRef]

45. Seong, K.; Seo, E.; Witek, K.; Li, M.; Staskawicz, B. Evolution of NLR resistance genes with noncanonical N-terminal domains in wild tomato species. *New Phytol.* **2020**, *227*, 1530–1543. [CrossRef]

46. Kanzaki, H.; Yoshida, K.; Saitoh, H.; Fujisaki, K.; Hirabuchi, A.; Alaux, L.; Fournier, E.; Tharreau, D.; Terauchi, R. Arms race co-evolution of Magnaporthe oryzae AVR-Pik and rice Pik genes driven by their physical interactions. *Plant J.* **2012**, *72*, 894–907. [CrossRef] [PubMed]

47. Cesari, S.; Thilliez, G.; Ribot, C.; Chalvon, V.; Michel, C.; Jauneau, A.; Rivas, S.; Alaux, L.; Kanzaki, H.; Okuyama, Y.; et al. The rice resistance protein pair RGA4/RGA5 recognizes the Magnaporthe oryzae effectors AVR-Pia and AVR1-CO39 by direct binding. *Plant Cell* **2013**, *25*, 1463–1481. [CrossRef] [PubMed]
48. Sarris, P.F.; Duxbury, Z.; Huh, S.U.; Ma, Y.; Segonzac, C.; Sklenar, J.; Derbyshire, P.; Cevik, V.; Rallapalli, G.; Saucet, S.B.; et al. A Plant Immune Receptor Detects Pathogen Effectors that Target WRKY Transcription Factors. *Cell* **2015**, *161*, 1089–1100. [CrossRef]

Essay

RNA-Seq Based Transcriptomic Analysis of Bud Sport Skin Color in Grape Berries

Wuwu Wen [2,†], Haimeng Fang [1,†], Lingqi Yue [1,†], Muhammad Khalil-Ur-Rehman [3], Yiqi Huang [1], Zhaoxuan Du [1], Guoshun Yang [1] and Yanshuai Xu [1,*]

1   College of Horticulture, Hunan Agricultural University, Changsha 410128, China
2   School of Agriculture and Biology, Shanghai Jiao Tong University, Shanghai 200240, China
3   Horticultural Sciences, The Islamia University of Bahawalpur, Bahawalpur 62300, Pakistan
*   Correspondence: yx56@hunau.edu.cn
†   These authors contributed equally to this work.

Abstract: The most common bud sport trait in grapevines is the change in color of grape berry skin, and the color of grapes is mainly developed by the composition and accumulation of anthocyanins. Many studies have shown that *MYBA* is a key gene regulates the initiation of bud sport color and anthocyanin synthesis in grape peels. In the current study, we used berry skins of 'Italia', 'Benitaka', 'Muscat of Alexandria', 'Flame Muscat', 'Rosario Bianco', 'Rosario Rosso', and 'Red Rosario' at the véraison stage (10 weeks post-flowering and 11 weeks post-flowering) as research materials. The relative expressions of genes related to grape berry bud sport skin color were evaluated utilizing RNA-Seq technology. The results revealed that the expressions of the *VvMYBA1/A2* gene in the three red grape varieties at the véraison stage were higher than in the three white grape varieties. The *VvMYBA1/A2* gene is known to be associated with *UFGT* in the anthocyanin synthesis pathway. According to the results, *VvMYBA1/A2* gene expression could also be associated with the expression of *LDOX*. In addition, a single gene (gene ID: Vitvi19g01871) displayed the highest expressions in all the samples at the véraison stage for the six varieties. The expression of this gene was much higher in the three green varieties compared to the three red ones. GO molecular function annotation identified it as a putative metallothionein-like protein with the ability to regulate the binding of copper ions to zinc ions and the role of maintaining the internal stable state of copper ions at the cellular level. High expression levels of this screened gene may play an important role in bud sport color of grape berry skin at the véraison stage.

Keywords: grape; bud sport; RNA-Seq; *MYB*

Citation: Wen, W.; Fang, H.; Yue, L.; Khalil-Ur-Rehman, M.; Huang, Y.; Du, Z.; Yang, G.; Xu, Y. RNA-Seq Based Transcriptomic Analysis of Bud Sport Skin Color in Grape Berries. *Horticulturae* **2023**, *9*, 260. https://doi.org/10.3390/horticulturae9020260

Academic Editor: Rossano Massai

Received: 5 November 2022
Revised: 5 February 2023
Accepted: 8 February 2023
Published: 15 February 2023

1. Introduction

1.1. Bud Sport

Many of the fruits we eat every day are extremely heterozygous in nature [1]. The genomes of fruit trees or vines are highly heterozygous, and in order to adapt to the natural environment, some fruit trees gradually develop inbred incompatibility; as a result, some of the excellent characteristics of fruit trees are lost. Most varieties of fruit trees, such as peach, grape, and citrus, are self-compatible. Most varieties of apple, pear, sweet cherry, and other fruit trees are self-incompatible, while male sterility sometimes occurs in grapes. The VviINP1 gene was identified as related to male sterility in grapes [2]. In order to maintain the excellent properties of fruit during production, asexual propagation (cuttings, strips, and grafting) is used to maintain the exceptional characteristics of fruit [3]. Among cultivation processes, some different mutative traits are observed in similar plants [4], and some mutations are stable to inherit and are called bud sport [5].

Plant bud sport is related to somatic cell mutation that occurs in the cells of the meristem of plant buds, usually expressed on branches, leaves, flowers, and fruits. The phenotype displayed by the bud sport is significantly different from that of the rest of the plant [6]. In general, bud sport is produced by cell division in the apical meristem of plants, which is triggered by mutations in the stable somatic cells of the first single cell and then fills the cell layer and forms a stable chimera [7,8]. Mutation in this cell gradually fill some or all of the meristem tissue during later stages of growth, and the mutation can be transferred to offspring and can enable mutants to reproduce asexually [9]. Bud sport brings certain types of new traits in the plant itself, while the original qualities of the plant parents are retained, which shape a new mechanism of genetic mutation [10]. Different quantitative genetic studies have located the SDI 119 quantitative trait locus (QTL) on linkage group (LG) 18, explaining up to 70% of phenotypic variance in the 120 seed content parameters. Looking into the potentials of grape varieties for table purposes, mutation-breeding programs have started for other characteristics using chemical and physical mutagens. This is very important for plants because not only the quality of plants can be improved but also more economic value can be generated [11–13].

At present, researchers and growers have selected bud sport varieties that are related to the early ripening, peel color, fruit size, and disease resistance of fruit trees according to different needs [14]. For example, through natural selection, radiation, or colchicine treatment, bud sports varieties related to early fruit ripening and peel color have been found in apples and grapevines [15,16]. Bud sport varieties with enhanced disease resistance have been found in peaches, plums, strawberries, and citrus [17–20], and varieties with enlarged fruit and doubled chromosomes have been found in bananas and kiwifruit [21,22].

1.2. Fruit Color

In fruit trees or vines, especially in apples and grapevines, peel color acts as one of the criteria for judging the ripeness of fruit, which is an important indicator and quality parameter of fruits. Numerous examples of fruit berry skin and flesh types of bud sports were reported [23]; the most common type of bud sport changes the color of the flesh or berry skin. The color change in fruit is mainly related to the change in anthocyanin content. Anthocyanins are secondary metabolites of flavonoids. In plants, flavonoids are believed to have a variety of functions, including defense against light coercion. Anthocyanin compounds play an important reproductive role as attractants in plant–animal interactions [24]. Changes in the contents of anthocyanins and synthetic pathways have been fully studied through many plant experiments [25,26].

According to multifaceted verification, some key regulatory genes in the anthocyanin synthesis pathway were analyzed [27]. In the early stages of the flavonoid biosynthesis process, CHS generates chalcone from the 4-coumarinyl-CoA and malonyl-CoA substrates. Chalcone isomerase catalyzes the formation of naringenin, which is the main metabolite of other synthetic branches of this pathway. Downstream of the flavonoid biosynthetic pathway, anthocyanins and leucine are common key substrates for the synthesis of anthocyanins and proanthocyanidins (PAs). Leucoanthocyanidin dioxygenase/anthocyanidin synthase (LDOX/ANS) can convert leucoanthocyanins to anthocyanidins, and anthocyanidins can be further glycosylated by uridine diphosphate (UDP)-glucose to forming flavonoid-O-glycosyltransferase (UFGT). O-methyltransferases (OMTs) catalyze the formation of O-methylated anthocyanins, such as petunidin, peonidin, and malvidin [28,29].

Most of the fruit and skin colors of different fruits, especially grape berries, are associated with the *MYB* gene regulation of anthocyanins [30–32]. The biosynthesis of fruit anthocyanins is controlled by a unique branching of R2R3 *MYB* transcription factors. Normally, the *MYB* gene interacts with the bHLH transcription factor and the WD40 complex protein to regulate the synthesis pathway of anthocyanins [33]. Studies related to grapes and apples have shown that the change in fruit color is due to the insertion of a reverse transcriptional transposon in the promoter region of *MYB* or is a deletion of the *MYB* gene and its upstream alleles that causes the fruit peel or flesh color change. When the *MYB* gene does not show expression or its related sequence alleles are missing, fruit color cannot change to red, blue, or purple [34].

1.3. Grape Bud Sport

Grapes (*Vitis vinifera* L.) are one of the most popular fruits in the world and are usually consumed fresh, as well as in the form of several value-added products. The varieties of grape are diverse, including color, fruit size, fruit type, aroma, and other characteristics that show difference in quality. Among them, color is one of the most important quality attributes for consumers. From the beginning, people have used fresh grapes and wine as a source of transmission to spread grapes all over the world. However, with the development of breeding technology, grape breeding started, and many somatic mutations associated with the quality of grapes have been discovered. Many new grape varieties have been developed through bud sport selection.

In the following figure, the color of line under a variety represents grape peel color: green represents green varieties, red represents red varieties, and black represents black and purple varieties.

The white grape 'Italia' could sport into red grapes of the 'Ruby Okuyama' and 'Benitaka' varieties. The red grape 'Okuyama Ruby' and the white grape 'Rosario Bianco' were crossed to produce the red grape 'Rosario Rosso'. The white grape 'Muscat of Alexandria' and the black-purple grape 'Schiava Grossa' were crossed to produce the black-purple grape 'Muscat Hamburg'. The hybridization of 'Bicane' white grapes and 'Muscat Hamburg' black-purple grapes produced the white grape 'Italia' (Figure 1).

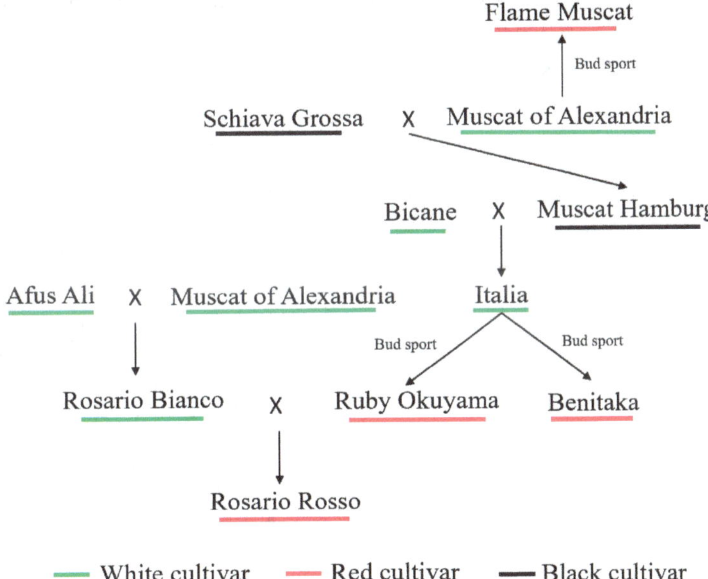

Figure 1. 'Italia' is associated with several grape bud sports and related relationship maps.

After thousands of years of natural hybridization and human selection, the color of the berry skins of grapes has become very diverse [35]. According to the presence or absence of anthocyanins in grape berry skin, which is divided into red and black or white varieties, this phenotype is controlled by a single gene locus [29]. There are four *MYBs* at this chromosome with two locations; at least two of these *MYBs* are mutated in white grapes. Either *VvMYBA1* or *VvMYBA2* (or both) can regulate berry peel color. For white grape, two mutations in the coding region of the *VvMYBA2* allele cause its inactivation, while it is not transcribed in white grapes due to the presence of retrotransposons in the promoter region of *VvMYBA1* [36,37]. This results in no accumulation of anthocyanins or very minute accumulation, and the berry skins and flesh color change from dark to light eventually. However, in some grape bud sport varieties, the deletion of the *Gret1* retransposon restores the function of *VvMYBA1*, and this deletion makes the color of grape berry skins and flesh white to black or purple [38]. However, some studies have shown that, in yellow-green or white bud sports of 'Cabernet Sauvignon' [39], with the exception of *VvMYBA1*, its homologous genes of *VvMYBA2r*, *VlMYBA1-1*, *VlMYBA1-2*, and *VlMYBA2* also regulate the synthesis of anthocyanins. In addition, there are functional and nonfunctional genes among these homologous genes and alleles [26]. Researchers found that, in white grapes, the allele of *VvMYBA1* is homozygous, while the alleles of *VvMYBA1* in red or black grapes are heterozygous [40]. It can be seen in many *MYB*-related genes in berries that play an important role in anthocyanin biosynthesis that the content of anthocyanins and the color of berry flesh and peels might be regulated by these genes.

1.4. Transcriptome Sequencing

Bud sport has been studied in many fruits; however, the mechanism of bud sport in grapes remains unclear. In order to understand the mechanism of berry peel color in relation bud sport, we utilize RNA-Seq technology to compare the 'Italia', 'Benitaka', 'Muscat of Alexandria', 'Flame Muscat', 'Rosario Bianco', and 'Rosario Rosso' varieties by selecting samples at 10 wpf (weeks post-flowering) and 11 wpf (12 samples in total). We conclude that, in addition to *UFGT*, the expression of the *LDOX* gene may also correlate with the expression of *VvMYBA1/A2*, and a new gene (gene ID: Vitvi19g01871) that exhibits the highest expression of all the detected genes in white varieties might play an important role at the véraison stage in 'green-red' bud sport berries.

2. Materials and Methods

2.1. Plant Materials

The research material (berries) used in this study was collected from the vineyard at the Zhengzhou fruit research institute (China) during 2020. The varieties used in the present research were 'Italia', 'Benitaka', 'Muscat of Alexandria', 'Flame Muscat', 'Rosario Bianco', and 'Rosario Rosso'. The vines were 10 years old with 'Y'-shaped tree forms. The berries of each cultivar were in the véraison stage, from 10 wpf (weeks post flowering) to 11 wpf. The red varieties showed notable change in berry color at 11 wpf (Figure 2). Three berries from the upper, middle, and lower parts of each cluster were selected from six uniform clusters. The berry skins were peeled off quickly and frozen in liquid nitrogen immediately. All frozen samples were stored at −80 °C for further analysis.

All the samples were allotted numbers as follows: It10 ('Italia' 10 wpf berries), It11 ('Italia' 11 wpf berries), Be10 ('Benitaka' 10 wpf berries), Be11 ('Benitaka' 11 wpf berries), Ma10 ('Muscat of Alexandria' 10 wpf berries), Ma11 ('Muscat of Alexandria' 11 wpf berries), Fm10 ('Flame Muscat' 10 wpf berries), Fm11 ('Flame Muscat' 11 wpf berries), Rb10 ('Rosario Bianco' 10 wpf berries), Rb11 ('Rosario Bianco' 11 wpf berries), Rb11 ('Rosario Rosso' 10 wpf berries), and Rr11 ('Rosario Rosso' 11 wpf berries) (as shown in Figure 2, respectively).

2.2. RNA Extraction and RNA-Seq

RNA was extracted from grape berry skins of different varieties using an RNA extract kit (Solebao Biotechnology Co., Ltd., Shanghai, China). The integrity of sample RNA was detected with agarose gel, the purity and concentration of RNA were detected with a NanoDrop-2000 instrument (Thermo Scientific, Waltham, MA, USA), and the RQN value was tested with Agilent5300 software. Follow-up experiments could be carried out when the RNA was not contaminated by impurities, such as pigment, protein, sugar, etc. The RQN $\geq$ 7, the brightness of 28/23S was greater than 18/16S, the RNA concentration $\geq$ 100 ng/uL, the OD260/280 = 1.8~2.2, the OD260/230 $\geq$ 2, and the total yield of RNA (>1 μg) met the requirements of two RNA libraries.

A Takara RT reagent kit (Takara, Shanghai, China) was used for cDNA and double-strand cDNA synthesis. RNA-Seq libraries were constructed using a TruSeq RNA sample prep kit v2 (Illumina, San Diego, CA, USA). The sequencing process was performed with an Illumina HiSeq 4000 SBS kit (300 cycles) system (Shanghai Majorbio Bio-pharm Biotechnology Co, Shanghai, China).

Figure 2. Fruit clusters of six varieties during véraison period (10 wpf and 11 wpf). (**A**) 'Italia' berries at 10 and 11 wpf; (**B**) 'Benitaka' berries at 10 and 11 wpf; (**C**) 'Muscat of Alexandria' berries at 10 and 11 wpf; (**D**) 'Flame Muscat' berries at 10 and 11 wpf; (**E**) 'Rosario Bianco' berries at 10 and 11 wpf; (**F**) 'Rosario Rosso' berries at 10 and 11 wpf.

2.3. Transcriptome Sequencing and Analysis

SeqPrep (https://github.com/jstjohn/SeqPrep, accessed on 12 February 2022) and Sickle (https://github.com/najoshi/sickle, accessed on 12 February 2022) were used for trimming the adaptors of raw reads and quality control of the raw reads to obtain high-quality reads. The clean reads were aligned to a reference genome (reference genome version: 12X.v2, website source: https://urgi.versailles.inra.fr/Species/Vitis/Data-Sequences/Genome-sequences, accessed on 14 February 2022) with HISAT2 (http://ccb.jhu.edu/software/hisat2/index.shtml, accessed on 14 February 2022) software, and the mapped reads of each sample were assembled with StringTie (https://ccb.jhu.edu/software/stringtie/index.shtml?t=example, accessed on 15 February 2022). To identify DEGs (differential expression genes) between two different samples, the expression level of each transcript was calculated according to the transcripts per million reads (TPM) method. RSEM (http://deweylab.biostat.wisc.edu/rsem/) was used to quantify gene abundances. The DEG analysis was performed using DESeq2/DEGseq/EdgeR with Q values (adjusted p-value ≤ 0.05, DEGs with $|\log_2 \mathrm{FC}| > 1$ and Q value ≤ 0.05 (DESeq2 or EdgeR)/Q value ≤ 0.001 (DEGseq) that were considered to be significantly different expressed genes. The output of normalized TPM values and the DEG analysis were performed using the Majorbio cloud platform (Shanghai Majorbio Bio-Pharm Technology Co., Ltd.).

2.4. Statistical Analysis

A correlation analysis was performed among *VvMYBA1*, *VvMYBA2*, *VvMYB5a*, *VvMYB5b*, *VvMYBPA1*, and grape pericarp anthocyanin synthesis genes using the transcription group TPM value at the level of $|r| > 0.7$ and $p < 0.05$. Expression level was significantly related to the genes. Pearson's correlation coefficient was used to measure the correlation between two random variables. The closer the Pearson value to 1, the higher the similarity of gene expression between samples, and the better the correlation between the samples.

SPSS v26.0 (Chicago, IL, USA) was used for the significance and correlation analysis of *MYB*-related regulatory genes related to anthocyanin synthesis structural genes data and correlation between anthocyanin synthesis structural genes and *VvMYBA1* and *VvMYBPA1* regulatory genes in two bud sport groups data.

3. Results

3.1. Quality Control Data Statistics

The total number of raw sequencings reads of each sample ranged from 41,748,704 to 48,476,130 among all the samples. After removing the low-quality reads, the average error rate of the sequencing bases of the clean reads after quality control was less than 0.026%. The percentage of the samples reaching Q20 quality reads was more than 97.74%, and the Q30 percentage was more than 93.32% among all the sequence data. The G and C base ratios were 45.96% and 47.01% of the total bases, respectively. The sequence alignment rates of clean reads matched with the reference genome ranged from 78.27% to 93.11% (Table 1).

Table 1. RNA-Seq data quality of all 12 varieties.

Sample Name	Raw Reads	Clean Reads	Error Rate (%)	Q20 (%)	Q30 (%)	GC Content (%)	Total Mapped
Be10	42,647,782	42,347,510	0.0245	98.26	94.66	46.27	34,536,653 (81.56%)
Be11	45,278,732	44,781,572	0.0254	97.85	93.73	47.01	35,050,465 (78.27%)
Fm10	45,595,408	45,132,012	0.0252	97.96	93.93	46.53	41,466,312 (91.88%)
Fm11	42,127,462	41,834,580	0.0248	98.11	94.32	46.74	38,000,195 (90.83%)
It10	41,804,432	41,371,026	0.0247	98.14	94.45	46.51	38,519,065 (93.11%)
It11	48,476,130	48,148,376	0.025	98.06	94.16	45.96	44,559,329 (92.55%)
Ma10	46,362,636	46,058,214	0.0248	98.1	94.28	46.4	41,828,638 (90.82%)
Ma11	47,352,198	46,978,274	0.0251	97.98	94	46.36	42,739,595 (90.98%)
Rb10	41,748,704	41,378,660	0.0252	97.95	93.95	46.35	37,569,526 (90.79%)
Rb11	43,902,394	43,576,488	0.0249	98.06	94.18	46.04	40,136,365 (92.11%)
Rr10	43,522,364	43,177,256	0.0249	98.06	94.22	46.48	39,821,925 (92.23%)
Rr11	42,739,842	42,458,930	0.0246	98.19	94.51	46.57	38,767,709 (91.31%)

(1) Raw reads: the total number of the raw sequencing data; (3) clean reads: the total number of clean sequencing data after quality filtering; (4) error rate (%): the average error rate of the sequencing base corresponding to the quality-filtered data, usually below 0.1%; (5) Q20 (%) and Q30 (%): base or read quality assessment parameters, Q20 and Q30 refer to the percentage of total bases with sequencing qualities of 99% and 99.9% above, respectively. Q20 is usually above 85% and Q30 is above 80%; (6) GC content (%): the percentage of G and C bases corresponding to the quality control data as a percentage of the total bases; (7) total mapped: the number of clean reads that can be matched on the genome.

3.2. Differentially Expressed Gene (DEG) Analysis

Through the differential expression analysis of the RNA-Seq data, 3124 DEGs were selected between It11 wpf ('Italia' grape skin samples at 11 weeks post-flowering) and It10 wpf. Compared to Be10 wpf, a total of 2707 DEGs were selected in Be11 wpf. In addition, 1766 DEGs were found between Ma11 wpf and Ma10 wpf, with 1716 DEGs between Fm11 wpf and Fm10 wpf. Rb11 wpf showed a total of 1640 DEG compared with Rb10 wpf, and Rr11 wpf showed 1579 DEGs compared with Rr10 wpf. The number of upregulated DEGs at the véraison stage (10 wpf to 11 wpf) was greater than the number of downregulated DEGs among the three white varieties of 'Italia', 'Muscat of Alexandria', and 'Rosario Bianco', while in red-colored varieties, 'Benitaka' and 'Flame Muscat' both showed lower upregulated DEG numbers at the véraison stage. Be10 wpf showed 1731 DEGs compared to It10 wpf, Fm10 wpf displayed 2790 DEGs compared to Ma10 wpf, and Rr10 wpf had 2962 DEGs compared to Rb10 wpf. For bud sport varieties in 'Benitaka', 'Italia', 'Flame Muscat', and 'Muscat of Alexandria', more upregulated DEG numbers were found at 10 wpf. Be11 wpf had a total of 2074 DEGs compared to It11 wpf, while 2000 DEGs were screened between Fm11 wpf and Ma11 wpf. Rr11 wpf had 3282 DEGs compared to Rb11 wpf. Among three comparisons of 'Benitaka' versus 'Italia', 'Flame Muscat' versus 'Muscat of Alexandria', and 'Rosario Rosso' versus 'Rosario Bianco', more downregulated DEG numbers were found at 11 wpf (Table 2).

Table 2. The numbers of DEGs among difference comparison groups.

Difference Comparison Group	Total DEG Number	Upregulated DEG Number	Downregulated DEG Number
It10_vs_It11	3124	1941	1183
Be10_vs_Be11	2707	1114	1593
It10_vs_Be10	1731	1095	636
It11_vs_Be11	2074	925	1149
Ma10_vs_Ma11	1766	1090	676
Fm10_vs_Fm11	1716	505	1211
Ma10_vs_Fm10	2790	1531	1259
Ma11_vs_Fm11	2000	551	1449
Rb10_vs_Rb11	1640	1152	488
Rr10_vs_Rr11	1579	865	714
Rb10_vs_Rr10	2962	931	2031
Rb11_vs_Rr11	3282	936	2346

3.3. Correlation Analysis among Each Sample

The Pearson correlation coefficient between It10 wpf and It11 wpf was close to 1, and It10 showed a positive correlation with It11. The Pearson correlation coefficients between It11 and Ma11 and between It11 and Rb11 wpf were also close to 1. The three white varieties of 'Italia', 'Muscat of Alexandria', and 'Rosario Bianco' showed good correlation (>0.8) at 11 wpf as well. The correlation coefficients between Ma10 wpf and Ma11 wpf and between Rb10 wpf and Rb11 wpf were close to 1, with 'Muscat of Alexandria' and 'Rosario Bianco' closely correlated. The correlation between the three red varieties of 'Benitaka', 'Flame Muscat', and 'Rosario Rosso' was low between 10 wpf to 11 wpf (Figure 3).

3.4. Gene Expression Level of VvMYBA1 in Berry Skins

The log2FC value was used to compare the expression levels of *VvMYBA1* in the comparisons of the GC10_vs_RC10 group and the GC11_vs_RC11 group. The results showed that log2FC >7, which means that the expression levels of the *VvMYBA1* gene in the three red varieties were much higher than those in three green varieties. In the comparison of the GC10_vs_GC11 group, the log2FC value was only 1.26, and *VvMYBA1* just reached the differential expression level (if the screening parameter was log2FC > 1.5, then it was not significant). In the comparison of the RC10_vs_RC11 group, the log2FC value was 1.95, and the expression of the *VvMYBA1* gene was significantly different (Figure 4).

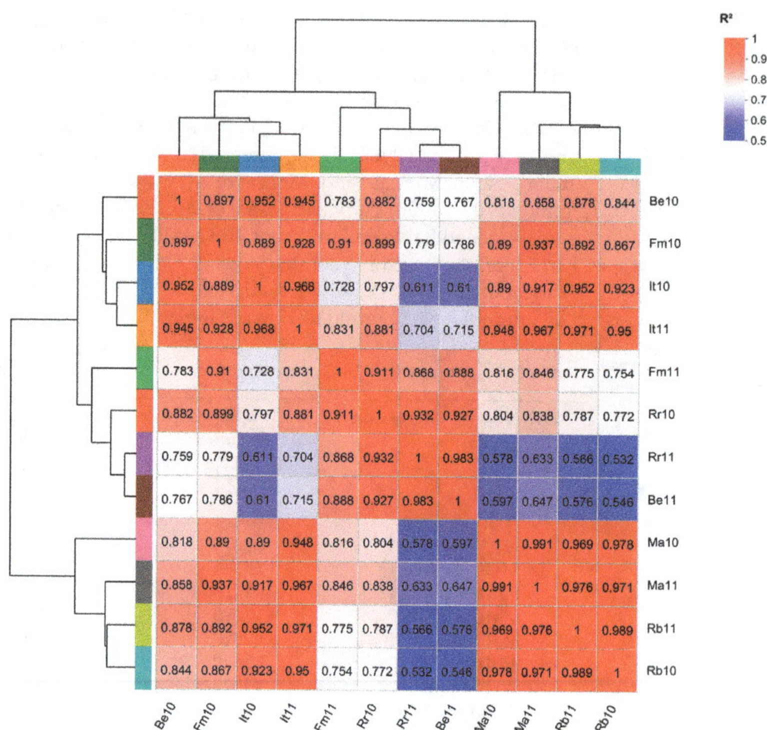

Figure 3. Correlation heatmap of six varieties (three groups).

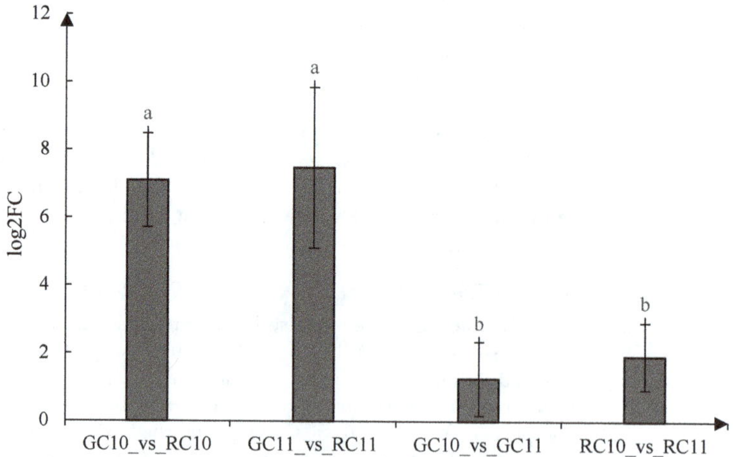

Figure 4. Differences in *VvMYBA1* expressions in red and white grape berry skins. 'GC' represents three green cultivars; 'RC' represents three red cultivars. a, b represent the significant level between the data ($p < 0.05$).

3.5. Anthocyanin-Synthesis-Related Gene Expression Analysis

Charenone synthase (CHS) is the first key enzyme of the flavonoid pathway. The gene expression in 'Benitaka' was higher at 10 wpf than at 11 wpf, and the gene expression level of Rr was lower at 10 wpf than 11 wpf. The expression level of the CHS-encoding gene

VvCHS was significantly different at the véraison stage for 'Benitaka' and 'Rosario Rosso' compared with other varieties. The TPM values of *VvCHS* in berry skins at the véraison stage during the transition period of 'Benitaka' were higher than those of 'Italia' (Figure 5A). Chalcone isomerase (CHI) catalyzed the isomerization of chalcone rings to form colorless flavonoids, and there was no significant difference in the expression of the coding gene *VvCHI* between 10 wpf and 11 wpf for each cultivar (Figure 5B). Flavanone 3-hydroxylase (F3H) is one of the key enzymes in the biosynthetic pathway of anthocyanins, while F3H, F3'H, and F3',5'H participate in the regulation of two branches of anthocyanin biosynthesis and the F3'H-controlled pathway for the synthesis of red anthocyanins. F3',5'H, on the other hand, regulates the synthesis of blue-violet delphinidin. The expression level of the F3'H-encoding gene *VvF3'H* was low in each sample, and there was no significant difference between 10 wpf and 11 wpf (Figure 5C). The F3',5'H-encoding genes of *VvF3'* and *5'H* were not expressed in It 10 wpf and Rr10 wpf, and the expressions of *VvF3', 5'H* in the grape berry skins of the two mutated red varieties, 'Benitaka' and 'Flame Muscat', were higher than in 'Italia' and 'Rosario Bianco' (Figure 5D). In addition, the expression level of *VvF3H* in 'Benitaka' was obviously higher than that in 'Italia', and the expression of the F3H-encoded gene *VvF3H* at 10 wpf and 11 wpf for each sample was very low and displayed no difference in each cultivar (Figure 5E). The expression level of the FLS-encoding gene *VvFLS* showed greater variation in the skin of the 'Benitaka' during véraison.

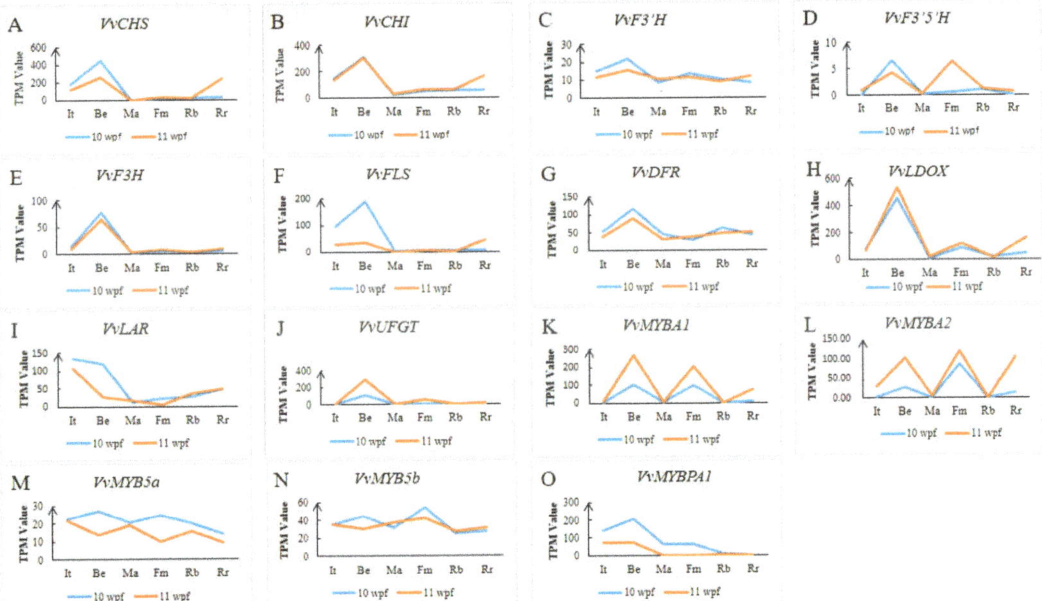

Figure 5. Anthocyanin synthesis structure gene and regulatory gene expression analysis of six varieties. (**A**) *CHS*, chalcone synthase; (**B**) *CHI*, chalcone isomerase; (**C**) *F3'H*, flavonoid 30-hydroxylase; (**D**) *F3'5'H*, flavanone3',5'-hydroxylase; (**E**) *F3H*, flavanone 3-hydroxylase; (**F**) *FLS*, flavonol synthase; (**G**) *DFR*, dihydroflavonol 4-reductase; (**H**) *LDOX*, leucoanthocyanidin dioxygenase; (**I**) *LAR*, leucoanthocyanidin reductase; (**J**) *UFGT*, anthocyanidin 3-*O*-glucosyltransferase; (**K–O**) *MYBA1, MYBA2, MYB5a, MYB5b, MYBPA1*, transcription factor encode genes, belonging to the R2R3 Myb family, which controls the last steps in the anthocyanins biosynthesis pathway.

Leucoanthocyanidin dioxygenase (*LDOX*) and *UFGT* successively catalyzed the oxidation of colorless proanthocyanidins to form colored delphinidin or anthocyanins and the glycosylation of catalytically unstable anthocyanins to form various stable anthocyanins.

The expression levels of the LDOX-encoding gene *VvLDOX*, the UFGT-encoding gene *VvUFGT*, and the regulatory genes *VvMYBA1* and *VvMYBA2* in the pericarps of the three red varieties were higher than those of the white varieties. Both the *VvUFGT* and *VvMYBA* genes were hardly expressed in the three white varieties during the véraison period (Figure 5H,J–L). The expressions of regulatory genes *VvMYBA5a* and *VvMYBPA1* in the 10 wpf grape berry skins of each cultivar were higher than at 11 wpf (Figure 5M,O).

3.6. Correlation Analysis between Anthocyanin-Synthesis-Related Structural Genes and VvMYBA1 and VvMYBPA1 Regulatory Genes among Bud Sport Varieties

In the 'Italia' vs. 'Benitaka' bud sport group, *VvMYBA1* and *VvUFGT* showed a significant positive correlation, and *VvMYBA1* may directly regulate *VvUFGT* expression to regulate anthocyanin synthesis. *VvMYBA1* was positively correlated with *VvCHS*, *VvCHI*, *VvF3H*, and *VvLDOX* in the 'Muscat of Alexandria' vs. 'Flame Muscat' bud sport group, while *VvMYBA1* was not significantly correlated with *VvUFGT* (Figure 6A,B). The mechanism of *VvMYBA1* regulation of the anthocyanin synthesis pathway in the pericarp was different between the 'Italia' vs. 'Benitaka' bud sport group and the 'Muscat of Alexandria' vs. 'Flame Muscat' bud sport group. The gene expressions of *VvMYBPA1* and *VvFLS* in the 'Italia' vs. 'Benitaka' bud sport group were positively correlated, and *VvMYBPA1* may directly regulate the expression of *VvFLS*. There was no significant correlation between *VvMYBPA1* and anthocyanin synthesis structural genes in the bud sport group of 'Muscat of Alexandria' vs. 'Flame Muscat', which may not be directly involved in the regulation of anthocyanin synthesis (Figure 6C,D).

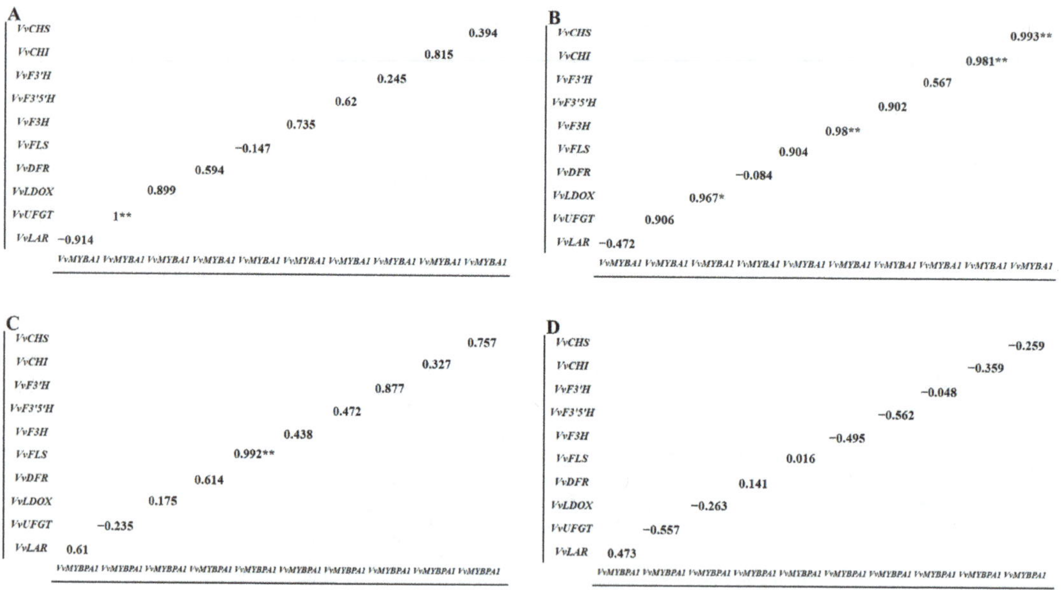

Figure 6. Correlation between anthocyanin synthesis structural genes and *VvMYBA1* and *VvMYBPA1* regulatory genes in two bud sport groups. (**A**) Correlation between anthocyanin synthesis structural genes and *VvMYBA1* regulatory gene in 'Italia' vs. 'Benitaka' bud sport group; (**B**) 'Muscat of Alexandria' vs. 'Flame Muscat' bud sport group anthocyanin synthesis structural genes and *VvMYBA1* regulatory gene correlation; (**C**) correlation between anthocyanin synthesis structural genes and *VvMYBPA1* regulatory gene in 'Italia' vs. 'Benitaka' bud sport group; (**D**) 'Muscat of Alexandria' vs. 'Flame Muscat' bud sport group anthocyanin synthesis structural genes associated with *VvMYBPA1* regulatory gene. '*,**' represents the two groups of data reached a significant level, '*' represents $p < 0.05$, '**' represents $p < 0.01$.

3.7. Correlation Analysis of MYB-Related Regulatory Genes and Anthocyanin Synthesis Structure

VvMYB5a and *VvMYB5b* were not significantly associated with structural genes in the anthocyanin synthesis pathway, which may not be directly involved in regulating the synthesis of anthocyanins. *VvMYBPA1* showed significant correlations with *VvCHS*, *VvF3'H*, *VvF3H*, *VvFLS*, and *VvLAR*, which may directly regulate the flavonoid pathway, anthocyanin synthesis, flavonol synthesis, and catechol synthesis in the anthocyanin synthesis pathway. *VvMYBA1* was positively correlated with *VvF3'5'H*, *VvLDOX*, and *VvUFGT*, which may be directly involved in regulating the synthesis of anthocyanins and regulating the *UFGT* catalytic formation of stable anthocyanin pathways. *VvMYBA2* was not significantly associated with structural genes in the anthocyanin synthesis pathway (Table 3, Figure 7). Among these, the regulation of *VvMYBA2* and *VvMYB5a* was not clear, while synthetic genes regulated by *VvMYBPA1* and *VvMYBA1* were clearly known.

Table 3. MYB-related regulatory genes related to anthocyanin synthesis structural genes. ** represents $p < 0.01$.

Gene	VvMYB5a	VvMYB5b	VvMYBPA1	VvMYBA1	VvMYBA2
VvCHS	0.194	0.154	0.731 **	0.384	0.231
VvCHI	−0.109	0.075	0.665	0.544	0.306
VvF3'H	0.459	0.487	0.839 **	0.431	0.22
VvF3'5'H	−0.09	0.33	0.349	0.747 **	0.481
VvF3H	0.24	0.184	0.709 **	0.585	0.222
VvFLS	0.468	0.263	0.873 **	0.121	−0.059
VvDFR	0.231	-0.09	0.663	0.374	0.007
VvLDOX	0.029	0.187	0.548	0.756 **	0.459
VvUFGT	−0.152	0.007	0.304	0.831 **	0.468
VvLAR	0.5	0.078	0.752 **	−0.246	−0.295

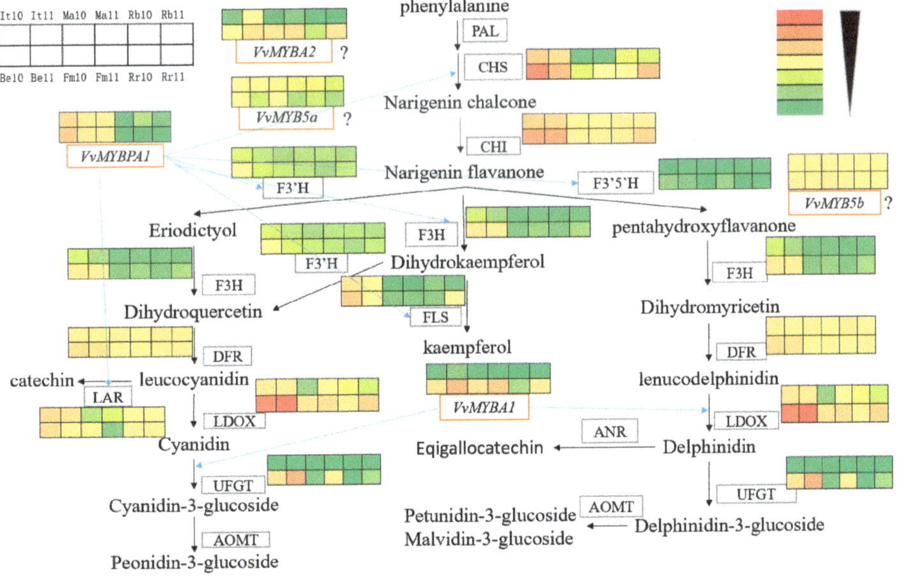

Figure 7. The relationships between the expression color scale of anthocyanin synthesis structural genes and regulatory genes in grape peels and the regulation of the *MYBA* gene. The blue dotted arrows represent the structural genes in the anthocyanin synthesis pathway regulated by the regulatory gene; the '?' indicates that the regulatory mechanism of the regulatory gene is not yet clear; the green-to-red color scale means the TPM values showed an increasing trend.

3.8. Screening of Genes Involved in the Regulation of Metal Ion Binding

When screening by sorting all gene expression levels (TPM values) in the three white grape varieties, a gene located on chromosome 19 (gene ID: Vitvi19g01871) was found to show the highest expression level. Its expression level was much higher than those of other genes, and the gene was highly expressed (almost the maximum) in the three red varieties. Interestingly, the expression levels of this gene in the pericarps of the three white the three red varieties during the same period were also different (Figure 8). From the comparison of 10 wpf and 11 wpf, this gene was upregulated at 11 wpf compared with 10 wpf in the white cultivar 'Italia', while the opposite was found in 'Muscat of Alexandria' and 'Rosario Bianco' (Figure 8). In the red varieties of 'Benitaka' and 'Rosario Rosso', the expression levels at 11 wpf were downregulated compared with 10 wpf, while the expression in 'Flame Muscat' was higher (Figure 8). According to the functional annotation, it was inferred that this gene encodes a metallothionein-like protein, which regulates the binding of copper ions and zinc ions. Copper ions are related to the synthesis of chlorophyll, which may have a certain impact on the change in peel color at the véraison stage.

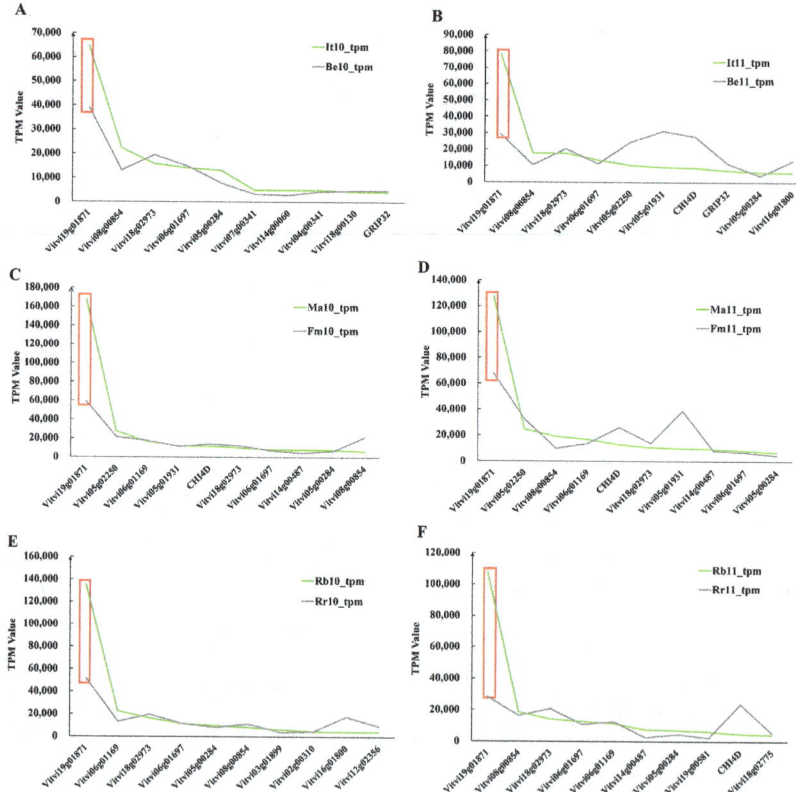

Figure 8. (**A**) 'Italia' 'Benitaka' TPM value almost top 10 of gene expression 10 wpf; (**B**) 'Italia' 'Benitaka' TPM value almost top 10 of gene expression 11 wpf; (**C**) 'Muscat of Alexandria' 'Flame Muscat' TPM value almost top 10 of gene expression 10 wpf; (**D**) 'Muscat of Alexandria' 'Flame Muscat' TPM value almost top 10 of gene expression 11 wpf; (**E**) 'Rosario Bianco' 'Rosario Rosso' TPM value almost top 10 of gene expression 10 wpf; (**F**) 'Rosario Bianco' 'Rosario Rosso' Top TPM value almost top 10 of gene expression 11 wpf. The green line represents the TPM values of white cultivars; the black line represents the TPM values of red and black cultivar.

4. Discussion

In recent years, it has been observed that many important fruit varieties are selected by bud sport [41,42]. According to statistics, there have been thousands of bud sport types on fruit trees, and some fruit trees can form a variety of bud sports. Due to the particularity of each bud sport, it brings certain characteristics in germplasm resources during fruit production and breeding. Therefore, this is an important approach used in the breeding of fruit crops.

The present study revealed that *VvMYBA1* showed elevated expression levels in the three red sport varieties at 10 wpf compared to three white varieties. In addition, after 11 wpf there were significantly higher *VvMYBA1* gene expression levels compared with the white cultivar grapes (Figure 4B). The *VvMYBA1* gene was proved to be a key transcription factor regulating color change in grape berry skins [43]. The *VvMYBA1* gene was expressed only in red berries, while it was hardly expressed in white berries (Figure 5K).

According to a correlation analysis, the majority of genes or enzymes related to the anthocyanin synthesis pathway were significantly correlated with *VvMYBPA1* and *VvMYBA1* (Table 3). Among them, five genes showed significant correlations with various genes, such as *VvMYBPA1*, *VvCHS*, *VvF3'H*, *VvF3H*, *VvFLS*, and *VvLAR*. Significant correlations of *VvF3'5'H*, *VvLDOX*, and *VvUFGT* with *VvMYBA1* were observed in our study.

The expression of the flavonoid 3-*O*-glucosyltransferase (UFGT) gene is essential for anthocyanin biosynthesis in grapes [44]. The *VvMYBA1* gene normally regulates the expression of *VvUFGT*, a key upstream gene of anthocyanin synthesis [45] considered to be the last step for catalyzing anthocyanin synthesis in the anthocyanin biosynthesis pathway [46], and both are very important in the formation of grape skin color. The RNA-Seq results indicated that the expression trends of *VvUFGT* in the three red varieties were consistent; among them, the expression level in 'Benitaka' was significantly higher than in the other two varieties and was not expressed in white grape varieties (Figure 5J). The above results are consistent with the results of a previous study conducted on 'Italia', 'Benitaka' and 'Flame Muscat' [5]. The Pearson's correlation analysis showed that *VvMYBA1* and *VvUFGT* were highly correlated with the same expression trend (Figure 5J,K). The results also indicated that *VvMYBA1* positively regulated the *VvUFGT* gene and played an important role in the biosynthesis of anthocyanins.

According to previous reports on anthocyanin synthesis in apples and bilberries, it was found that *MYBPA1* could also regulate the expression of *UFGT* [47]. In this experiment, the correlation between these two genes was not high. This may explain why, among the three groups of varieties (the 'Italia' vs. 'Benitaka' group, the 'Muscat of Alexandria' vs. 'Flame Muscat' group, and the 'Rosario Bianco' vs. 'Rosario Rosso' group), *MYBPA1* was not a key transcription factor regulating *UFGT* and the anthocyanin biosynthesis pathway. Its specific regulatory mechanism still needs further study.

MYBPA1 plays an important role in the anthocyanin biosynthesis pathway, and the expression of *MYBPA1* is positively correlated with anthocyanin accumulation [48]. In blue bilberries, the *MYBPA1* and *MYBA* transcription factors can activate the expression of DFR and ANS genes in the anthocyanin biosynthesis pathway, which are considered key genes for anthocyanin biosynthesis [49]. In this study, the expression levels of the *VvMYBPA1* gene in the two groups of bud sport varieties of 'Italia' vs. 'Benitaka' and 'Muscat of Alexandria' vs. 'Flame Muscat' were higher at 10 wpf compared with 11 wpf, while this gene was not expressed in the 'Rosario Bianco' vs. 'Rosario Rosso' group (Figure 5O). The expression trends of five structural genes (*VvCHS*, *VvF3'H*, *VvF3H*, *VvFLS*, and *VvLAR*, Table 3) related to *VvMYBPA1* were different in the three groups of tested varieties (Figure 5A–E,I). The phylogenetic analysis depicted that the flavonoid-related R2R3 MYBs of *VmMYBPA1* and *VvMYBPA1* belonged to the same group. *VmMYBPA1* could regulate the expression of *CHS* and significantly regulated the expression of the *F3'5'H* gene, while *VmMYBPA1* expression was significantly decreased in white mutant berries compared with blueberries [50], which indicates that it was related to anthocyanin biosynthesis. The expression level of *MYBPA1* is associated with the accumulation of proanthocyanidins (PA) during the early development of grape berries. The expression level of *MYBPA1* was lower before the véraison stage and peaked at two weeks following the véraison stage, later showing a low expression level. *MYBPA1* activates the promoters of *LAR* and *ANR* in grapes [51]. The expression of *VvMYBPA1* was opposite to that of *VmMYBPA1*, as expressed in 'Italia'. This is in contrast to previous studies showing no expression observed in white grape varieties. A previous study also found that *VvMYBPA1* could also be expressed in seeds [52]. The above results might indicate that the pathway or regulation mechanism of the *MYB* gene in anthocyanin synthesis is different in diverse species.

The gene expression analysis showed that the expression of *VvLDOX* was consistent with the expression trends of *VvMYBA1* and *VvUFGT* in other test materials, except for in 'Italia' (Figure 5H,J–K). *LDOX* has a unique expression pattern in the biosynthesis of anthocyanin in grape peels, and its expression levels were very high in red or black peels, which was related to the content of anthocyanin. *UFGT* is present in many tissues of grape, as well as in the skins of white and red grape varieties, while the expression of *LDOX* is not as absolute as *UFGT* [53]. *VvMYBPA1* was found to activate *VvLDOX* expression in grapes [49], and this result suggested that the expression of *VvLDOX* in 'Italia' may be related to the regulation of *VvMYBPA1*, while there was no significant correlation between *VvMYBPA1* and *VvLDOX*.

In addition to the above results, an interesting point found in this study was the gene located on chromosome number 19 (Gene ID: Vitvi19g01871). The gene expression levels (TPM values) of green varieties at 10 weeks and 11 weeks post-flowering were between 64,711–168,489 and 78,173–127,381, respectively. The expression levels of red grape varieties at 10 weeks and 11 weeks post-flowering were between 39,130–59,249 and 28,319–67,849, respectively. The expression levels of this gene in green varieties were much higher than those in red varieties, as well as much higher than all the other differentially expressed genes (Figure 8). The gene was annotated by GO molecular gene function as a metallothionein-like protein that regulates the binding of copper ions to zinc ions. Copper ions play an important role in the redox of plant respiration and are related to chlorophyll synthesis, which is important for photosynthesis. Increased photosynthesis and chlorophyll lead to excessive chlorophyll accumulation in grape peel cells. However, its mechanism of action in the process of bud sport peel color and anthocyanin synthesis is still unclear, and the function of this gene needs to be further verified at molecular or cellular levels.

5. Conclusions

In this study, it was found that *MYBA1/2* and *MYBPA1*, the key genes involved in anthocyanin synthesis in grapes, were highly expressed in red grape varieties, and their expression levels in white grapes were significantly lower than in red grapes. The expressions of *UFGT* and *LDOX* genes were positively correlated with the key peel-color-related gene of MYBA. A newly discovered gene (gene ID: Vitvi19g01871) in this study may play a key regulatory role in grape skin coloration.

Author Contributions: Experimental design, Y.X. and W.W.; data curation, H.F. and L.Y.; figures and table making, Y.H. and Z.D.; writing, H.F. and W.W.; manuscript review and editing, M.K.-U.-R. and G.Y. All authors have read and agreed to the published version of the manuscript.

Funding: This work was supported by grants from the National Natural Science Foundation of China Youth Fund (No. 32002015), the Natural Science Foundation Youth Fund of Hunan Province (2021JJ40252), and the Hunan Provincial Department of Education Scientific Research Project—General Project (19C0926).

Data Availability Statement: Not applicable.

Conflicts of Interest: The authors declare no conflict of interest.

References

1. Hansche, P.E.; Beres, W. Genetic remodeling of fruit and nut trees to facilitate cultivar improvement. *HortScience* **1980**, *15*, 710–715. [CrossRef]
2. Massonnet, M.; Cochetel, N.; Minio, A.; Vondras, A.M.; Lin, J.; Muyle, A.; Garcia, J.F.; Zhou, Y.; Delledonne, M.; Riaz, S.; et al. The genetic basis of sex determination in grapes. *Nat. Commun.* **2020**, *11*, 2902. [CrossRef] [PubMed]
3. Foster, T.M.; Aranzana, M.J. Attention sports fans! The far-reaching contributions of bud sport mutants to horticulture and plant biology. *Hortic. Res.* **2018**, *5*, 44. [CrossRef] [PubMed]
4. Wu, B.; Li, N.; Deng, Z.; Luo, F.; Duan, Y. Selection and evaluation of a thornless and HLB-tolerant bud-sport of pummelo citrus with an emphasis on molecular mechanisms. *Front. Plant Sci.* **2021**, *12*, 739108. [CrossRef]
5. Du, X.; Wang, Y.; Liu, M.; Liu, X.; Jiang, Z.; Zhao, L.; Tang, Y.; Sun, Y. The assessment of epigenetic diversity, differentiation, and structure in the 'Fuji' mutation line implicates roles of epigenetic modification in the occurrence of different mutant groups as well as spontaneous mutants. *PLoS ONE* **2020**, *15*, e0235073. [CrossRef]
6. Azuma, A.; Kobayashi, S.; Goto-Yamamoto, N.; Shiraishi, M.; Mitani, N.; Yakushji, H.; Koshita, Y. Color recovery in berries of grape (*Vitis vinifera* L.) 'Benitaka', a bud sport of 'Italia', is caused by a novel allele at the *VvmybA1* locus. *Plant Sci.* **2009**, *176*, 470–478. [CrossRef]
7. Szymkowiak, E.J.; Sussex, I.M. What chimeras can tell us about plant development. *Annu. Rev. Plant Biol.* **1996**, *47*, 351–376. [CrossRef]
8. Stewart, R.N.; Meyer, F.G.; Dermen, H. Camellia 'Daisy Eagleson' a graft chimera of *Camellia sasanqua* and *C. japonica*. *Am. J. Bot.* **1972**, *59*, 515–524. [CrossRef]
9. Darwin, C. *The Variation of Animals and Plants under Domestication (Cambridge Library Collection—Darwin, Evolution and Genetics, pp. I–Ii)*; Cambridge University Press: Cambridge, UK, 2010. [CrossRef]
10. Franks, T.; Botta, R.; Thomas, M.R.; Franks, J. Chimerism in grapevines: Implications for cultivar identity, ancestry and genetic improvement. *Theor. Appl. Genet.* **2002**, *104*, 192–199. [CrossRef]
11. Liu, G.; Li, B.; Hu, P.; Zhou, G. Analysis of differences in mineral element content of early maturation of navel orange and its parent seedlings. *Trop. Agric. Sci.* **2017**, *37*, 21–25.
12. Carolina, R.; Rafael, T.P.; Nuria, M.; Nieves, D.; José, A.V.; Cécile, M.; Thierry, L.; Javier, I.; Manuel, T.; Juan, V.; et al. The major origin of seedless grapes is associated with a 4 missense mutation in the MADS-box gene VviAGL11. *Plant Physiol.* **2018**, *177*, 1234–1253.
13. Tetali, S.; Karkamkar, S.P.; Phalake, S.V. Mutation breeding for inducing seedlessness in grape variety ARI 516. *Int. J. Minor Fruits Med. Aromat. Plants* **2020**, *6*, 67–71.
14. Visser, T.; Verhaegh, J.J.; De Vries, D.P. Pre-selection of compact mutants induced by X-ray treatment in apple and pear. *Euphytica* **1971**, *20*, 195–207. [CrossRef]
15. Kuksova, V.B.; Piven, N.M.; Gleba, Y.Y. Somaclonal variation and in vitro induced mutagenesis in grapevine. *Plant Cell Tissue Organ Cult.* **1997**, *49*, 17–27. [CrossRef]
16. Hamill, S.D.; Smith, M.K.; Dodd, W.A. In vitro induction of banana autotetraploids by colchicine treatment of micropropagated diploids. *Aust. J. Bot.* **1992**, *40*, 887–896. [CrossRef]
17. Predieri, S.; Gatti, E. Effects of gamma radiation on microcuttings of plum (*Prunus salicina* Lindl.)'Shiro'. *Adv. Hortic. Sci.* **2000**, *14*, 7–11.
18. Hashmi, G.P.; Hammerschlag, F.A.; Huettel, R.N. Growth, development, and response of peach somaclones to the root-knot nematode, *Meloidogyne incognita*. *J. Am. Soc. Hortic. Sci.* **1995**, *120*, 932–937. [CrossRef]
19. Wu, J.H.; Ferguson, A.R.; Murray, B.G. Manipulation of ploidy for kiwifruit breeding: In vitro chromosome doubling in diploid Actinidia chinensis Planch. *Plant Cell Tissue Organ Cult.* **2011**, *106*, 503–511. [CrossRef]
20. Takahashi, H. Breeding of strawberry resistant to Alternaria black spot of strawberry varieties (Alternaria althernata strawberry pathotype). *Bul. Akita Pref. Coll. Agri.* **1993**, *19*, 1–44.
21. Ge, H.; Li, Y.; Fu, H.; Fu, H.; Luo, L.; Li, R.; Deng, Z. Production of sweet orange somaclones tolerant to citrus canker disease by in vitro mutagenesis with EMS. *Plant Cell Tissue Organ Cult.* **2015**, *123*, 29–38. [CrossRef]
22. Shamel, A.D.; Pomeroy, C.S. Bud mutations in horticultural crops. *J. Hered.* **1936**, *27*, 487–494. [CrossRef]
23. Lamo, K.; Bhat, D.J.; Kour, K.; Pratap, S. Mutation studies in fruit crops: A review. *Int. J. Curr. Microbiol. Appl. Sci.* **2017**, *6*, 3620–3633. [CrossRef]
24. Allan, A.C.; Hellens, R.P.; Laing, W.A. *MYB* transcription factors that colour our fruit. *Trends Plant Sci.* **2008**, *13*, 99–102. [CrossRef] [PubMed]
25. Winkel-Shirley, B. Flavonoid biosynthesis. A colorful model for genetics, biochemistry, cell biology, and biotechnology. *Plant Physiol.* **2001**, *126*, 485–493. [CrossRef] [PubMed]
26. Grotewold, E. The genetics and biochemistry of floral pigments. *Annu. Rev. Plant Biol.* **2006**, *57*, 761–780. [CrossRef]
27. Jaakola, L. New insights into the regulation of anthocyanin biosynthesis in fruits. *Trends Plant Sci.* **2013**, *18*, 477–483. [CrossRef]
28. Tohge, T.; de Souza, L.P.; Fernie, A.R. Current understanding of the pathways of flavonoid biosynthesis in model and crop plants. *J. Exp. Bot.* **2017**, *68*, 4013–4028. [CrossRef]
29. Jun, J.H.; Xiao, X.; Rao, X.; Dixon, R. Proanthocyanidin subunit composition determined by functionally diverged dioxygenases. *Nat. Plants* **2018**, *4*, 1034–1043. [CrossRef]

30. Gu, C.; Liao, L.; Zhou, H.; Wang, L.; Deng, X.; Han, Y. Constitutive activation of an anthocyanin regulatory gene *PcMYB10. 6* is related to red coloration in purple-foliage plum. *PLoS ONE* **2015**, *10*, e0135159. [CrossRef]

31. Sun, B.; Zhu, Z.; Cao, P.; Chen, H.; Chen, C.; Zhou, X.; Mao, Y.; Lei, J. Purple foliage coloration in tea (*Camellia sinensis* L.) arises from activation of the R2R3-MYB transcription factor CsAN1. *Sci. Rep.* **2016**, *6*, srep32534. [CrossRef]

32. Zhou, Y.; Zhou, H.; Lin-Wang, K.; Vimolmangkang, S.; Espley, R.; Wang, L.; Allan, A. Transcriptome analysis and transient transformation suggest an ancient duplicated MYB transcription factor as a candidate gene for leaf red coloration in peach. *BMC Plant Biol.* **2014**, *14*, 388. [CrossRef] [PubMed]

33. Matus, J.T.; Aquea, F.; Arce-Johnson, P. Analysis of the grape *MYB R2R3* subfamily reveals expanded wine quality-related clades and conserved gene structure organization across Vitis and Arabidopsis genomes. *BMC Plant Biol.* **2008**, *8*, 83. [CrossRef] [PubMed]

34. Zhang, L.; Hu, J.; Han, X.; Li, J.; Gao, Y.; Zhang, G.; Tian, Y. A high-quality apple genome assembly reveals the association of a retrotransposon and red fruit colour. *Nat. Commun.* **2019**, *10*, 1494. [CrossRef] [PubMed]

35. Boss, P.K.; Davies, C.; Robinson, S.P. Expression of anthocyanin biosynthesis pathway genes in red and white grapes. *Plant Mol. Biol.* **1996**, *32*, 565–569. [CrossRef] [PubMed]

36. Kobayashi, S.; Goto-Yamamoto, N.; Hirochika, H. Retrotransposon-induced mutations in grape skin color. *Science* **2004**, *304*, 982. [CrossRef] [PubMed]

37. Walker, A.R.; Lee, E.; Bogs, J.; McDavid, D.; Robinson, S.; Thomas, M. White grapes arose through the mutation of two similar and adjacent regulatory genes. *Plant J.* **2007**, *49*, 772–785. [CrossRef] [PubMed]

38. Lijavetzky, D.; Ruiz-García, L.; Cabezas, J.A.; Andres, M.T.; Bravo, G.; Lbanez, A.; Cabello, F. Molecular genetics of berry colour variation in table grape. *Mol. Genet. Genom.* **2006**, *276*, 427–435. [CrossRef]

39. Walker, A.R.; Lee, E.; Robinson, S.P. Two new grape varieties, bud sports of Cabernet Sauvignon bearing pale-coloured berries, are the result of deletion of two regulatory genes of the berry colour locus. *Plant Mol. Biol.* **2006**, *62*, 623–635. [CrossRef]

40. This, P.; Lacombe, T.; Cadle-Davidson, M.; Owens, C.L. Wine grape (*Vitis vinifera* L.) color associates with allelic variation in the domestication gene *VvmybA1*. *Theor. Appl. Genet.* **2007**, *114*, 723–730. [CrossRef]

41. Granhall, I. Spontaneous and induced bud mutations in fruit trees. *Acta Agric. Scand.* **1954**, *4*, 594–600. [CrossRef]

42. Alexander, V. *Breeding for Ornamentals: Classical and Molecular Approaches*; Springer Science & Business Media: Rehovot, Israel, 2002.

43. Kobayashi, S.; Ishimaru, M.; Hiraoka, K.; Honda, C. Myb-related genes of the Kyoho grape (*Vitis labruscana*) regulate anthocyanin biosynthesis. *Planta* **2002**, *215*, 924–933.

44. He, P.; Li, L.; Wang, H.; Chang, Y. An RNA-Seq analysis of the peach transcriptome with a focus on genes associated with skin colour. *Czech J. Genet. Plant Breed.* **2019**, *55*, 166–169. [CrossRef]

45. Azuma, A.; Kono, A.; Sato, A. Simple DNA marker system reveals genetic diversity of *MYB* genotypes that determine skin color in grape genetic resources. *Tree Genet. Genomes* **2020**, *16*, 29. [CrossRef]

46. Bogs, J.; Jaffé, F.W.; Takos, A.M.; Walker, A.R.; Robinson, S.P. The grapevine transcription factor *VvMYBPA1* regulates proanthocyanidin synthesis during fruit development. *Plant Physiol.* **2007**, *143*, 1347–1361. [CrossRef]

47. Wang, N.; Qu, C.; Jiang, S.; Chen, Z.; Xu, H.; Fang, H.; Su, M.; Zhang, J.; Wang, Y.; Liu, W.; et al. The proanthocyanidin-specific transcription factor *MdMYBPA1* initiates anthocyanin synthesis under low-temperature conditions in red-fleshed apples. *Plant J.* **2018**, *96*, 39–55. [CrossRef] [PubMed]

48. Primetta, A.K.; Karppinen, K.; Riihinen, K.R.; Jaakola, L. Metabolic and molecular analyses of white mutant Vaccinium berries show down-regulation of *MYBPA1*-type R2R3 MYB regulatory factor. *Planta* **2015**, *242*, 631–643. [CrossRef] [PubMed]

49. Günther, C.S.; Dare, A.P.; McGhie, T.K.; Deng, C.; Lafferty, D.J.; Plunkett, B.J.; Grierson, E.R.; Turner, J.L.; Jaakola, L.; Albert, N.W.; et al. Spatiotemporal modulation of flavonoid metabolism in blueberries. *Front. Plant Sci.* **2020**, *11*, 545. [CrossRef] [PubMed]

50. Karppinen, K.; Lafferty, D.J.; Albert, N.W.; Mikkola, N.; McGhie, T.; Allan, A.C.; Afzal, B.M.; Häggman, H.; Espley, R.V.; Jaakola, L. *MYBA* and *MYBPA* transcription factors co-regulate anthocyanin biosynthesis in blue-coloured berries. *New Phytol.* **2021**, *232*, 1350–1367. [CrossRef] [PubMed]

51. Liu, Y.; Shi, Z.; Maximova, S.N.; Payne, M.J.; Guiltinan, M.J. *Tc-MYBPA* is an Arabidopsis TT2-like transcription factor and functions in the regulation of proanthocyanidin synthesis in Theobroma cacao. *BMC Plant Biol.* **2015**, *15*, 160. [CrossRef]

52. Terrier, N.; Torregrosa, L.; Ageorges, A.; Vialet, S.; Verriès, C.; Cheynier, V.; Romieu, C. Ectopic expression of *VvMybPA2* promotes proanthocyanidin biosynthesis in grapevine and suggests additional targets in the pathway. *Plant Physiol.* **2009**, *149*, 1028–1041. [CrossRef] [PubMed]

53. Boss, P.K.; Davies, C.; Robinson, S.P. Anthocyanin composition and anthocyanin pathway gene expression in grapevine sports differing in berry skin colour. *Aust. J. Grape Wine Res.* **1996**, *2*, 163–170. [CrossRef]

MDPI

St. Alban-Anlage 66

4052 Basel

Switzerland

www.mdpi.com

Horticulturae Editorial Office

E-mail: horticulturae@mdpi.com

www.mdpi.com/journal/horticulturae